# Manual of
## FIRE
## Safety

Manual of
**FIRE**
Safety

O Fire, may we be those
Who have the right thought
And the divine vision
And through all the days
Pass safe and beyond the danger.

So, be of easy access to us
Agni, as a father to his son
Abide with us for our well-being.

— *Rig Veda*
Translation by Sri Aurobindo

# Manual of
# FIRE
# Safety

**N. Sesha Prakash** BE, MS

Consultant in
Safety and Environmental Engineering and Management

Visiting Faculty
Mangalore Institute of Fire and Safety Management

# CBS Publishers & Distributors Pvt Ltd

New Delhi • Bengaluru • Chennai • Kochi • Kolkata • Mumbai
Bhopal • Bhubaneswar • Hyderabad • Jharkhand • Nagpur • Patna • Pune
Uttarakhand • Dhaka (Bangladesh) • Kathmandu (Nepal)

Manual of
FIRE
Safety

ISBN: 978-81-239-1990-4

Copyright © Publishers

First Edition: 2011
Reprint: 2013, 2014, 2017, 2020, 2022

Published by Satish Kumar Jain and produced by Varun Jain for

**CBS Publishers & Distributors** Pvt Ltd

4819/XI Prahlad Street, 24 Ansari Road, Daryaganj, New Delhi 110 002, India.
Ph: 23289259, 23266861, 23266867  Website: www.cbspd.com
Fax: 011-23243014       e-mail: delhi@cbspd.com; cbspubs@airtelmail.in.
*Corporate Office:* 204 FIE, Industrial Area, Patparganj, Delhi 110 092
Ph: 4934 4934       Fax: 4934 4935   e-mail: publishing@cbspd.com; publicity@cbspd.com

**Branches**

• Bengaluru: Seema House 2975, 17th Cross, K.R. Road,
  Banasankari 2nd Stage, Bengaluru 560 070, Karnataka
  Ph: +91-80-26771678/79       Fax: +91-80-26771680       e-mail: bangalore@cbspd.com
• Chennai: 7, Subbaraya Street, Shenoy Nagar, Chennai 600 030, Tamil Nadu
  Ph: +91-44-26680620, 26681266       Fax: +91-44-42032115       e-mail: chennai@cbspd.com
• Kochi: 68/1534, 35, 36, Power House Road, Opp. KSEB, Kochi 682018, Kerala
  Ph: +91-484-4059061-65       Fax: +91-484-4059065       e-mail: kochi@cbspd.com
• Kolkata: 6/B, Ground Floor, Rameswar Shaw Road, Kolkata-700 014, West Bengal
  Ph: +91-33-22891126, 22891127, 22891128       e-mail: kolkata@cbspd.com
• Mumbai: 83-C, Dr E Moses Road, Worli, Mumbai-400018, Maharashtra
  Ph: +91-22-24902340/41       Fax: +91-22-24902342       e-mail: mumbai@cbspd.com

*Representatives*

| | | | |
|---|---|---|---|
| • Bhopal | 0-8319310552 | • Bhubaneswar | 0-9911037372 |
| • Hyderabad | 0-9885175004 | • Jharkhand | 0-9811541605 |
| • Nagpur | 0-9421945513 | • Patna | 0-9334159340 |
| • Pune | 0-9623451994 | • Uttarakhand | 0-9716462459 |
| • Dhaka (Bangladesh) | 01912-003485 | • Kathmandu (Nepal) | 977-9818742655 |

*Printed at* Rashtriya Printer, Dilshad Garden, Delhi, India

*to*

_____

*My Father*

Since ages, fire *(Agni)* has been worshipped by many religions. In fact, the dawn of civilization was with the discovery of fire and the harnessing of the potential of fire energy. We cannot think of life without fire.

Fire safety, the aspect of protecting life, property and the environment from the dangers of an uncontrolled fire, is, of course, the most important element of the concept of safety and reliability. Safety is a subset of reliability. Safety leads to reliability. For safety, saving human life is the primary concern. Reliability aims to improve the overall system performance by increasing safety and reducing failures and breakdowns.

We take safety for granted unless and until disaster strikes close to us. The agony and trauma of a fire tragedy is unbearable. A perfectly safe system does not exist. All systems are unsafe to some degree. But how much of risk is acceptable is the question.

Safety attitude is the approach of an individual or a group to a hazard and the consequent risk. Individual safety attitudes are conditioned by the society's safety culture.

Safety culture influences the perceived risk levels which is the ability to perceive hazards.

Safety culture is determined by

Individual : Difficult to observe and evaluate.
Situation and behavior : These two can be observed and assessed.

If the individual cannot be changed, the behavior can be changed, by imparting safety values and beliefs, which is possible through Safety education. Safety education is the principle method to promote self-protection.

Learning only by experience may be very dear and too late. Knowledge of safety management will help in improving the safety attitude and making systems safe, both for humans and the environment. Regulations alone cannot provide the requisite change.

Therefore, safety awareness may be enhanced by

- Introducing safety regulations and penalising violators
- Providing safety information and disseminating knowledge by education and training

Economic conditions may force us to compromise on safety but it will turn out to be more expensive in the long run. Safety issues get underscored by the much larger damage caused by natural calamities such as earthquakes, floods and volcanic eruptions.

Improving safety may mean simple housekeeping, which is not as simple as we think, or enhancing the reliability of the system. Technology, being a problem-solving and a dynamic process, provides greater reliability. Technology represents a more practical aspect of knowledge. Technology increases safety.

As important as the technology are the skills necessary to operate technological systems. Technological development is the only way to raise the standard and quality of life. True and meaningful development should improve the quality of life. In addition to indicators of quality of life such as per capita income, infant mortality, delivery deaths, life expectancy, education, health facilities, communication; accident rate, health and safety hazards and risk level (safety of human life) should also be considered.

*Manual of Fire Safety* is an attempt to enhance the awareness for safety and, in particular, the ways and means of safe utilization and controlling of fire in all its forms.

I acknowledge the contributions of the various organizations where I have worked, being the source of information and knowledge, and to my family, for enduring the long hours that were required to complete this book.

This book is dedicated to my father, who bestowed me with all the inspiration.

N. Sesha Prakash

# Contents

# 1

# *Fire Science*

> *"O Agni, God, we kindle thee, refulgent, wasting not away, that this more glorious fuel may send forth for thee its shine to heaven. Bring food to those who sing thy praise! The Son of Strength; for is he not our gracious Lord? Let us serve him who bears our gifts! In battles may he be our help and strengthener, be the saviour of our lives!"*
>
> *"Rig Veda"*

## WHAT IS FIRE?

Fire is a chemical reaction in which a combustible material combines with oxygen in the atmosphere to give out heat and flame. Fire is combustion accompanied by flame or glow, which escapes its normal confines to cause damage. Fire is the heat and light of burning.

Burning or combustion can be described as a chemical reaction involving a fuel and oxygen. We can classify burning into controlled combustion and uncontrolled combustion.

- Controlled burning/combustion–which can be beneficially used (e.g. cooking, lighting, engines for motor vehicles and aircraft).
- Uncontrolled burning/combustion–which can get started unintentionally and can spread (e.g. house fires, wild fires, forest fires, explosions).

Combustion reactions are exothermic, i.e. they release heat.

In case of combustion reactions involving gases and vapours, the combustible must be present at levels of concentration lying between values known as 'flammable limits'.

With the exception of spontaneous ignition, the process can only begin when an external source of heat is applied to the fuel.

## Effects of Fire

Fire is a force, a hazard, and a potential annihilator, which can at no stage be ignored or be treated with anything but the highest priority. It can prove to be a powerful destroyer if neglected and has been considered foremost in the list of hazardous conditions. Fire is

> *Fire is uncontrolled burning. Fire is uncontrolled combustion*

1

heralded by smoke and it is necessary to react in the initial stages in order to gain control over the fire before it can reach a large and devastating magnitude. Damages in a fire accident are enormous and catastrophic.

## SMOKE

Smoke is the combined product of fire, consisting of:
- The products of combustion
- Any unburnt fuel
- Any other gases released by the combustible material
- Any excess air entrained in the combustion products.

Before a fire reaches the flame stage, smoke is generated by the smouldering fire. Thereby, smoke gives an alert alarm about the fire.

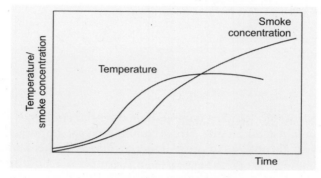

**Fig. 1.1:** Development of smoke.

As far as deaths from fire incidents are concerned, reports show that smoke has caused more fatal incidents than the fire itself. Even when there have been deaths in a fire tragedy due to apparent burns, death would have occurred due to smoke inhalation and the burns and charring happening after the death. Since smoke can spread much more easily than the fire, smoke can cause harm even at a distance away from the fire.

Smoke also obstructs visibility, which prevents or hinders escape from a fire situation. Emergency lights provided for fire exigencies should be sufficiently bright to improve the visibility. To reduce the toxic effect of smoke inhalation, a first aid support can be to cover the nose/mouth with a wet cloth.

### Effects of Smoke

Smoke from a fire has killed more people than the fire itself.

In smaller enclosed spaces, as the hot gases produced by a fire descend from the ceiling, the temperature of the interior rises rapidly. The temperature can easily exceed 100° C which is the maximum short time sustainable temperature for living beings and the threat of temperature is more immediate than that of the smoke.

In larger spaces, however, the temperature rise is much slower and also since the hot gases of the fire spread out, the descending gases are cooler. In such cases, the toxic properties of the smoke are of greater concern.

So, the primary threat at distances away from a fire is smoke and not heat.

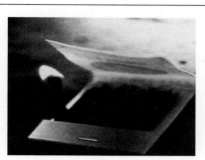

**A safety match**

> **Safety matches**
>
> Need to be kept out of reach of children. Many fires have been started with matches and lighters.

### How do safety matches work?

When a safety match is struck, phosphorus sesquisulphide in the rough "striking" strip reacts with potassium chlorate in the head to give an initial spark. This sets off a solid-state reaction in the match head, which raises its temperature and starts to decompose the organic components to create a cloud of smoke. The sulphur vapour, which has a very low ignition temperature, bursts into flame and sets fire to the cloud of gas, in turn igniting the wood. Otherwise, the match would simply flash like a sparkler.

Smoke from fires contains heat and the products of combustion.

The products of combustion include gases such as $CO$, $CO_2$ and water vapour. $CO$ and $CO_2$ are poisonous gases. Hence smoke is poisonous. The products of combustion are also irritating and cause coughing or eye burning. It is also difficult to see through smoke thereby affecting visibility. Smoke contains particles of unburnt carbon as a result of incomplete combustion.

Smoke is also hot. Because smoke is usually hotter than the surrounding air it tends to rise. So the best way to escape from the smoke from a fire is to crawl on the floor.

To drive away smoke in an enclosed space such as a room, it is necessary to allow entry of fresh air by openings at a lower level so that when the smoke rises, it draws the fresh air, and cools and gets diluted. It is also necessary to consider providing ventilators at the top level, during building designs, to enable the hot smoke to escape from the top.

*If you play with fire, you get burnt–proverb*

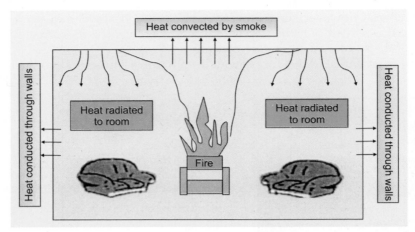

**Fig. 1.2:** Compartmental fire.

## Effect of Carbon Monoxide

Occupants and fire-fighting personnel should not be exposed to concentrations of carbon monoxide (CO) in excess of 55 parts per million (ppm) or 55 mg/m$^3$.

Carbon monoxide is particularly dangerous in that it is odourless, colourless, and tasteless. Its effects are cumulative; doses that may be tolerable by individuals over brief periods may prove to be dangerous to them when repeated or prolonged over several hours. Carbon monoxide combines with blood to form carboxyhemoglobin (COHb). CO accumulates rapidly in the blood; however, the body is extremely slow in reducing the COHb level which may account for its toxic action. Maximum COHb levels have been set at 5% for all system design objectives and aviation system performance limits and at 10% for all other system performance.

> *Smouldering fires–more smoke; flaming fires–more heat*

## Smoke Toxicity

A measurement used to estimate the toxicity of a smoke from a fire is L (CL) 50, which is the concentration-time product required for death to occur in 50% of the animals exposed to the smoke.

*Movement of smoke:* Smoke like all fluids moves wherever there is difference in pressure between two points. The most obvious force which acts on the smoke is the buoyancy created by heat which makes it travel upwards. The smoke within a building also travels to other parts of the building due to air currents caused by air conditioner systems, operating fans or due to temperature difference between the inside and outside areas of the building.

> *We do not inherit this world from our ancestors; we borrow it from our children–*
> *sustainable development thought*

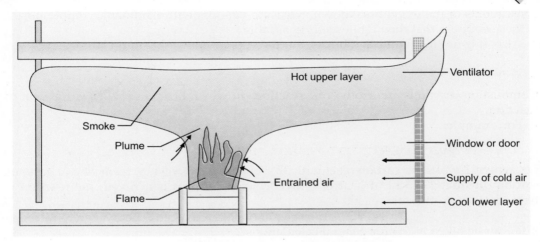

**Fig. 1.3:** Ventilation in a compartmental fire.

## Burning of plywood
### (As also some textiles, foams, etc.)

Burning of plywood releases formaldehyde.

Formaldehyde, used in the adhesives required for manufacturing of plywood, produces a colourless, pungent smelling gas which can (when more than 0.1 ppm) cause watery eyes, burning sensation in the throat and eye and nausea. Higher concentrations can trigger asthma attacks, can also cause cancer and skin diseases.

Burning of wood, as reported by Douglas Drysdale (An Introduction to Fire Dynamics), requires heating, by radiation to 600° C for it to ignite spontaneously. In case the heating is by conduction, a temperature of 490° C is sufficient for the spontaneous ignition to take place.

Also, given the same ignition source, natural teak wood takes longer time to ignite than plywood.

## COMBUSTION

Fire means combustion, which may be defined as a chemical reaction involving rapid oxidation accompanied by the evolution of light and heat.

Combustion is a continuous process, requiring continuous supply of new fuel and new oxidizer in the flame zone. In addition, the new fuel and new oxidizer must be brought up to ignition temperature before they react by high speed collision of their molecules to start the chain reaction between their free radicals which bring about the release of available energy mainly in the form of heat. This heat is fed back from the flame zone and the hot products of combustion to the new supplier of fuel.

Flame is actually a vapour phase chemical reaction between fuel vapour and oxygen. Even solids do not burn as such. In fact, when heat is applied to wood, the chemical constituents of the wood breakdown, releasing vapours with flammable compounds. It is these vapours that then ignite to form a flame.

Also, it is not the liquid that burns, but the vapour above the liquid surface.

## Combustion Reactions

Combustion reactions are exothermic reactions in which heat is evolved as the heat of reaction.

For example:

$$C \text{ (solid)} + O_2 \text{ (gas)} \longrightarrow CO_2 \text{ (gas)} + \text{(heat)}$$

i.e., when 12 g of solid carbon react with 32 g of gaseous oxygen to form 44 g of gaseous carbon dioxide, 393.5 kJ of heat is evolved. Since net heat is evolved, the reaction is termed as an exothermic reaction.

*Composition of air:* Air is 79.2% nitrogen and 20.8% oxygen. The oxygen to nitrogen ratio can be taken as 1:4 for general calculations.

*Combustion reactions:* The combustion reactions at normal temperatures and pressures are given below.

*Combustion of carbon:* Carbon is one of the principal constituents of most of the fuels such as wood, petroleum products, etc.

$$C + O_2 \longrightarrow CO_2 + 94 \text{ kcal of heat}$$

with air, without air

$$2C + O_2 \longrightarrow 2CO + 52.8 \text{ kcal of heat}$$

CO can further burn as

$$CO + \tfrac{1}{2} O_2 \longrightarrow CO_2 + 67.6 \text{ kcal of heat}$$

*Combustion of hydrogen:* In most of the combustible materials, hydrogen is associated with carbon. In the combustion reaction, hydrogen produces water (steam).

$$H_2 + \tfrac{1}{2} O_2 \longrightarrow H_2O + 68.4 \text{ kcal of heat}$$

If the steam is not condensed,

$$H_2 + \tfrac{1}{2} O_2 \longrightarrow H_2O + 57.7 \text{ kcal of heat}$$

## FLAMES

By flames we mean the products of combustion which are hot. The heat in the fire causes the fuel to decompose into simpler components. This process is known as pyrolysis. The pyrolysis process generates some sort of molecules. These molecules are known as free radicals. They combine with oxygen to generate combustion products that escape. The reaction also generates light and heat. The light is visible and appears as flame. There are a large number of free radicals in a flame.

*Forewarned is forearmed*

## Types of Flames

### Premix Flames

When the fuel is deliberately mixed with the air before combustion takes place, the flame usually burns very efficiently as there is plenty of air within the flame itself. This is called a premixed flame. This type of flame is very hot and with very little soot, e.g. a gas stove flame.

### Diffused Flames

In this fire, the combustible vapour is generated by the hot fuel and has to mix with the oxygen present in the air. This type of flame is not as efficient and the flame may be very sooty and not as hot as a premixed flame. The oxygen and the fuel must diffuse together in the flame.

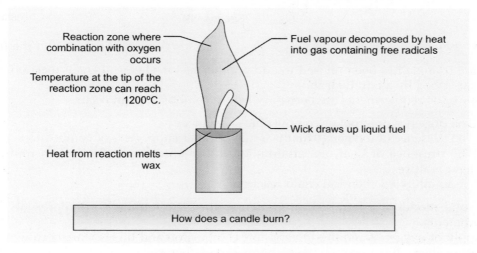

Fig. 1.4: Burning of a candle.

The heat and light in the flame are generated in the reaction zone. The flame appears yellowish if there are a large number of soot particles, which get hot and radiate heat. When the flame cools, it becomes smoke, while the carbon particles become soot. The flame appears blue, if there is little or no soot. Flames in premixed fires are blue with very little soot.

In diffusion flames, the rate of burning is determined by the rate of mixing of the fuel and oxygen, which is controlled by the degree of ventilation. In enclosed spaces such as a room, the ventilation depends on the room configuration. This aspect is very important in the design of buildings. Solid and liquid fuels produce diffused flames as the mixing occurs only during combustion when the fuel gives off flammable vapours.

In premixed flames, ventilation does not impose a restriction on the rate of burning as the oxygen is already mixed with the fuel. Hence premixed flames burn faster, which is the case with gaseous flames.

## THE FIRE TRIANGLE

Fire results from combination of fuel, heat and oxygen which in combination is called fire triangle.

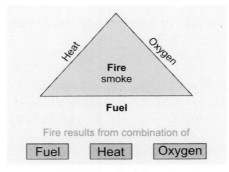

**Fig. 1.5:** Fire triangle.

## FIRE TETRAHEDRON

The fire triangle has been revised to add another factor–chemical chain reaction, to be then described by a tetrahedron.

Four basic requirements are necessary before combustion can occur.

- The presence of a fuel or combustible substance.
- The presence of oxygen (usually as air) or other supporters of combustion.
- The presence of heat, i.e. attainment and maintenance of a certain minimum temperature.
- An uninhibited chemical chain reaction.

A solid model of a tetrahedron known as the "fire tetrahedron" represents these requirements.

Supply of oxygen is required to support combustion and this is usually drawn from the atmosphere. But in some cases, the materials themselves (e.g. explosives) contain sufficient oxygen to maintain combustion.

Most solids and liquids require to be heated above their normal temperature before they can emit flammable vapours, though some liquids (e.g. petrol or methylated spirit) emit flammable vapours at normal temperatures.

Fire risk of a fuel is mainly dependent on its volatility or tendency to vapourize and so develop an explosive or combustible mixture with air. The flash point is usually taken as a guide to volatility. Flash points are usually taken as a measure of fire safety in storage, handling and transport of all the flammable liquids.

### Explanation of the Fire Tetrahedron

*Fuel*

A fuel is any substance that can undergo combustion. Anything that burns can act as a fuel. The majority of fuels encountered are organic and contain carbon and combinations of hydrogen and oxygen in varying ratios. In some cases, nitrogen will be present;

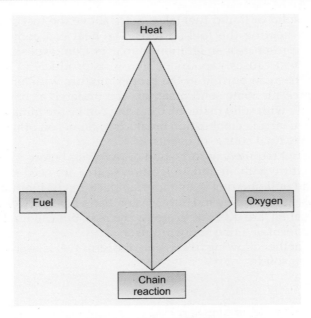

**Fig. 1.6:** Fire tetrahedron.

*A tetrahedron is a regular polyhedron with four faces; each face is a triangle with the three meeting at each vertex. The tetrahedron has 4 vertices and 6 edges. The tetrahedron is also called a triangular pyramid.*

examples include wood, plastics, gasoline, alcohol, and natural gas. Inorganic fuels contain no carbon and include combustible metals, such as magnesium or sodium.

All matter can exist in one of three phases—solid, liquid, or gas. The phase of a given material depends on the temperature and pressure and can change as conditions vary. If cold enough, carbon dioxide, for example, can exist as a solid (dry ice). The normal phase of a material is that which exists at standard conditions of temperature [21° C] and pressure [101.6 kPa or 1 atmosphere at sea level].

Fuels exist in many different forms. Some are gases, e.g. methane and propane. Some are liquids, e.g. petrol and kerosene, which are burnt in engines and some are solids, e.g. wood and coal, which are burnt in the fireplace. Many plastics, paper and cotton are also fuels. So nearly everything can burn, given ignition, and sufficient oxygen and combustion support.

Gases burn more easily because the gas and oxygen can mix easily and the heat generated encourages further burning. For a liquid to burn, the heat from the fire must boil off some of the fuel so that it becomes a vapour and can mix with the oxygen in the air. For a solid to burn, the heat from the fire has to decompose the fuel into vapours that can mix with oxygen in the air.

*FOHC–fuel, oxygen, heat, chain reaction*

Combustion of a solid or liquid fuel takes place above the fuel surface in a region of vapours created by heating the fuel surface. The heat can come from the ambient conditions, from the presence of an ignition source, or from exposure to an existing fire. The application of heat causes vapours or pyrolysis products to be released into the atmosphere where they can burn, if in the proper mixture with air, and if a competent ignition source is present. Some solid materials can undergo a charring reaction where oxygen reacts directly with solid material. Charring can be the initial or the final stage of burning. Sometimes charring combustion breaks into flame; on other occasions charring continues through the total course of events.

Gaseous fuels do not require vapourization or pyrolysis before combustion can occur. Only the proper mixture with air and an ignition source are needed.

The form of a solid or liquid fuel is an important factor in its ignition and burning rate, e.g., a fine wood dust ignites easier and burns faster than a block of wood. Some flammable liquids, such as diesel oil, are difficult to ignite in a pool but can ignite readily and burn rapidly when in the form of a fine spray or mist.

For the purposes of the following discussion, the term fuel is used to describe vapours and gases rather than solids.

## FUEL LOAD OR FIRE LOAD

Calorific value of a fuel is the total heat released by combustion of unit weight of the fuel.

### Calorific Value of Fuel Mixtures

The calorific value of fuel mixtures (e.g. natural gas) can be calculated from the calorific values of the constituent gases.

Natural gas composition is–$CH_4$–95.9%, $C_2H_6$–3.8%, $N_2$–0.3%.

If the gross calorific value of $CH_4$ is 9530 kcal/$Nm^3$, of $C_2H_6$ is 16860 kcal/$Nm^3$, (nitrogen is not combustible and hence need not be considered).

Natural gas calorific value = (0.959 × 9530) + (0.0038 × 16860)
= 9819 kcal/$Nm^3$.

### Fire Load

Fire load is the amount of heat in kilocalories liberated per sq.m of floor area by the combustion of the contents and any combustible parts of the building itself.

Therefore, fire load is given by the formula:

Weight of all contents × average calorific value of the contents/floor area.

*Temperatures of fumes:* The temperature of fumes originating from a burning fuel can be calculated from the empirical relation.

$$T_f = T_o + (PCI/(V_f \times C_m))$$

Where $T_f$–theoretical temperature of the flame or fumes,° C
$T_o$–initial temperature of the fuel,° C
$V_f$–volume of fumes, $Nm^3$
$PCI$–lower calorific value of the fuel–kcal/$Nm^3$
$C_m$–Specific heat of the fumes.

For example, for hydrogen, burning in air,

1 volume of hydrogen gives 3 volumes of fumes.
The calorific value of hydrogen is 2610 kcal/Nm$^3$.
If $T_o$ is 0° C, and $C_m = 0.429$ kcal/Nm$^3$

$$T_f = 0 + 2620/(3 \times 0.429) = 2025° \text{ C}.$$

In actual combustion conditions in a fire, due to incomplete combustion and radiation losses, the fume temperature will somewhat lower, which may be taken as 70–80% of the above.

## OXIDIZING AGENT

In most fire situations, the oxidizing agent is the oxygen in the earth's atmosphere. Fire can occur in the absence of atmospheric oxygen when fuels are mixed with chemical oxidizers. Many chemical oxidizers contain readily released oxygen. Ammonium nitrate fertilizer ($NH_4NO_3$), potassium nitrate ($KNO_3$), and hydrogen peroxide ($H_2O_2$) are examples.

Anaerobic combustion of ammonium nitrate –

$$NH_4NO_3 \longrightarrow N_2O + 2H_2O$$

Normal air contains 21% oxygen. Oxygen is a colourless gas. (Oxygen is essential to human life and to most plants and animal lives as well.) In oxygen-enriched atmospheres, such as in areas where medical oxygen is in use or in high-pressure diving or medical chambers, combustion is greatly accelerated. Materials that resist ignition or burn slowly in air can burn vigorously when additional oxygen is present. Combustion can be initiated in atmospheres containing very low percentages of oxygen, depending on the fuel involved. As the temperature of the environment increases, the oxygen requirements are further reduced. While flaming combustion can occur at concentrations as low as 14 to 16% oxygen in air at room temperatures of 21° C, flaming combustion can continue at close to 0% oxygen under post-flashover temperature conditions. Also, smouldering combustion once initiated can continue in a low-oxygen environment even when the surrounding environment is at a relatively low temperature. The hotter the environment, the less oxygen is required. This later condition is why wood and other materials can continue to be consumed even though the fire is in a closed compartment with low oxygen content. Fuels that are enveloped in a layer of hot, oxygen-depleted combustion products in the upper portion of a room can also be consumed.

It should be noted that certain gases can form flammable mixtures in atmospheres other than air or oxygen. One example is a mixture of hydrogen and chlorine gas.

For combustion to take place, the fuel vapour or gas and the oxidizer should be mixed in the correct ratio. In the case of solids and liquids, the pyrolysis products or vapours disperse from the fuel surface and mix with the air. As the distance from the fuel source increases, the concentration of the vapours and pyrolysis products decreases. The same process acts to reduce the concentration of a gas as the distance from the source increases.

Fuel burns only when the fuel/air ratio is within certain limits known as the flammable (explosive) limits. In cases where fuels can form flammable mixtures with air, there is a minimum concentration of vapour in air below which propagation of flame does not occur. This is called the lower flammable limit. There is also a maximum concentration above which flame will not propagate called the upper flammable limit. These limits are generally expressed in terms of percentage by volume of vapour or gas in air.

The flammable limits reported are usually corrected to a temperature of 0° C and 1 atmosphere. Increases in temperature and pressure result in reduced lower flammable limits possibly below 1% and increased upper flammable limits. Upper limits for some fuels can approach 100% at high temperatures. A decrease in temperature and pressure will have the opposite effect. Caution should be exercised when using the values for flammability limits found in the literature. The reported values are often based on a single experimental apparatus that does not necessarily account for conditions found in practice.

The range of mixtures between the lower and upper limits is called the flammable (explosive) range. For example, the lower limit of flammability of gasoline at ordinary temperatures and pressures is 1.4%, and the upper limit is 7.6%. All concentrations by volume falling between 1.4 and 7.6% will be in the flammable (explosive) range. All other factors being equal, the wider the flammable range, the greater the likelihood of the mixture coming in contact with an ignition source and thus the greater the hazard of the fuel. Acetylene, with a flammable range between 2.5 and 100%, and hydrogen, with a range from 4 to 75%, are considered very dangerous and very likely to be ignited when released.

Every fuel/air mixture has an optimum ratio at which point the combustion will be most efficient. This occurs at or near the mixture known as the Stoichiometric ratio. When the amount of air is in balance with the amount of fuel (i.e. after burning there is neither unused fuel nor unused air), the burning is referred to as Stoichiometric. This condition rarely occurs in fires except in certain types of gas fires.

| Flammability limits of some common fuels with air | | |
|---|---|---|
| *Fuel gas* | *Lower limit (%)* | *Upper limit (%)* |
| Hydrogen | 4.0 | 75.0 |
| CO | 12.5 | 74.0 |
| Methane | 5.3 | 14.9 |
| Ethane | 3.0 | 12.5 |
| Propane | 2.2 | 9.5 |
| Butane | 1.9 | 8.5 |
| Pentane | 1.5 | 7.8 |
| Ethylene | 3.1 | 32.0 |
| Acetylene | 2.5 | 100.0 |
| Petrol (gasoline) | 1.4 | 7.6 |

The flammability limits get widened when fuel-oxygen mixtures are involved.
The flammability limits for multi-component fuel mixtures can be predicted approximately from Le Chatelier's rule.

Fires usually have either an excess of air or an excess of fuel. When there is an excess of air, the fire is considered to be fuel controlled. When there is more fuel present than air, a condition that occurs frequently in well-developed room or compartment fires, the fire is considered to be ventilation controlled.

In a fuel-controlled compartment fire, all the burning will take place within the compartment and the products of combustion will be much the same as burning the same material in the open. In a ventilation-controlled compartment fire, the combustion inside the compartment will be incomplete. The burning rate will be limited by the amount of air entering the compartment. This condition will result in unburnt fuel and other products of incomplete combustion leaving the compartment and spreading to adjacent spaces. Ventilation-controlled fires can produce massive amounts of carbon monoxide.

If the gases immediately vent out of a window or into an area where sufficient oxygen is present, they will ignite and burn when the gases are above their ignition temperatures. If the venting is into an area where the fire has caused the atmosphere to be deficient in oxygen, such as a thick layer of smoke in an adjacent room, it is likely that flame extension in that direction will cease, although the gases can be hot enough to cause charring and extensive heat damage.

## HEAT

The heat component of the tetrahedron represents heat energy above the minimum level necessary to release fuel vapours and cause ignition. Heat is commonly defined in terms of intensity or heating rate (kilowatts) or as the total heat energy received over time (kilojoules). In a fire, heat produces fuel vapours, causes ignition, and promotes fire growth and flame spread by maintaining a continuous cycle of fuel production and ignition. Normally, the heat required is initially supplied by a source of heat and is then provided by the combustion process itself. The amount of heat required to cause ignition depends on the nature of the material. A gas or vapour may be ignited by a small spark, whereas a solid may require a greater heat source.

### Heat Transfer

The transfer of heat is a major factor in fires and has an effect on ignition, growth, spread, decay (reduction in energy output), and extinction. Heat transfer is also responsible for much of the physical evidence used by investigators in attempting to establish a fire's origin and cause.

It is important to distinguish between heat and temperature. Temperature is a measure that expresses the degree of molecular activity of a material compared to the activity at a reference point such as the freezing point of water. Heat is the energy that is needed to maintain or change the temperature of an object.

When heat energy is transferred to an object, the temperature increases. When heat is transferred away, the temperature decreases.

In a fire situation, heat is always transferred from the high-temperature mass to the low-temperature mass. Heat transfer is measured in terms of energy flow per unit of time (kiolwatts). The greater the temperature difference between the objects, the more energy is transferred per unit of time and the higher the heat transfer rate is. Temperature can be compared to the pressure in a fire hose and heat or energy transfer to the waterflow in gallons per minute.

We must therefore understand that heat transfer is accomplished by three mechanisms—conduction, convection, and radiation. When carrying a fire investigation or analyzing a fire, the role of the three modes of heat transfer, and an understanding of each is necessary.

## Conduction

The transfer of heat through a solid, like the heat we feel when we touch the outside of a hot stove. Conduction is the form of heat transfer that takes place within solids when one portion of an object is heated. Energy is transferred from the heated area to the unheated area at a rate dependent on the difference in temperature and the physical properties of the material. The properties are the thermal conductivity, the density, and the heat capacity. The heat capacity (specific heat) of a material is a measure of the amount of heat necessary to raise its temperature (kcal/kg/degree of temperature rise).

If thermal conductivity is high, the rate of heat transfer through the material is high. Metals have high thermal conductivities, while plastics or glass have low thermal conductivity values. Other properties being equal, high-density materials conduct heat faster than low-density materials. This is why low-density materials make good insulators. Similarly, materials with a high heat capacity require more energy to raise the temperature than materials with low heat capacity values.

Generally, conduction heat transfer is considered between two points with the energy source at a constant temperature. The other point will increase to some steady temperature lower than that of the source. This condition is known as steady state. Once steady state is reached, thermal conductivity is the dominant heat transfer property. In the growing stages of a fire, temperatures are continuously changing, resulting in changing rates of heat transfer. During this period, all three properties, thermal conductivity, density, and heat capacity play a role. Taken together, these properties are commonly called the thermal inertia of a material.

The impact of the thermal inertia on the rise in temperature in a space or on the material in it, is not constant throughout the duration of a fire. Eventually, as the materials involved reach a constant temperature, the effects of density and heat capacity become insignificant relative to thermal conductivity. Therefore, thermal inertia of a material is most important at the initiation and early stages of a fire (pre-flashover).

Conduction of heat into a material, as it affects its surface temperature, is an important aspect of ignition. Thermal inertia is an important factor in how fast the surface temperature will rise. The lower the thermal inertia of the material, the faster the surface temperature will rise. Conduction is also a mechanism of fire spread. Heat conducted through a metal wall or along a pipe or metal beam can cause ignition of combustibles in contact with the heated metals. Conduction through metal fasteners such as nails, nail plates, or bolts can result in fire spread or structural failure.

## Convection

Convection is the transfer of heat by moving particles of liquids or gases, like the heat that flows from hot water to the rice grains when we boil them or the heat that flows out of the kettle in a flow of steam.

Convection is the transfer of heat energy by the movement of heated liquids or gases from the source of heat to a cooler part of the environment.

Heat is transferred by convection to a solid when hot gases pass over cooler surfaces. The rate of heat transfer to the solid is a function of the temperature difference, the surface area exposed to the hot gas, and the velocity of the hot gas. The higher the velocity of the gas, the greater the rate of convective transfer.

In the early history of a fire, convection plays a major role in moving the hot gases from the fire to the upper portions of the room of origin and throughout the building. As the room temperatures rise with the approach of flashover, convection continues, but the role of radiation increases rapidly and becomes the dominant heat transfer mechanism. Even after flashover, convection can be an important mechanism in the spread of smoke, hot gases, and unburned fuels throughout a building. This can spread the fire or toxic or damaging products of combustion to remote areas.

### Radiation

Radiation is the transfer of heat by infrared electromagnetic radiation—heat we can feel without touching, like the heat from the sun or from a room heater.

Radiation is the transfer of heat energy from a hot surface to a cooler surface by electromagnetic waves without an intervening medium, e.g. the heat energy from the sun is radiated to earth through the vacuum of space. Radiant energy can be transferred only by line-of-sight and will be reduced or blocked by intervening materials. Intervening materials do not necessarily block all radiant heat, e.g. radiant heat is reduced on the order of 50% by some glazing materials.

The rate of radiant heat transfer is strongly related to a difference in the fourth power of the absolute temperature of the radiator and the target. At high temperatures, small increases in the temperature difference result in a massive increase in the radiant energy transfer. Doubling the absolute temperature of the hotter item without changing the temperature of the colder item results in a 16-fold increase in radiation between the two objects.

The rate of heat transfer is also strongly affected by the distance between the radiator and the target. As the distance increases, the amount of energy falling on a unit of area falls off in a manner that is related to both the size of the radiating source and the distance to the target.

## UNINHIBITED CHEMICAL CHAIN REACTION

Combustion is a complex set of chemical reactions that results in the rapid oxidation of a fuel producing heat, light, and a variety of chemical by-products. Slow oxidation, such as rust or the yellowing of newspaper, produces heat so slowly that combustion does not occur. Self-sustained combustion occurs when sufficient excess heat from the exothermic reaction radiates back to the fuel to produce vapours and cause ignition in the absence of the original ignition source.

Flame reactions involve radicals and free atoms, which participate in chain reactions. The chain reaction theory states that an exothermic reaction is the result of a chain of reactions involving reactive species. The start of this chain reaction is by the presence of certain atoms, which are produced by chain-initiating reactions. These atoms then become chain carriers. The reaction of a chain carrier may lead to the formation of new chain carriers (chain branching) and this tends to accelerate the reaction rate.

## Inhibiting

Checking or stopping the exothermic reaction of substances, which contain oxygen within themselves is known as inhibiting. Certain materials have oxygen in themselves. Such materials when on fire, cannot be extinguished by smothering as they give out oxygen from themselves, e.g. gunpowder, TNT, RDX, etc. Such fires are due to self-sustaining, continuous heat producing (exothermic) chemical reactions. Such fires can be extinguished only by breaking the chemical chain reaction. This type of extinguishing method is called inhibiting.

Combustion of solids can occur by two mechanisms—flaming and smouldering. Flaming combustion takes place in the gas or vapour phase of a fuel. With solid and liquid fuels, this is above the surface. Smouldering is a surface-burning phenomenon with solid fuels and involves a lower rate of heat release and no visible flame. Smouldering fires frequently make a transition to flaming after sufficient total energy has been produced or when airflow is present to speed up the combustion rate.

## Rate of Combustion

Rate of combustion varies from substance to substance and is described as slow, rapid or spontaneous.

1. *Slow combustion:* A chemical reaction accompanied by slow evolution of heat, but not light, is slow combustion, e.g. cotton-waste burning in an ill-ventilated space.
2. *Rapid combustion:* A chemical reaction accompanied by rapid evolution of heat and in many cases by an appreciable amount of light is called rapid combustion, e.g. petroleum products.
3. *Spontaneous combustion:* It is combustion occurring as a result of heat by the absorption of atmospheric oxygen at ambient temperature without the application of external heat, provided a supporter of combustion is present, e.g. phosphorus in contact with wood and sawdust when exposed to a steam pipe.

Flammable liquids–have a flash point of less than 37.8° C

Combustible liquids—have a flash point of 37.8° C or greater.

**Fig. 1.7:** Flame.

## Types of Combustion

Fire combustion can be of two types:

* Flame combustion (volume burning).
* Surface burning.

### Flame Combustion

Flame combustion is the direct burning of a gaseous fuel. Flame is a relatively fast exothermic gas-phase reaction distinguished by its ability to propogate subsonically through space. It is usually, but not exclusively, accompanied by the emission of visible light. These are open flames. The rate of burning is usually high and a high temperature is produced.

Flames can be of two types—premixed flames and diffusion flames.

* *Premixed flames:* The reactants are mixed before entering the reaction zone. This condition exists in a gas burner or stove or in a laboratory Bunsen-burner and is relatively controlled.
* *Diffusion flames:* The air for combustion is supplied by diffusion from the surrounding air. This refers to gases burning on mixed vapour and air. Such burning is difficult to control.

### Surface Burning

Fire occurring on the surface of a solid fuel. This is also sometimes called as a glow or deep-embered seated fire. However, this fire also takes place at the same temperature as open flames.

Always think before you bite!

**Fig. 1.8:** While fighting a fire—"think before acting".

*Prevention is better than cure*

## THEORY OF FIRE EXTINGUISHMENT

Fires can be suppressed by controlling or removing one or more sides of the fire tetrahedron.

### Methods of Fire Extinction

Fire extinction consists in the limitation of one or more of the factors mentioned in the fire tetrahedron. Fires can be prevented or suppressed by controlling or removing one or more sides of the tetrahedron. Methods of fire extinction may, therefore, be conveniently classified under the following headings:

- Starvation or limitation of fuel.
- Smothering or limitation of oxygen or blanketing.
- Cooling or limitation of temperature.
- Stopping the chemical chain reaction (also known as chemical intervention or inhibition).

### Starvation (Removal of Fuel)

Means to deprive the burning fuel of more fuel, whereby once the fuel is exhausted, the fire will automatically die out.

a. By removing combustible material from the neighbourhood of the fire.
b. By removing the fire from the neighbourhood of combustible material.
c. By sub-dividing the burning material.

### Smothering (Removal of Oxygen/Suffocation)

Suffocation means to deprive the burning fuel of oxygen which is the main supporter of combustion. Absence of oxygen will not allow combustion of any sort. (If the burning process is slow, without any flames, it can be termed as smouldering). If the oxygen content of the atmosphere in the immediate neighbourhood of the burning material can be sufficiently reduced, (say less than 15%), combustion will cease. Following are the methods of smothering.

a. Application of viscous coating on burning material, i.e. use of foam.
b. Application of cloud of finely divided particles of dry powder from a pressurizer.
c. Application of an inert gas, steam or vapourizing liquids from a fire extinguisher or extinguishing systems.

### Cooling (Removal of Heat)

Cooling means reducing the temperature of the fuel below its ignition point. For cooling principle in fire extinction, water is the one most commonly employed as it is one of the best extinguishing media available in large quantity at a very low cost.

### Stopping the Chemical Chain Reaction

Loss of chain carriers has a quenching effect on the fire. Use of inhibitors reduces reaction rates by combining with and thereby destroying the chain carriers. Such methods are effective on only solid fires as in the case of a gaseous fuel fire, the radicals required by the chain reaction are continuously supplied by diffusion from areas of high concentration of the radicals.

## DEVELOPMENT OF FIRE

### Four Stages of Fire

1. *Incipient stage:* Invisible products of combustion are given off, no visible smoke, flame or heat is still not present.
2. *Smouldering stage:* If the burning process is slow, without any flames, it can be termed as smouldering. Combustion products are now visible as smoke. Flame or heat is still not present.
3. *Flame stage:* Actual fire now exists. Appreciable heat is not present, but follows almost instantaneously.
4. *Uncontrolled stage:* (Heat stage).

At this point, large amounts of heat, and smoke is produced. There is uncontrolled heat and the fire is rapidly expanding in space.

### *Time Available for Fighting a Fire*

- Class A — stages 1 and 2 — 0 to 10 minutes
- Class B — stages 1 and 2 — 0 to 1 minutes
- Class C — stages 1 and 2 — 0 to 30 seconds

All fires start with an ignition. For ignition to occur, all of the following conditions must exist simultaneously:

- Sufficient heat must be available to provide the required energy for the chemical reaction to start.
- There has to be enough fuel vapour in the air, not too much, not too little.
- There has to be sufficient air (oxygen).

If there is insufficient energy, the molecules do not react with each other. The energy to cause ignition might be in the form of the heat from a match or a spark or a lighted cigarette.

The energy makes the molecules of fuel and that of oxygen nearby, to move faster. If they do not move fast enough, they will bounce off each other when they collide. If the speed is sufficient, the molecules strike and start off the combustion reaction.

All fuels require different amounts of energy to cause an ignition. Some are easier to ignite than others. It is also easier to ignite a fuel if it has a lot of surface area. A pile of wooden sticks is easier to ignite than a wooden log. The sticks have much more surface area than the log.

Once started, a fire becomes a self-sustaining "heat engine". A fire is the burning of vapourized fuel, which when burnt produces heat, which in turn converts more fuel to vapour, thus continuing the burning cycle.

An incipient fire produces heat which is then radiated to the fuel surface to vapourize more of the fuel to keep up the fire. The fire also draws in oxygen (air) from the surroundings to complete the combustion reaction. Volatile flammable liquids require very little energy to vapourize and hence can develop very fast.

### Growth of Fires

A fire spreads by the transfer of heat from the fire source to the surroundings. Most of the heat transfer is by convection and radiation. At lower temperatures in the affected space, convection plays a major role, with the heat being carried by the flames, which being of lesser density, tend to move upward. This heats up the roof (in an interior fire) and from the roof, the heat is radiated back to the floor.

In case of fires of spilled material and bulk material, heat is conducted from the ignition point to the rest of the material and the spread of the fire is very rapid.

The rate of growth of a fire depends on the fuel, which is burning, and other conditions such as supply of air to the fire. Fires in liquids and gases can grow very fast. The size of a fire is measured by its heat release rate.

Practically every fire can be extinguished within the first few minutes, if proper equipment is brought into service by trained persons. The essentials are the right kind of equipment, kept in good condition, in the right place, with persons trained to use the equipment and to bring it into service immediately after the fire is discovered.

A fire which is allowed to spread creates panic amongst people, and panic, in many cases, has caused more loss of life than the fire itself (Fig. 1.10).

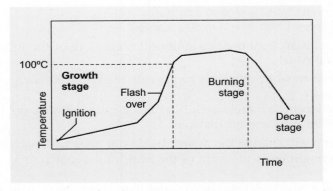

**Fig. 1.9:** Growth of fire.

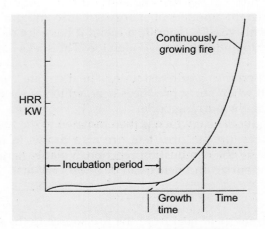

**Fig. 1.10:** Growth and HRR of a fire.

## Test for Growth of an Interior Fire

In the test conducted, combustible material in a chamber is ignited and the fire spread and temperature rise are monitored. The exposure lasts for 2 hours. The temperature rises to 1000° C after a few minutes and peaks at 1350° C after one hour. This test is depictive of the rise of temperature in a fire in enclosed spaces such as residences, storage rooms, tunnels, etc. The temperature–time data is given in Fig. 1.11.

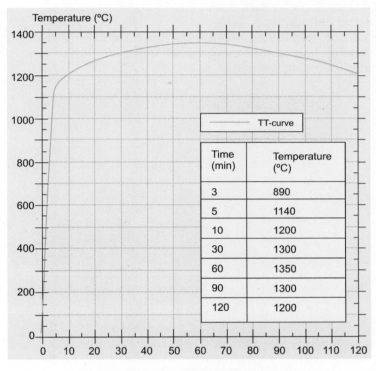

| Time (min) | Temperature (°C) |
|------------|------------------|
| 3 | 890 |
| 5 | 1140 |
| 10 | 1200 |
| 30 | 1300 |
| 60 | 1350 |
| 90 | 1300 |
| 120 | 1200 |

**Fig. 1.11:** Time–temperature curve of a fire.

### Petrol

A mixture of more than 200 volatile hydrocarbons in the range of C4 to C12, suitable for use in spark ignited internal combustion engine. Regular automotive petrol has a flash point of –40° C.

*The great fire of London happened in 1966*

### Reasons for Major Spread of Fire

- Delayed discovery of fire
- Missing, damaged or uncharged fire fighting equipment.
- Lack of fire separation and compartmentalization.
- Lack of effective fixed fire protection systems.
- Leaks or spillage of oils, greases, hydraulic fluid and fuels.
- Sub-standard level of house-keeping.
- Restricted or difficult access for fire fighters and fire engines.
- Unmanned or unmonitored areas.
- Overstocking or over storing of products, raw materials or engineering equipment.
- Lack of suitably trained personnel available to deal with the fire in the vital first few minutes following ignition.

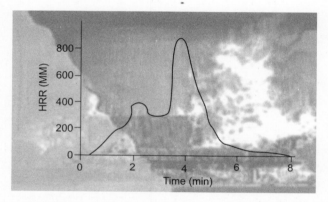

**Fig. 1.12:** HRR in an interior environment with burning upholstery.

### Heat Release Rate

Heat release rate (HRR) is the primary fire hazard indicator. The rate of heat release, especially the peak amount, is the primary characteristic determining the size, growth, and suppression requirements of a fire environment.

### TYPES OF FIRES

Fire types based on the size and the manner in which the fire spreads are—pool fires, jet fires, flash fires, fire balls, and anaerobic fires.

Pool fires and jet fires are less likely to give rise to consequences which will affect people over considerable distances. Flash fires and fire balls are fires which spread instantaneously and therefore cause extreme damage to life and property. Anaerobic fires are those that arise from solid substances used as propellants in guns and rockets and other substances as certain nitrated celluloses. The fires they cause are extremely violent and hard to extinguish.

> *Smoke and toxic gases have caused more deaths than the fire itself*

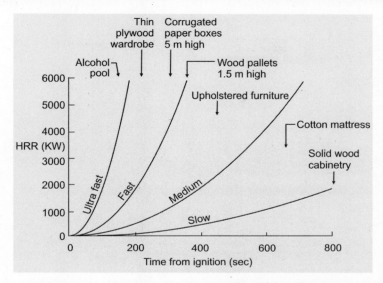

**Fig. 1.13:** Time taken to reach growth stage of different materials in an interior fire.

### Pool Fires

A poolfire may arise from the ignition of a spillage of a flammable liquid on the ground or a fire in an open storage tank. Pool fires will have a height of about 1.75 to 2.5 times the diameter of the spill or tank.

An estimate of the power or heat content of a pool fire = $10^5$ watts (100 kW) per $m^3$ of the flame volume.

### Jet Fires

A combustible material escaping from a container under pressure entraps air to produce, on ignition, a jet flame. As an example, a jet fire can occur when a leak of a flammable gas from a joint in a pipeline gets ignited.

### Flash Fire

A flash fire results from the ignition of a cloud of combustible vapours or dust which has drifted some distance away from the point of release and has become mixed with atmospheric air to form a flammable mixture.

Flash fires can easily propagate back to the source, which then causes a flash over.

*Behold, how great a matter a little fire kindleth —James Bible*

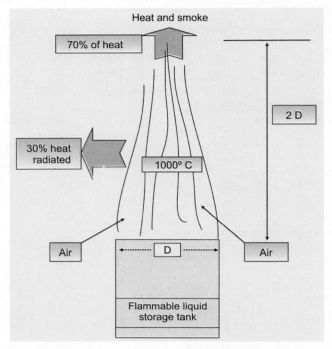

Fig. 1.14: Pool fires.

## Fire Balls

Fire balls are created from the ignition of flammable vapour clouds. Such clouds burn with great intensity, with burn-up rates measured in tonnes of fuel per second.

Fire balls occur when a large quantity of fuel stored under pressure is suddenly released and then gets ignited. Fire balls can rise up in the air in the form of a cloud or a ball.

The radius of a fire ball

$$R = 28 \times m^{0.33} \text{ metres.}$$

Where $m$ = mass of the fuel ignited (tons).

## Boiling Liquid Expanding Vapour Explosion (BLEVE)

A BLEVE can occur when a storage tank containing liquefied vapours receives thermal energy of great intensity. This can happen when a storage tank is subjected to the heat from a fire in its surrounding. The heating-up causes the vapour pressure inside the tank to rise–leading to a rupture of the tank and consequent release and ignition of the flammable liquid. This leads to a large and rapidly expanding "fire ball" or a blast wave.

## Flashover

As a fire builds up in an enclosed space, the smoke produced becomes hotter and it starts radiating more and more heat to the adjacent areas. If the temperature reaches above the ignition temperature of the adjacent lying materials, the materials ignite suddenly and flashover occurs. In enclosed spaces, the whole space starts to burn and becomes very hot. Flashovers occur when the temperature of the hot smoky gases reaches about 600° C. This is a very hot temperature. After flash over, the temperature will suddenly rise to above 1000° C. Such fires are very difficult to bring under control. When a fire reaches the flashover stage in an interior fire, fighting the fire with portable fire extinguishers or even with the limited resources of a fire brigade is futile and many a time, the fire fighting may be called off, as the situation can be endangering to the fire fighting personnel.

## EXPLOSIONS

Explosions involving fires are in fact extremely fast fires. Explosions, such as a pressure vessel explosion, not involving fires are known as cold fires.

Explosion means a rapid or violent process in which a large amount of energy is released suddenly.

Explosions have very great disaster potential and cause great loss of life and damage to property.

**Fig. 1.15:** Explosion.

### National Electric Code (NEC) Explosive Atmosphere Classifications

Certain locations are hazardous because the atmosphere does or may contain gas vapour or dust in explosive quantities. These locations are classified according to the type of explosive agent which may be present.

Class I
Group A–Acetylene
Group B–Butadiene, hydrogen, manufactured gases containing more than 30% hydrogen by volume.
Group C–Acetaldehyde, cyclopropane, ethylene.
Group D–Acetone, ammonia, benzene, butane, gasoline, methane (natural gas), toluene, xylene.

Class II
> Group E–Aluminum, magnesium and other dusts with similar characteristics.
> Group F–Carbon black, coke, coal dust.
> Group G–Flour, starch or grain dust.

Class III
> Easily ignitable such as rayon, cotton and other materials of similar nature.

*Explosive: A chemical that causes a sudden, almost instantaneous release of pressure, gas, and heat when subjected to sudden shock, pressure or high temperature, e.g. dynamite.*

## Gas Explosions

### Example of Hydrogen—Nitrogen Mixture Explosion

A process uses a mixture of hydrogen and nitrogen, which together provide a neutral and reducing atmospheric envelope over the product, which is being processed at a high temperature.

*System features*

- The mixture contains about 10% hydrogen and the balance nitrogen.
- The process is carried on at a high temperature of about 900° C
- The mixture of gases is held in the process space at a pressure of 20–30 pascal.

Considering the flammability limits of hydrogen, has a lower flammability limit of 12%. Therefore, hydrogen forms an explosive mixture at a concentration above 12%.

*Control design:* The control design methodology to prevent the formation of an explosive mixture is:

- In case the nitrogen flow to the system fails, the hydrogen flow should also cut-off immediately.
- Flow limitation is installed to limit hydrogen flow to 12%.

### Minimizing Risk of Explosion

Risk of explosion can be minimized by isolating hazardous substances from heat sources and by using spark arrestors, vents, drains, or other safety techniques.

## COMPRESSED GASES

### Acetylene

Acetylene is a highly flammable hydrocarbon which, with oxygen, produces the hottest flame (3200° C). Acetylene is very unstable and can become dangerously explosive, if compressed above 1 bar in the free state. Acetylene forms an explosive mixture with copper and alloys containing more than 67% copper.

### Hydrogen

The lightest known gas, hydrogen is highly inflammable and burns in air with an almost invisible pale blue flame. Keep flames and sparks away from hydrogen, as with other fuel gases. Do not "crack" the valve of a hydrogen cylinder to blow out dirt, etc. as it could be dangerous.

### Nitrogen and Helium

Nitrogen and helium gases do not present any great hazard. Both are non-flammable and generally considered to be inactive. Chief precaution is to make sure that there is good ventilation whenever they are stored or in confined spaces, otherwise, high concentrations could cause suffocation.

### Oxygen

Oxygen is non-flammable, but in combination with fuel gases (acetylene, hydrogen, propane) and combustible substances, will burn fiercely at great speed. Safety precautions must be strictly adhered to when using oxygen. Carelessness can even lead to serious accidents. Pure oxygen, under pressure, can cause spontaneous combustion when in contact with oil or grease.

Liquid oxygen, too, can be very dangerous if mishandled. For example, if spilled on asphalt paving, the oxygen soaked asphalt may explode if stepped-on.

Oxygen should not be used as a substitute for compressed air to dust off materials or to operate pneumatic tools.

- Oxygen and other flammable gases should be stored separately in well-ventilated areas, away from excessive heat and physical hazards.
- When stored inside buildings, cylinders of oxygen should not be in close proximity to fuel gases.
- Compressed gas cylinders are not designed for temperatures in excess of 55° C.

Grease and oil must be kept away from oxygen cylinders since oxygen under pressure can cause spontaneous combustion when in contact with them.

### Liquid Petroleum Gas (LPG)

LPG is used as a domestic fuel. LPG is considered safe for domestic use as it has a short flammability range, can be easily liquefied, has high calorific value or heat content, and burns with very little smoke. LPG is used at home in portable cylinders. As such extra precautions are necessary in the design and use of LPG.

### *Properties of LPG*

LPG is a mixture of propane and butane. LPG for domestic use is odourized (to different levels) to enable easy detection of gas leaks.

### General

| | |
|---|---|
| Indian standard specification | IS: 4576 |
| Vapour pressure at 65° C kg/cm$^2$ | 16.87 |
| Volatility-evaporation temp (for 95% volume at NTP) | 2° C |
| Copper-strip corrosion at 38° C not worse than | 1 |
| Dryness | No free entrained water |
| Odour | Level 2 |

| Chemical composition | | |
|---|---|---|
| Ethane | % | 1 max |
| Propane | % | 38 max |
| Iso-butane | % | 19 max |
| Nor-butane | % | 41 min |
| Iso-pentane | % | 1 max |
| Volatile pulphur | % | 0.003 max |

*Liquid*

| | |
|---|---|
| Density at 15° C kg/liter | 0.557 |
| Volume of liquid per kg at 15° C liters | 1.85 |
| Vapour pressure at 15° C bar | 5.3 |
| Gross calorific value kcal/kg | 11840 |
| Net calorific value kcal/kg | 10920 |
| Boiling point at atmospheric pressure ° C | 0 |

*Vapour*

| | |
|---|---|
| Density at 15° C kg/m$^3$ | 2.21 |
| Volume of gas per kg at 15° C m$^3$ | 0.48 |
| Latent heat of vapourization at 15° C kcal/kg | 86 |
| Gross calorific value kcal/Nm$^3$ | 26200 |
| Net calorific value kcal/Nm$^3$ | 24100 |
| Air required for combustion m$^3$/m$^3$ | 29 |

*Safety Precautions in use of LPG Cylinders at Home*

Residential homes are special with respect to the occupancy and the nature of activities carried on therein. Presence of children and the aged, activities such as cooking and festivities, resting and sleeping make a residence multifarious and diverse. A set of minimum safety precautions are to be exercised in using LPG at home, which can result in a safe environment.

- The hose conveying the gas from the cylinder to the stove needs to be replaced periodically (ISI marked).
- The LPG cylinder should not be exposed to heat.
- The regulator should be switched off when not in use.
- The sealing O-ring in the cylinder should be replaced periodically. Soap solution can be used to check for any leaks.
- The cylinder should not be subjected to shocks and vibrations.

*Be prepared–scout motto–Baden Powel*

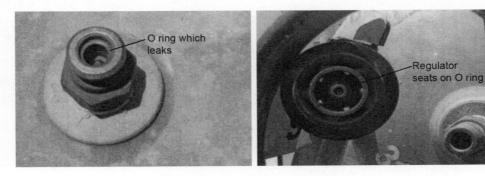

**Fig. 1.16:** LPG cylinder leak source.

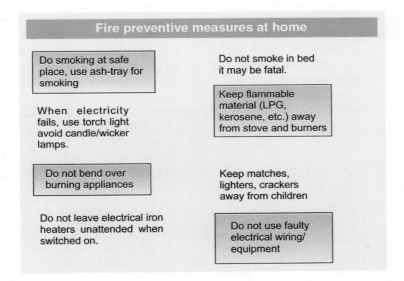

**Fire preventive measures at home**

Do smoking at safe place, use ash-tray for smoking

Do not smoke in bed it may be fatal.

When electricity fails, use torch light avoid candle/wicker lamps.

Keep flammable material (LPG, kerosene, etc.) away from stove and burners

Do not bend over burning appliances

Keep matches, lighters, crackers away from children

Do not leave electrical iron heaters unattended when switched on.

Do not use faulty electrical wiring/ equipment

## CAUSES OF FIRE

It is an established fact that most of the fires are due to human failure to comply with established procedures and failure to recognize safe work practices or carelessness, malicious intent or simple incompetence.

Statistics show that most fires result from a very small number of causes.

Major causes in non-industrial premises are arson, careless use or disposal of cigarettes and matches or electrical source of ignition.

*Spontaneous combustion can occur in damp charcoal/coal heaps*

## Main Ignition Causes

- Faults or misuse of electrical equipment.
- Burning and faulty generating equipment.
- Faulty oil, gas and electrical heating, boiler and drying equipment.
- Smoking, matches, cigarettes, etc.
- Uncontrolled rubbish burning.
- Hot products or waste products.
- Friction generation–bearings, conveyors, drive belts.
- Static electricity.
- Spontaneous chemical ignition.
- Arson and deliberate ignition.

## IGNITION

Ignition causes initiation of combustion. Ignition is a source of intense heat which provides the heat necessary for the fire-combustion to be initiated. For a fire to start at temperatures below the auto-ignition temperature of the combustible material, ignition is necessary.

In order for most materials to be ignited, they should be in a gaseous or vapour state. A few materials may burn directly in a solid state or glowing form of combustion including some forms of carbon and magnesium. These gases or vapours should then be present in the atmosphere in sufficient quantity to form a flammable mixture. Liquids with flash points below ambient temperatures do not require additional heat to produce a flammable mixture.

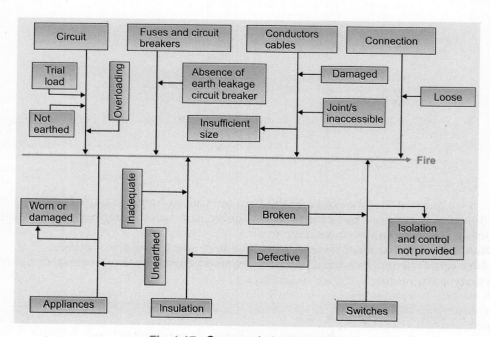

**Fig. 1.17:** Causes of electrical fires.

The fuel vapours produced should then be raised to their ignition temperature. The time and energy required for ignition to occur is a function of the energy of the ignition source, the thermal inertia of the fuel, and the minimum ignition energy required by that fuel and the geometry of the fuel.

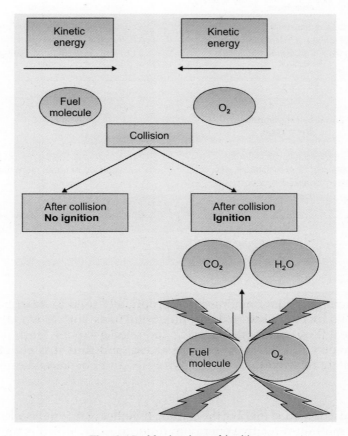

**Fig. 1.18:** Mechanism of ignition.

If the fuel is to reach its ignition temperature, the rate of heat transfer to the fuel should be greater than the conduction of heat into or through the fuel and the losses due to radiation and convection. A few materials, such as cigarettes, upholstered furniture, sawdust, and cellulose insulation, are permeable and readily allow air infiltration. These materials can burn as solid phase combustion, known as smouldering. This is a flameless form of combustion whose principal heat source is char oxidation. Smouldering is hazardous, as it produces more toxic compounds than flaming combustion per unit mass burned, and it provides a chance for flaming combustion from a heat source too weak to directly produce flame.

| Factors developing heat in the electric circuit | |
|---|---|
| **Overload**<br><br>• Melting of fuse wire<br>• Breakdown of insulation<br>(If fuse wire is up-rated by user.) | **Short circuit**<br><br>Damage to insulation at the point of short circuit electrical sparks<br>(120 amp. current at 240 volt. raises the temp. over 1000°C) |
| **Localized resistance in the circuit**<br><br>If the circuit is not securely connected. High resistance path causes localized heating at the poor junctions. | **In line ARC**<br><br>Breakage in the conductor, when current is flowing, causes sparking of temp. more than 1000°C. |

The term smouldering is sometimes inappropriately used to describe a non-flaming response of a solid fuel to an external heat flux. Solid fuels, such as wood, when subjected to a sufficient heat flux, will degrade, gasify, and release vapors. There usually is little or no oxidation involved in this gasification process, and thus it is endothermic. This is more appropriately referred to as forced pyrolysis, and not smouldering.

## Collision Theory

A bimolecular reaction must involve the coming together of two molecules. The molecules must possess a minimum relative energy in order to react, which is represented by the activation energy. The rate of reaction is markedly influenced by the temperature, since a rise in temperature favours the formation of active molecules and also increases the molecular velocity. Therefore, the reaction rate is dependent on:

- The activation energy.
- The temperature.
- The frequency factor, which represents the total frequency of encounters between reactant molecules.

## Ignition of Solid Fuels

For solid fuels to burn with a flame, the substance should either be melted or vapourized (like thermoplastics) or be pyrolyzed into gases or vapours (i.e. wood or thermo set

| Static electricity—the cause of fire |
|---|
| The ability of electrically insulating material to build an accumulation of electric charges on their surfaces, gives rise to the term "static electricity". |
| If surface of the material is not connected to the earth or to an oppositely charged object, the accumulation and density of the charges increases, resulting in a spark when the electrical potential is high. |
| Lightning is also an electrostatic discharge. Several million volts potential between the storm clouds and earth are discharged and current of an average capacity about 20000 amp. Having energy $10^{10}$ joules is present in each flash. |
| The electrostatic discharge and lightning striking combustible material causes ignition and striking human beings causes electrocution and burns. |
| Earthing, bonding, humidification and ionization are the preventive measures. |

plastic). In both examples, heat must be supplied to the fuel to generate the vapours. When solid materials are heated to a high enough temperature, their molecules breakdown to produce flammable vapours that can chemically react with oxygen in the air.

High-density materials of the same generic type (woods, plastics) conduct energy away from the area of the ignition source more rapidly than low-density materials, which act as insulators and allow the energy to remain at the surface. For example, given the same ignition source, oak takes longer to ignite than a soft pine. Low-density foam plastic, on the other hand, ignites more quickly than high-density plastic.

The amount of surface area for a given mass (surface area to mass ratio) also affects the quantity of energy necessary for ignition. It is relatively easy to ignite one kilo of thin wood shavings with a match, while ignition of a one-kilo solid block of wood with the same match is very unlikely.

Because of the higher surface area to mass ratio, corners of combustible materials are more easily burned than flat surfaces.

In the presence of an already existing fire, the fire acts as a pilot flame to ignite any combustible matter in the surrounding. The absence of the pilot flame requires that the fuel vapours of the first item ignited be heated to their auto-ignition temperature. In "An Introduction to Fire Dynamics", Dougal Drysdale reports two temperatures for wood to auto-ignite or spontaneously ignite. These are heating by radiation to above 600° C, and heating by conduction to above 490° C.

For spontaneous ignition to occur as a result of radiative heat transfer, the volatiles released from the surface should be hot enough to produce a flammable mixture above its auto-ignition temperature when it mixes with unheated air. With convective heating on the other hand, the air is already at a high temperature and the volatiles need not be as hot. When exposed to their ignition temperature, thin materials ignite faster than thick materials (e.g. paper vs. plywood).

### Ignition of Liquids

In order for the vapours of a liquid to form an ignitable mixture, the liquid should be at or above its flash point. The flash point of a liquid is the lowest temperature at which it gives off sufficient vapour to support a momentary flame across its surface based on an appropriate ASTM (American Society for Testing and Materials) test method. The value of the flash point may vary depending on the type of test used. Even though most of a liquid may be slightly below its flash point, an ignition source can create a locally heated area sufficient to result in ignition.

Atomized liquids or mists (those having a high surface area to mass ratio) can be more easily ignited than the same liquid in the bulk form. In the case of sprays, ignition can often occur at ambient temperatures below the published flash point of the bulk liquid provided the liquid is heated above its flash point and ignition temperature at the heat source.

### Ignition of Gases

Combustible substances in the gaseous state have extremely low mass and require the least amount of energy for ignition.

Gas leaks are dangerous and potential fire hazards, since they can be ignited very easily. In the case of a fire due to a large gas leak, the gas supply must be brought to a minimum positive pressure or completely cut-off before trying to extinguish the flame. Equipment surrounded by flame can be cooled with water until the flame is completely extinguished.

**Fig. 1.19:** Explosion–proof lamp, explosion–proof plug.

### Precautions Against Ignition

- Where there is a danger of fire or explosion from presence of flammable or explosive substances, electrical equipment such as switches should be of flame proof type.
- Adequate and effective measures are to be taken to prevent accumulation of static charges.

- Safety shoes used in such areas should not have metal nails in the soles, which is likely to cause sparks for ignition.
- Smoking and lighting of matches are prohibited in such areas.

## Spontaneous Ignition and Combustion

Examples of spontaneous ignition and combustion are discussed below.

Certain metals such as sodium (Na), potassium (K), lithium (Li), zirconium (Zr), titanium (Ti) and cesium (Cs) when left exposed to air ignite spontaneously because of their high order of reactivity with moist air. Therefore, sodium and potassium are always kept immersed in kerosene.

Chemicals like yellow phosphorus ignite spontaneously when left exposed to air. Therefore, from a point of view of safety, yellow phosphorus is always kept inside water in a completely close container and away from any heat.

Chemicals like potassium permanganate and glycerin, when they come in contact with each other—first fumes are produced (slow combustion), afterwards they burst into flame (rapid combustion), without application of external heat. It is an example of spontaneous combustion.

Materials like iron sulphide, oil soaked rags, vegetable fibre soaked in oil and coal are subject to slow oxidation reaction, when left exposed to air at normal temperatures. If large amount of these materials are stored in the form of heap and allowed to stand for a long period of time and the heat liberated by the oxidation reaction in the interior of the mass is not allowed to escape, the temperature rise in the interior leads to their ignition and results in spontaneous combustion.

Bacteriological decomposition of organic materials and oxidation of carbonaceous organic materials at ordinary temperature evolve heat, causing a rise in temperature. If the ventilation is inadequate and the heat thus evolved cannot be dissipated away, the rate of decomposition and oxidation increases which result in further increase in temperature. When the temperature reaches the ignition temperature, spontaneous combustion takes place. Haystack fire provides a familiar example of this effect. If the moisture content of the haystack is just right, then a whole range of micro-organisms such as bacteria, yeast, fungi can digest it and produce sufficient heat to raise the temperature of the haystack to its ignition temperature causing spontaneous combustion. The ignition temperature of a haystack is 172° C.

The precautionary measure to prevent spontaneous ignition is to prevent formation of air pockets and provide adequate ventilation in areas where materials susceptible to spontaneous ignition is present.

## ELECTRICAL FIRES

If the statistics of fire causes is examined, electrical sources are seen as the cause of majority of fires. Correct design of electrical systems is therefore very important.

## Causes of Electrical Fires

a. Overloading of cables by currents–cables overheat and the life of insulation is shortened.
b. Short-circuit of conductors (reason, e.g. mechanical damage to insulation.)

c. Leakage of current to earth (e.g. due to failure of cable insulation).
d. Loose connections–resulting in localized overheating.
e. Arcs and sparks from electrical faults.

## Preventing Electrical Fires

Electrical wiring should be regularly checked and any worn or weak wires should be replaced, including cords on appliances.

Cords or wiring should be not run under carpets, where they might become damaged and set the carpet on fire.

Frequently electrical power outlets are overloaded and such abuse can lead to cable damage and consequent fires.

*Protecting cables:* Power cables, when exposed to heat due to the operating conditions or due to heat evolved during a fire, can undergo further damage due to the loss of insulation of the conductors. Electrical conductors then need to be protected thermally.

> **Explosion proof motor:** A totally enclosed motor designed to withstand an internal explosion of specified gases or vapours and not allow the internal flame or explosion to escape. Such motors are required to be used in hazardous locations.

## CLASSIFICATION OF FIRES

The different types of fire risks are classified according to the nature of the fuel. The purpose of classification is to help in deciding upon the suitability of the types of extinguishing agents available for fire fighting.

The comprehensive classification defines 5 types of fires. A more concise definition can classify fires into only A, B, and C categories.

*Class 'A' fire:* Fire due to ordinary combustible materials like wood, textiles, paper, rubbish and the like. Class A fires are the most common. The effective agent to extinguish this type of fire is generally water in the form of a jet or spray.

Light hazard—Occupancies like offices, assembly halls, canteens, restrooms, etc.

Ordinary hazard—Occupancies like saw mills, carpentry shops, book binding shops, small timber yards, engineering workshops, etc.

Extra hazard— Occupancies like large timber yards, godowns storing fibrous materials, flour mills, cotton mills, jute mills, large wood working factories, etc.

*Class 'B' fire:* Fire in flammable liquids or liquefied solids like oils, petroleum products, solvents, grease, paint, etc. This type can be divided into two groups, those that are miscible and immiscible with water. Depending on the types, different extinguishing agents such as water spray, foam, dry chemical powder, carbon dioxide, etc. can be used. For fires involving alcohol, alcohol resistant foam can be used.

*Class 'C' fires:* Fires from flammable gases and for fire involving electrically energized equipment (under NFPA classification) and delicate machinery and the like. This type of fire can be extinguished by smothering with the help of foam, dry chemical powder, carbon dioxide, halon, etc.

*Class 'D' fires:* Fire from reactive chemicals, active and combustive metals and the like. Examples are magnesium, sodium, potassium. Extinguishing agents containing water are ineffective and sometimes dangerous too. Carbon dioxide and the bicarbonate classes of dry chemical powders are also hazardous, if applied to most metal fires. Powdered graphite, powdered talc, soda ash, limestone and dry sand are normally suitable for class D fires. This type of fire can be extinguished by smothering with the help of special type of dry chemical powder (ternary eutectic chloride).

*Class 'K' fires:* Fires in cooking appliances that involve combustible cooking media (vegetable or animal oils and fats). [Class K fires are under NFPA (National Fire Protection Association) USA classification only].

Many fires involve a combination of the types of materials.

**Carbon monoxide:** A gaseous molecule having the formula CO, which is the product of incomplete combustion of organic materials. Carbon monoxide has an affinity for haemoglobin approximately 200 times stronger than oxygen and is highly poisonous. CO is a flammable gas which burns with a blue flame and has explosive limits of 12 to 75%. Carbon monoxide has approximately the same vapour density as air, 0.97 (air = 1.00).

### Flammable and combustible liquids
### (NFPA definition and classification)

Flammable liquids include any liquids having a flash point below 37.8° C (100° F) and having a vapour pressure not exceeding 276 kPa (40 psi) (absolute) at 37.8° C (100° F). Flammable liquids shall be subdivided as follows: (1) Class I liquids shall include those having flash points below 37.8° C (100° F) and are subdivided as follows: (a) Class IA liquids shall include those having flash points below 22.8° C (73° F) and having a boiling point below 37.8° C (100° F). (b) Class IB liquids shall include those having flash points below 22.8° C (73° F) and having a boiling point above 37.8° C (100° F). (c) Class IC liquids include those having flash points at or above 22.8° C (73° F) and below 37.8° C (100° F). Combustible liquids shall be or shall include any liquids having a flash point at or above 37.8° C (100° F). They shall be subdivided as follows: (1) Class II liquids shall include those having flash points at or above 37.8° C (100° F) and below 60° C (140° F); (2) Class IIIA liquids shall include those having flash points at or above 60° C (140° F) and below 93.3° C (200° F); (3) Class IIIB liquids include those having flash points at or above 93.3° C (200° F).

## CLASS OF FIRE AND SUITABLE FIRE EXTINGUISHERS

The classification is for the purpose of effective fire fighting.

### European and Indian Classification

 **Class A fires:** Fires of combustible organic materials like wood, paper, textiles, etc.

 **Class B fires:** Fires of flammable liquids or liquefiable solids like petroleum, alcohol, oil, tar, etc.

 **Class C fires:** Fires of flammable gases or liquefied gases such as propane, hydrogen, acetylene, etc.

 **Class D fires:** Fires of combustible metals and their alloys like magnesium, aluminium, potassium, lithium, zinc.

**Caution Note for Electrical fires:** Any fires involving or started by electrical equipment can be of the class A to D. Flow of electricity must be stopped and disconnected and then the extinguishant suitable for the material involved in the fire should be employed.

## American (NFPA) classification

| Class of fire | | Suitable fire extinguisher |
|---|---|---|
|  A |  Ordinary combustibles | **Ordinary combustibles fires** Chemical extinguishers of soda-acid type, gas/expelled water and antifreeze types and water buckets and foam |
|  B |  Flammable liquids | **Flammable liquid gas fires** Chemical extinguishers of carbon dioxide, dry-power and sand buckets, foam |
|  C |  Electrical equipment | **Electrical fires** Chemical extinguishers of carbon dioxide, inert gases and halon |
|  D |  | **Combustible metal fires** Special dry-powder and sand buckets. |
|  K |  | **Liquid cooking media fires** Special chemical agent. |

### Fire extingnisher use guide

**Carbon dioxide:** A molecule consisting of one atom of carbon and two atoms of oxygen which is a major combustion product of the burning of organic materials. Carbon dioxide ($CO_2$) is the result of complete combustion of carbon. In the gaseous form, $CO_2$ is used as a fire extinguisher. In the solid form, $CO_2$ is known as dry ice. $CO_2$ is heavier than air, with a vapour density of 1.53 (air = 1.00).

## Fire extinguisher ratings

The rating of a fire extinguisher indicates the amount of fire the extinguisher can put off or extinguish, when used by a non-expert person (expressed numerically).

**Class A extinguisher rating:** The amount of water the fire extinguisher holds and the amount of fire it will extinguish. The test fire consists of wood cribs of 36 sq mm cross-section and for a fire of 3A rating (smallest fire), fifty-six sticks are used.

**Class B extinguisher:** States the number of square meters of a flammable liquid fire that can be extinguished with the extinguisher. The class B fire for the test consists of a circular tray with a fuel floating on a layer of water. The smallest class B fire is made by a tray of 720 mm diameter with a total fuel/water content of 13 litres.

**Class C, D and K extinguishers:** These types of extinguishers generally have no rating.

**Multipurpose rating:** Means for use on many types of fires. Extinguishers with multipurpose ratings also show labels for all the types of fires on which it can be used.

Additionally, class A and class B fire extinguishers have a numerical rating which is based on tests conducted by Underwriter's laboratories that are designed to determine the extinguishing potential for each size and type of extinguisher.

### Extinguisher Rating

*Rating indication on a fire extinguisher:* The higher the numerical value on the extinguisher, the more fire it will put out.

In India, there is no rating system for portable fire extinguishers, since there is no testing and certifying agency.

# 2

# *Fire Protection*

*And the Angel of the Lord appeared unto him in a flame of fire out of the midst of a bush; and he looked, and, behold, the bush burned with fire, and the bush was not consumed.*

*—Exodus 3.2*

## FIRE PRECAUTIONS

Fire precautions are defined by BS-4422 as "the measures taken and the fire protection provided in a building or other fire risk to minimize the risk to occupants, contents and structure from an outbreak of fire."

Fire precaution has three phases:

1. *Fire prevention:* Fire prevention means fire precautions that are intended to prevent occurrence of fire. Measures to prevent the outbreak of a fire and/or to limit its effects include procedures as well as physical measures to reduce the probability of the occurrence of fire, e.g. fitting an earth leakage circuit breaker to an electrical installation may reduce the chance of a fire of electrical origin.

   Procedure for routine inspection can also lessen the risk by leading to early identification and rectification of faults.

2. *Fire protection:* Fire protection means fire precautions that afford a degree of protection if fire does occur.

3. *Fire plans:* Fire plans consist of measures such as emergency plans, escape routes, etc.

Fire prevention is the concern of everyone. Fire protection reduces the high fire loss. Fire fighting is the last resort adopted on the failure of fire prevention and fire protection. Fire plans help manage and mitigate the fire risk.

Fire prevention involves adopting measures to ensure that harmful fires do not start.

## Measures to Prevent Fire

- Fire legislation (statutes and regulations)
- Inspection plans and checklist

- Fire proof design (heat and fire-resistant materials) (passive fire protection measures)
- Promotion of fire safety by fire safety education

A well written fire safety plan shall provide for the following:
- Use of alarms
- Transmission of alarm to control department
- Response to alarm
- Isolation of fire
- Evacuation of immediate area
- Evacuation of smoke compartment
- Preparation of floors and building for evacuation
- Extinguishment of fire

## Measures to Prevent Fire Growth

A fire, as we know, is characterized by ignition, growth and spread.
- Removal of waste
- Security of flammable material storage
- Avoidance of unnecessary storage
- Avoiding introducing ignition sources

## Fire Legislation

Standards and codes related to fire safety for buildings, manufacturing facilities, storage facilities and for other fields such as transport have been developed. These standards are for fire protection as well as for design standards for products and fire performance. Rules and regulations made by legislation must be accepted and appreciated by the society, if it is to be enforced and provide benefit.

In India, such standards, codes and practices are framed and implemented by BIS, controller of explosives, factories inspectorate, Tariffs Advisory Committee (TAC), etc.

Internationally, the NFPA (National Fire Protection Association), ASTM (American Society for Testing and Materials), in the USA are the leading agencies involved in establishing the codes and standards.

## Fire Safety Education

Educating people—because people cause, and can prevent, almost all fires. Education helps people to be alert to fire dangers (Fig. 2.1).

## DESIGN FOR FIRE SAFETY

### Fire Load

*Considerations in determining the fire load in a building compartment* (NFPA method).
The approach is to define a fire that is confined to the object of origin as one whose heat release rate is sufficient to result in a steady state upper layer temperature of 100° C. This upper layer temperature would result in a radiant flux to other combustibles in the room of about 1 kW/m$^2$ which is insufficient to drive flame spread on most materials.

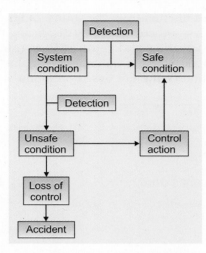

**Fig. 2.1:** Preventive action principle.

Similarly, a heat flux of 3 kW/m² would typically drive flame spread only near the object of origin where the flux from the flame provides additional drive for flame spread. A heat flux of 15 kW/m² may ignite other objects in the room but is below flashover. A heat flux of 25 kW/m² is characteristic of flashover that would result in flames out of the door and spread to the adjacent compartment. The energy needed to produce the target temperature is a function of the geometry, heat losses, and ventilation of the space.

## Fire Resistance

*Definition of fire resistance* (BS 4422 part 2): Ability of an element of building construction, component or structure to fulfill for a stated period of time, the required stability, fire integrity and/or thermal insulation and/or other expected duty in a standard fire resistance test.

Fire resistance is the means and methods used to protect materials such as wood, plastic, textiles, paper and other materials against fire. Even such incombustible materials such as steel and concrete can be affected by intense fire. Steel can melt and concrete can crack.

Fire retardants help prevent materials from burning or being severely damaged when exposed to fire. Some increase the time it takes for treated articles to burst into fire. Others cause a material to extinguish itself if it is ignited by a brief fire, thereby preventing the fire from spreading to surrounding objects.

## Use of Intumescent Materials

These materials expand upon exposure to heat or react with fire or heat to produce non-combustible carbon char, which thereby prevents the heat from regenerating. This

*Grease and oil should not be applied on fitting in oxygen pipelines,
as they can get ignited*

intumescing action separates the fuel from the source of ignition. Some of the materials also produce nitrogen, which smothers the fire by displacing the oxygen. Their insulating property prevents fire penetration and reduces heat transfer. They are capable of providing protection up to 1500° C. Such materials are applied as brush or spray coatings.

Typical water-based intumescent paint based on mono-ammonium phosphate.

| Ingredient | Content % |
| --- | --- |
| Water | 20.4 |
| Surfactant | 0.5 |
| Mono-ammonium phosphate | 25.6 |
| Chlorinated paraffin | 5.1 |
| Latex | 17.4 |
| $TiO_2$ | 8.5 |
| Others | 22.5 |

In textiles, nearly permanent fire resistance is obtained through processes that molecularly bond fire retardant materials to the fabric. These fire retardant materials help resist fire by interfering with the vapour–oxygen combustion reaction. Treatments consisting of chemicals such as antimony oxide and borate salts with halogenated polymer binder are employed for textiles used for tents. Such treatment is especially important for children's sleepwear. The cloth does not catch fire.

Temporary fire retardation may be obtained by soaking fabrics in solutions of such chemicals as boric acid and ammonium sulfate.

Materials such as ceramic fiber insulation and asbestos (asbestos is now being restricted) and asbestos substitutes such as refrasil/woven graphite provide protection from fire and heat (Fig. 2.2).

*Standard fire resistance test* [ISO 834 (1975) or BS 476]: Tests are conducted to determine the fire resistance rating for a given method of design of construction and also to assist in the development of new products with improved fire-resistant characteristics.

In the test, the structural element is loaded so as to produce the same stress as during a fire situation.

## Fire-resistant Materials

Fire resistant materials are materials with insulation properties. Examples of such materials are refrasil, glass fibre, aramid, kevlar, proban which are employed as fire-resistant materials.

*Refrasil:* Refrasil is a trade name for a material made of non-respirable fibres of amorphous silica products. Refrasil can withstand temperatures of up to 1000° C (special grades can withstand up to 1260° C). Their use presents no known health hazard and is

(a)     (b)

**Fig. 2.2:** (a) A refrasil cloth (silica fiber), (b) A woven graphite cloth.

the ideal substitute for asbestos. It is available in the form of cloth, tapes, sleeves, etc. They can be used for personal protection and also for equipment protection. They can act as fire walls and thermal insulation for equipment, cables, etc. which may be exposed to the high temperatures of a fire.

**A place for everything and everything in its place**

## WORK AREA CHECKLIST

Area _____ Inspected by _____

_____

_____

Elements for Inspection

*House Keeping*

- Are combustibles kept to a minimum?
- Combustible trash removed on regular basis.
- Oily rags kept in metal containers with lids.

*Smoking*

- Smoking areas are free of combustible material?
- Adequate cigarette disposal provided in smoking area (Fig. 2.3).

*Electrical*

- Extension cords are not used in place of permanent wiring.
- Surge suppressor used for computer circuits.
- Electrical lights are clear of combustible material.
- Combustibles are not stored in front of electrical outlets/panels.
- Emergency lights operate and are positioned correctly to illuminate the path.

*Heating*

- Heating units have ample clearance for ventilation.
- Heating units have been insulated.

*Fire*

- Sprinkler system control valves are in good condition.
- Fire hydrant accesses are clear.
- Sprinkler heads have a minimum 45 cm clearance.
- Fire exits are accessible and visible, access not blocked.
- Fire exits are checked periodically.
- Fire doors are free of obstruction
- Fire alarm stations are visible.

*Exits*

- All emergency exits are clear and unobstructed.
- Fire exits are adequately illuminated.
- Fire drills are conducted regularly
- Evacuation plans and staging are posted.

*Comments:* _____

_____

_____

_____

_____

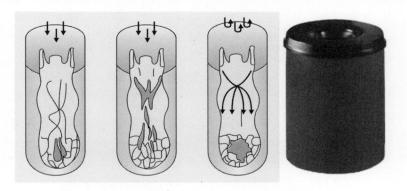

**Fig. 2.3:** Cigarette dispensing bins—principle of function.

## FIRE PROTECTION

Factors for determining the fire protection requirements

- Fire load—fire load is the amount of heat in kilocalories liberated per sq.m of floor area by the combustion of the contents and any combustible parts of the building itself.

    = Weight of contents × calorific value/floor area.
- Occupancy—nature and number.
- Class of fire and consequent hazard.
- Other factors—distribution of load, speed of burning, ease of ignition, nature of materials, etc.

| Typical fire loads in different occupancies | | |
|---|---|---|
| *Occupancy* | *kcal per sq m* | *Fire resistance required* |
| Offices | 45000 to 150000 | One hour |
| Residences | 200000 to 300000 | One hour |
| Shops, workshops | 300000 to 500000 | Two hours |
| Godown, warehouses | More than 500000 | Four hours |

Fire protection is achieved by two means:
- Fire suppression.
- Fire detection and alarm

Fire suppression means fire fighting using fire extinguishing agents.

*Unsatisfactory conditions should be corrected as soon as possible in order to maintain the highest degree of safety.*

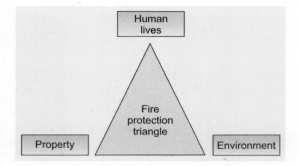

Fig. 2.4: Objectives of a fire protection system.

The two basic aspects of fire fighting systems are—extinguishing the fire and reducing the loss due to fire. These two systems are entirely dependent on the efficiency of the fire fighting systems installed and the awareness about the usage and principles of fire protection by the operating personnel. It is therefore imperative that the fire protection systems available be maintained in a state of peak performance at all times.

## EXTINGUISHING AGENTS

1. Water
2. Foam
3. Powder
4. Gaseous agent
5. Vapourizing liquids

The most commonly available and effective fire-fighting medium is water. The most versatile extinguisher is one filled with multi-purpose dry powder (sometimes referred to as ABC dry powder).

### Water

Water absorbs 85 calories/gram on being heated from room temperature to boiling point. So, water readily absorbs heat and therefore reduces the temperature of the fuel to below its ignition point. Water absorbs 540 calories in changing from liquid to steam (latent heat of vapourization). Therefore, steam production from water, on being applied as a mist, has a higher cooling effect and also steam acts by suffocation.

### Advantages of Water

• Water is easily available everywhere, and is hence the most commonly used fire extinguishing agent. Although all types of water can be utilized, it is preferable to use soft water. Sea water can cause corrosion of pumps, piping and fittings.

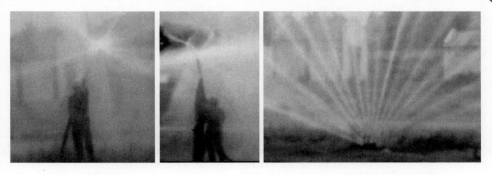

Mist or fog spray                    Curtain mist spray

**Fig. 2.5:** Mist or fog spray and curtain mist spray.

- Water can be easily stored and pumped to great distances.
- Water is non-toxic and does not pose any health hazard.
- Water can be applied as a jet and also as a mist or fog (Fig. 2.5).
- Water suppresses fires by cooling as well as by smothering. Steam produced by application of water has a cooling as well as smothering effect.
- Additives can be mixed with water to enhance its effect.

### Disadvantages of Water

Water has its own disadvantages:

- Water is not a clean agent. After application, the fire scene area needs clean-up.
- Water is a conductor of electricity. So water cannot be used where there is energized electricity. Electrical power has to be cut off before water can be applied.
- Water can damage sensitive equipment such as electronic and electrical equipment.
- Water can cause material damage since many materials dissolve in water.
- Water can cause run off.
- Water application has an efficiency of only 20 to 25%. Much of the water gets wasted.
- Caution is to be exercised in using water on flammable liquid fires, as the water run-off can spread the fire.
- Use of water can pollute the public water supply system.

### Foam

Foam used for fire fighting consists of tiny air or gas bubbles, which can form a compact fluid layer or blanket on the surface of flammable liquids and thus preventing combustion of the liquid surface.

### Types of Foam Compounds

There are two types of foam compounds in general use.

- Mechanical foam compound
- Chemical foam compound

## 1. Mechanical Foam Compound

These foams are produced by air entraption by mechanical means such as aspiration or air blowing.

Mechanical foam compounds are again classified into 3 types depending on the expansion ratio of the compound.

### Classification based on the Expansion Ratio

Fire-fighting foam is a stable aggregation of small bubbles of lower density than oil or water that exhibits a tenacity for covering horizontal surfaces. Air foam is made by mixing air into a water solution, containing a foam concentrate, by means of suitably designed equipment. It flows freely over a burning liquid surface and forms a tough, air-excluding, continuous blanket that seals volatile vapours from access to air. It resists disruption from wind and draft or heat and flame attack and is capable of resealing in case of mechanical rupture. Fire-fighting foams retain these properties for relatively long periods of time. Foams are also defined by expansion and are subdivided into three ranges of expansion. These ranges correspond broadly to three types of usage described below.

1. Low-expansion foam—expansion up to 20.
2. Medium-expansion foam—expansion from 20 to 200.
3. High-expansion foam—expansion from 200 to 1000.

*Low expansion foam:* These foam compounds can expand up to 20 times (normal rating between 5 and 15). The foam is typically produced by self-aspirating foam branch pipes.

> While dispensing flammable liquids from large containers the container should be grounded. Also, the drum should be bonded by using a bond between the dispensing and the receiving containers.

Since low expansion foams contain more water, they provide the cooling effect of the water. These compounds can be used for hydrocarbon liquid fuel fires.

*Medium expansion foam:* These foam compounds can expand up to 200 times their volume and are generally produced by self-aspirating foam branch pipes with nets. Since the water content is less, does not provide the cooling effect. So, they are generally used on hydrocarbon fuels with low boiling point. The foam has a tendency to breakdown due to the heat of the fire.

*High expansion foam:* These foam compounds have an expansion of up to 1000 times their volume and are generally produced by foam generators using air fans. These foam types are useful for hydrocarbon gases stored under cryogenic conditions and for warehouse protection.

> *Common salt can be used to extinguish some class D fires*

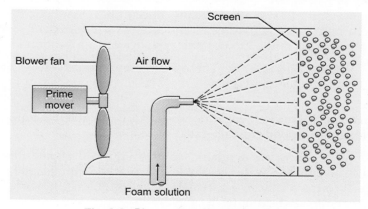

**Fig. 2.6:** Blower type foam generator.

## 2. Chemical Foam

When two or more chemicals are added, foam is generated due to chemical reaction. The most common ingredients used for chemical foam are sodium bicarbonate and aluminium sulfate with added stabilizers. These types of foam compounds can be used in portable fire extinguishers, where sometimes, mechanical formation of foam is not possible.

### Classification of Foams by Composition

*Protein-based foam:* These are made from hydrolyzed proteins with added stabilizers and preservatives. They form a thick foam blanket and is suitable for hydrocarbon liquid fires, but cannot be used on water miscible liquids. Their effectiveness is limited on low flash point fuels which have a lengthy pre-burn time.

*Fluoro-protein foam:* Fluoro-protein foam contains added fluoro-chemicals which increases its effectiveness on low flash point fuels.

---

### Water-miscible liquid

Liquids that are water-miscible include low molecular weight (3 carbons or less) alcohols, such as methyl alcohol, ethyl alcohol, n-propyl alcohol, isopropyl alcohol, and allyl alcohol. Acetone and tetrabutyl alcohol are also water-miscible. When water-miscible flammable liquids are mixed with water, a homogeneous solution is formed. The flash point, fire point, heat of combustion, and heat release rate for the solution will be different from the pure liquid. The flash point and fire point of the solution will increase as the water concentration increases. At a certain water concentration, which varies for different liquids, the fire point will no longer exist and the solution will no longer present a fire hazard.

---

### Synthetic Foams

Synthetic foam concentrate is based on foaming agents other than hydrolyzed proteins and include aqueous film-forming foam (AFFF) concentrates, medium- and high-expansion foam concentrates, and other synthetic foam concentrates. Synthetic foam concentrate contains hydrocarbon surface active agents and acts as a wetting agent, foaming agent, or both. In general, its use is limited to portable nozzle foam application for spill fires.

*Aqueous film forming foam (AFFF):* These foam compounds consist of fluorocarbon surfactants, a foaming agent and a stabilizer. These compounds can be used with all types of water.

This foam has quick knock down properties and is suitable for liquid hydrocarbon fuels, but the foam has a poor drainage rate.

*Multipurpose AFFF:* These are synthetic foaming liquids designed for fire protection of water soluble solvents and water insoluble hydrocarbon liquids, and can also use all types of water. It forms a cohesive polymeric layer on the liquid surface and thereby suppresses the vapours and extinguishes the fire.

| *Flammable/combustible hydrocarbon* | *Suitable foam type* |
|---|---|
| Hydrocarbons with flash point below 37.8° C | Protein, AFFF, fluoroprotein, FFFP, and alcohol-resistant AFFF or FFFP |
| Flash point at or above 37.8° C | All foams |
| Liquids requiring alcohol-resistant foams | Alcohol-resistant foams |

*Film forming fluoroprotein (FFFP):* This is a combination of the AFFF and the fluoroprotein foam and thereby provides a high level of knock down effect and post-fire security and burn back resistance. This foam is suitable for hydrocarbon liquid fuel fires including deep pool fires of low flash point fuels which have lengthy pre-burn time.

Bonding is an excellent safeguard against generation of static spark while transferring flammable liquids by free fall.

### Foam systems

The lowering of surface tension occurs when foam is used to fight oil, kerosene or petrol fires. Foamed water is an excellent fire fighting medium, but for it to be effective, it must spread across the burning fuel's surface. Fluorinated surfactants are used to reduce water's surface tension below 20 mN/m, and the foam produced spreads because this is lower than the surface tension of the fuel (petrol, for example, has a surface tension of 21.8 mN/m). This technique was pioneered in the 1960's by the US Navy researchers to fight fires aboard aircraft carriers. The foam can be further stabilized by adding partially hydrolyzed proteins and is then called as film forming fluoro-proteins.

## Powder (Dry Chemical and Wet Chemical)

### Regular Dry Chemical

These powders are primarily a mixture of sodium bicarbonate, silicates and stearates. It is also known as BC dry powder. The powder agent made of sodium bicarbonate ($NaHCO_3$) is suitable for use on all types of flammable liquid and gas fires (class B) and for fires involving energized electrical equipment (class C). It is particularly effective on fires in common cooking oils and fats. In combination with these materials, the sodium bicarbonate-based agent reacts to form a type of soap (soponification), which floats on a liquid surface, such as in deep-fat fryers, and effectively prevents re-ignition of the grease. Sodium bicarbonate-based dry chemical is not generally recommended for the extinguishment of fires in ordinary combustibles (class A), although it can have a transitory effect in extinguishing surface flaming of such materials.

Use of sodium bicarbonate as dry chemical powder:

- $NaHCO_3 + HEAT \longrightarrow CO_2 + OH + Na$
- $OH + H^* \text{ (active in flame)} \longrightarrow H_2O$
- $Na + H^* \text{ (active in flame)} \longrightarrow NaH$

The powders can also be made from potassium bicarbonate or potassium bicarbonate with urea base. Commercially available agents are essentially potassium bicarbonate ($KHCO_3$), potassium chloride ($KCl$), and urea-based potassium bicarbonate ($KC_2N_2H_3O_3$). All three agents are suitable for use on all types of flammable liquid and gas fires (class B) and also for fires involving energized electrical equipment (class C). It is generally recognized that salts of potassium are more effective in terms of chemical extinguishment mechanisms than sodium salts in extinguishing class B fires, except those in deep-fat fryers and other cooking equipment.

Dry chemicals based on the salts of potassium are not generally recommended for the extinguishment of fires in ordinary combustibles (class A), although they can have a transitory effect in extinguishing surface flaming of such materials.

Use of potassium chloride as dry chemical powder:

$$KCl \longrightarrow K^+ + Cl^-$$

Potassium chloride

$$K + H \longrightarrow KH$$
$$KH + OH^- \longrightarrow K^+ + H_2O$$
$$Cl + H \longrightarrow HCl$$

Potassium bicarbonate-based powders are more effective than the sodium bicarbonate based powders. BC dry powders are effective against class B and C fires. They are used in areas where flammable liquids are present.

### MSDS for BC dry chemical

Mixture of sodium bicarbonate (90%), silicates and stearates (9.8%), water (0.2%)
White powder, relative density (@20.4° C) = 2, apparent density = 1 gm/cc
Partially soluble in water.
NFPA hazard identification—health–1, fire–0, reactivity–0.
Emergency first aid—In case of eye contact, wash thoroughly with water for at least 15 minutes

*Specifications of dry chemical powders:* Dry chemicals should conform to IS 4308 standard.

| Dry powder specifications | |
|---|---|
| Specification | IS: 4308 |
| Apparent density | 0.98–1.04 |
| Sieve analysis | |
| % - 100 mesh | < 1% |
| % - 200 mesh | > 92% |
| Hygroscopicity % gain | 1.0% max |
| Wet caking test | No lumps |
| Moisture % | 0.2 max |
| Water repellency | > 1.5% |

BC dry powders are not suitable for deep fat fires or for situations in which it is necessary to penetrate deep confined spaces, e.g. within enclosures.

Dry chemical powders manufactured to IS: 4308 are compatible with foam and are suitable for B and C class fires.

The manufactured powders are generally packed in bags or metal containers.

## Multipurpose Dry Chemical

A multipurpose dry chemical powder is essentially made up of mono-ammonium phosphate (fluidized and siliconized mono-ammonium phosphate). Mono-ammonium phosphate is a non-toxic dry chemical agent. It is effective against class A, B and C fires. This agent has as its base, mono-ammonium phosphate ($NH_4H_2PO_4$) and is similar in its effect on class B and class C fires to the other dry chemicals. However, it does not possess a soponification characteristic and should not be used on fires in deep-fat fryers. Unlike the other dry chemicals, it does have a considerable extinguishing effect on class A materials. This material melts at 350° C. The agent, when heated, decomposes to form a molten residue that will adhere to heated surfaces. On combustible solid surfaces (class A), this characteristic excludes the oxygen necessary for propagation of the fire. In class B fires, it smothers and breaks the chain reaction and also since it is non-conductor of electricity, it prevents the conduction of electricity back to the fire fighter.

These powders are ideal for most applications, but are more expensive than regular dry chemical.

By multipurpose, we mean that the powder can be used on all the three types of fires—class A, class B and class C fires. Hence multipurpose dry chemical powders are also called as ABC powders.

Cleaning up of ABC chemical powders is easy as the chemical does not normally bake and stick to a hot metal surface.

Use of mono-ammonium phosphate (MAP) as dry chemical powder:

(MAP)
$$NH_4H_2PO_4 \xrightarrow{\text{Fire}} NH_3 + H_3PO_4$$

For class B and C fires:

$$NH_3 + H^* \longrightarrow NH_4$$
$$NH_4 + OH^* \longrightarrow NH_3 + H_2O$$

For class A fire:

$$H_3PO_4 \longrightarrow HPO_3 + H_2O$$

## Advantages of Dry Chemical Powders

- No detioration during storage
- Non-toxic and non-corrosive
- Harmless to skin
- Non-conductor of electricity
- Better than $CO_2$ on liquid fires
- Good for blast fires.

## Disadvantages of Dry Chemical Powders

Dry chemical powder based on sodium bicarbonate cannot be used on certain metal fires such as fires from sodium, magnesium, etc.

Dry chemical powders cannot be used to produce inert atmospheres to prevent gas/vapour formation. Re-ignition can take place in case of liquid fires since powder does not cool the burning substance below its ignition temperature.

Powders create dust–thereby cleanup is required after application. The dust can create breathing problems to people in the vicinity. Evacuation of people in the application area may be necessary before use. Dust masks need to be worn to prevent powder dust inhalation.

Dry chemical powders have a short discharge distance; thereby can only be applied from close range.

Dry chemical powders are prone to caking. Regular maintenance of powder systems is necessary to keep them in operating condition.

Although powders as a group are multipurpose, different types of powders are required for different class of fires.

## Powder Agent for Class D Fires

Class D fires are fires involving combustible metals which are generally found in many machining and milling operations. Sodium chloride-based dry powder is the most common extinguishing agent used for this purpose.

Metal fires pose special problems to the fire fighting operations as water or foam cannot be used as fire extinguishing agent because some metals react violently with water, even at room temperature. Ordinary dry chemical powders are not suitable for fighting metal fires.

## Combustion of Combustible Metals

Fire of high intensity can occur in certain metals. Ignition is generally the result of frictional heating, exposure to moisture, or exposure from a fire in other combustible materials.

The greatest hazard exists when these metals are in the molten state, in finely divided forms of dust, turnings, or shavings.

Ternary eutectic chloride (TEC) based dry powders are specially formulated for metal fires. It is treated with specialty chemicals to impart flow properties and excellent water repellency. TEC powder fuses at temperature of 600°–650° C and has high latent heat of fusion. The powder absorbs heat from the burning metal and forms a hard crust on the metal surface. This crust provides a blanketing effect thereby cutting off oxygen and the heat radiation from the burning metal surface. Chloride radicals of the powder react with the free radicals of metals responsible for the fire propagation and this results in a retardation of fire propagation chain reaction. The unique composition of ternary eutectic chloride of alkali metals gives TEC powders fast and effective fire knock down capability, which cannot be matched by ordinary mixtures of chloride salts of metals or phosphate base powders. TEC is extensively tested and found suitable for reactive metals like Li, Na, K, Mg, Al powders and paste, Ti, Fe and their alloys. The powder has also been tested with explosive composition of these metals as well as delay detonation compositions. It is also used for ammonium perchlorate and other such compounds. TEC can extinguish small oil fires and can also be used for electrical fires.

TEC powder can be applied gently through fire extinguisher or through an automatic powder dispensing system. Dry powder can also be manually applied on the burning metal with the help of a scoop, a shovel or a plastic bag. Care should be taken to cover the surface completely without pinholes. The covered surface should be left undisturbed for at least 30 to 45 minutes. Due care should be taken while dispensing dry powder through the fire extinguisher. High pressure may splash the burning metal and the fire may start at another place. Dry powder should be dispensed off gently on the burning surface at low pressure with the help of a diffuser.

> Ternary eutectic chloride (TEC)–typical composition:
> Potassium chloride –51%; barium chloride –29%;
> sodium chloride –20%

### Wet Chemical Powders

Wet chemical extinguishing systems use chemical in liquid form. They are ideally suited for class K fires.

An important characteristic of class K fires is auto-ignition. Its vapours are not easily ignited. However, when heated to high temperatures, they can auto-ignite. Even after extinguishing the fire, re-ignition can occur until the liquid is cooled below its auto-ignition temperature.

The most common wet chemical extinguishing agents is potassium carbonate based solution. Potassium acetate can also be used. This solution is discharged as fine droplets into a protected area. The main extinguishing action is by cooling caused by the heat of vapourization. Re-ignition is prevented by soponification, wherein the wet chemical agent combines with the oil or grease to form a soapy layer on the surface to seal off the fuel from the oxygen and also allowing the oil/grease to cool below the auto-ignition temperature. A fine mist spray is used, which can prevent splashing.

Generally, wet chemical systems are designed as fixed fire protection systems with piping and operating at a pressure of 12 bar. The spray is usually actuated for about 45 seconds to a minute.

## Gaseous Agent

Because $CO_2$ is denser than air and will not support combustion, it has become one of the most significant methods of suppressing and fighting electrical and grease fires. In addition, the application of $CO_2$ in firefighting is clean, non-toxic and leaves no residue.

> **Precautions in use** (Powders)
> Chemical type goggles to be used
> during application.
> In some applications, dust mask
> is required.

Carbon dioxide is the most common gaseous agent used for fire protection.

Carbon dioxide extinguishes fire by reducing the concentrations of oxygen, the vapour phase of the fuel, or both in the air to the point where combustion stops.

Sometimes a mixture of $CO_2$ and $N_2$ is used. For computer rooms only $CO_2$ is used. Gaseous agents are generally used for enclosed spaces such as rooms.

$CO_2$ produces rapid cooling of the air when introduced, leading to condensation of moisture on equipment. There can be severe cooling effect (thermal shock) which can affect some equipment.

About 30% $CO_2$ will extinguish almost any fire in a room. But more than 7% leads to respiratory asphyxiation. So, only if there is no occupant or after the occupants can safely evacuate, the gas should be discharged.

$CO_2$ is an excellent smothering agent for extinguishing fires. It is about 1.5 times heavier than air. It can easily be liquefied and bottled, normally under pressure of approximately 73.2 bars. When used on fire, the liquid $CO_2$ boils off rapidly into $CO_2$ gas taking away some heat. The gas extinguishes the fire by smothering. It is very quick, clean and non-conductor of electricity. It is non-toxic and does not cause any damage to most of the materials; it does not leave any residue. It can be stored for long period and it is easily available. $CO_2$ is very effective in enclosed areas. It can penetrate into places that cannot be reached by other means. It does not damage delicate machinery and instruments. It can be used safely on electrical equipment and panels. $CO_2$ is highly asphyxiating and cannot be detected by smell or colour. About 9% concentration causes unconsciousness. Any compartment, which has been flooded with $CO_2$, must be fully ventilated before entering without wearing breathing apparatus.

When liquid $CO_2$ is released from a pressurized storage cylinder there is an extremely rapid expansion from liquid to gas, which produces a refrigerating effect that converts part of $CO_2$ into snow. This snow, which has a temperature of about —79° C, soon sublimes (i.e. changes directly from solid to gas). This sublimation produces a cooling effect but it is the smothering effect of the gas, which is of primary importance in the extinguishing gas. The critical temperature of $CO_2$ is 31.35° C. If the liquid is heated above this temperature, it will pass into gaseous stage without any change in volume. It is therefore important that $CO_2$ is maintained in liquid form prior to use as a fire extinguishing agent. At normal temperature and pressure, the expansion ratio of $CO_2$ is 1:450.

### Special Gaseous Agents

A special gaseous system uses a mixture of $CO_2$, argon and nitrogen. This gas mixture extinguishes a fire by reducing the oxygen level below 15%, the point at which most combustibles will no longer burn. Simultaneously, the carbon dioxide in the mixture stimulates the uptake of oxygen by the human body, thereby protecting anyone who might be trapped in the fire area from the effects of the lowered oxygen level.

Concentration of extinguishing agent (gaseous and vapourizing) required to extinguish fires.

| | |
|---|---|
| Halon 1211 | 3.0 to 8.0 |
| Carbon dioxide | 18 to 26 |
| Nitrogen | 29 to 40 |

### Vapourizing Liquids

Halogenated hydrocarbons (halons) are a group of extinguishing agent, which are under pressure in liquid form. They are released in such a way as to vapourize readily in a fire zone. They extinguish fire mainly by interfering with the chemical reaction involved in the spreading of flame. They possess little cooling effect. Halon can be effectively used in class B and C fires especially in enclosed spaces. They are electrically non-conductive and safe to use on electrical or electronic equipment. They cause little damages. Halons are hydrocarbon compounds of a group of five non-metallic elements know as halogens. They are fluorine, chlorine, bromine, iodine and astatine. For various reasons, iodine and astatine are not suitable as extinguishing agents and can be ignored.

Halon is a clean alternative. But halon has been banned by the Montréal protocol on substances that deplete the ozone layer. By Halon, is meant halogenated hydrocarbons, which have been employed to extinguish fire. Two halogenated bromine containing agents are in common use—halon 1301($CF_3Br$) and halon 1211($CF_2BrCl$). These halogenated fire suppression agents have the advantage that they are clean in nature and therefore leave no residues following their use to extinguish a fire. Hence, they are generally employed for the protection of computer rooms, EDP facilities, museums, libraries and the like, where the use of water can often cause more secondary damage to the property being protected than is caused by the fire itself. Halons can be employed as total flooding or as streaming applications. Materials with low boiling point are more suitable for total flood applications, while those with high boiling points are better suited for streaming applications. Halon 1301 has a boiling point of $-58°$ C and is employed for total flooding applications, whereas, halon 1211 has boiling point of $-4°$ C is employed as a streaming agent. (Chemicals with boiling points lower than $-15°$ C are too gaseous for streaming applications.)

*Magnesium metal dust can ignite easily*

But these bromine and chlorine containing compounds are ascertained to be capable of the destruction of the earth's protective ozone layer, e.g. halon 1301 has an ozone depletion potential (ODP) rating of 10 and halon 1211 has an ODP of 3. Hence, these compounds are to be phased out of production and use.

When using halon 1211 or BCF, the maximum exposure is 1 min at 4–5% concentration. After release, the area should be well ventilated before human entry.

Use of halon 1301 (BTM):

$$CF_3Br \text{ (BTM)} \xrightarrow{\text{Fire}} CF_3 + Br$$

$$Br + H^* \text{ (active)} \xrightarrow{480° \text{ C}} HBr$$

$$HBr + OH^* \text{(active)} \longrightarrow Br + H_2O$$

## Characteristics of Halon

The criteria determining the suitability of halons for fire fighting are:

- Their effectiveness as extinguishing agents
- Toxicity, including that of their products on decomposition
- Physical properties, e.g. boiling and freezing points
- Effect on materials with which they come in contact

## Identification

The actual names of the halons are shortened to sets of initial letters as under:

| Name Initials | Halons | Number |
|---|---|---|
| Bromotrifluoromethane | BTM | 1301 |
| Bromochlorodifluoromethane | BCF | 1211 |
| Dibromotetrafluoroethane | DYE | 2402 |
| Chlorobromomethane | CBM | 1011 |
| Methyl bromide | MB | 1001 |

Halons 1301 and 1211 are the least toxic. The exposure of personnel to these agents during fire-fighting is likely to be well within the maximum safe values, except where a total flooding system is being discharged. In a fire, an important aspect to be considered is the effect of the decomposition products of halon. They are much more toxic than the halons themselves. However, it is the combustion products of the actual fire like smoke and carbon monoxide, which pose the greatest hazard. Halons must not be used on chemicals containing their own elements. They are not to be used on reactive metals and metal hydrates.

The halogenated hydrocarbons used for extinguishing fires have the special property of vapourizing readily when heated. They are generally known as vapourizing liquids. They form a dense, heavier than air, cloud of non-flammable vapour which not only blankets a fire but also interferes with the chemical reaction of flame propagation in the burning materials. Vapourizing liquids are non-conductors of electricity. They are normally harmless to delicate electronic equipment. Their worst drawback is their toxicity. Some famous extinguishing media such as CTC and methyl bromide have been withdrawn from use because of their toxicity. Freon type vapourizing liquids are currently used, as they are less toxic. The most common substances used are bromo-chlorodifluoromethane (BCF) and bromotrifluoromethane (BTM).

Halogenated extinguishing agents are hydrocarbons in which one or more hydrogen atoms are replaced by atoms from the halogen series fluorine, chlorine, bromine and iodine. The replacement confirms non-flammability and enhances flame extinguishing property to many of the resulting compounds. The use of halogenated hydrocarbons as fire extinguishing media is not permitted in accommodation areas and living spaces. It is allowed only in machinery spaces such as engine rooms, pump rooms, etc.

The three elements generally found in fire extinguishing media are fluorine, chlorine and bromine. Substitution of a hydrogen atom in a hydrocarbon with these elements influences the relative properties in the following manner: Fluorine provides stability to the compound, reduces toxicity, boiling points and increases thermal stability. Chlorine provides fire extinguishing effectiveness, increases toxicity, boiling point and reduces thermal stability. Bromine works almost similar to chlorine but to a greater degree.

Compounds containing combinations of fluorine, chlorine and bromine can possess varying degree of effectiveness, chemical and thermal stability, toxicity and volatility. They are very suitable for fires involved with electricity since they do not become electrically conductive in presence of water. As they vapourized easily, they do not leave any corrosive or abrasive residue after use. The extinguishing mechanism of the halogenated agent is a chain reaction, which interferes with the combustion process. On weight basis, halon is approximately 2.1/2 times more effective than $CO_2$. Inhalation of 4 to 5% of this gas is maximum that can safely be inhaled in one minute.

### Steam as a Fire Extinguishant

Steam in large quantities may be used to extinguish a fire. In situations where it is readily available, it can be used in fixed installations. Ships and industries having individual compartments are adapting steam to fill the compartment under pressure for smothering.

### FM 200

Another vapourizing liquid fire extinguishing agent is FM 200 (trade mark). It essentially consists of a chemical–Heptafluoropropane (manufactured by the Great Lakes Co,). FM 200 is a colourless gas which is liquefied under pressure for storage. Like halon 1301, it has a low toxicity. It is generally super-pressurized with nitrogen to 24.8 bar. It aridly extinguishes most of the fires through a combination of chemical and physical mechanisms. FM 200 does not have any ozone depletion effect as it does not contain halogens. It requires a concentration of only 7%. It is generally used as replacement for halon systems for fire protection for data processing rooms and for electronic equipment rooms.

| Comparison between halon 1301 and FM 200 | | |
|---|---|---|
| | *Halon 1301* | *FM 200* |
| Ozone depletion potential | 16 | 0 |
| Atmospheric life time | 87–110 years | 31–41 years |
| Class A extinguishing less than concentration | 4.1% | 5.8% |
| Class B fire extinguishing-concentration. | 4.1% | 5.8–6.6% |
| Chemical structure | $CF_3Br$ | $CF_3CHFCF_3$ |

Examples of extinguishant for protecting class A hazards are as follows:

1. Water type
2. Halogenated agent type
3. Multipurpose dry chemical type
4. Wet chemical type

Examples of extinguishant for protecting class B hazards are as follows:

1. Aqueous film forming foam (AFFF)
2. Film forming fluoroprotein foam (FFFP)
3. Carbon dioxide
4. Dry chemical type
5. Halogenated agent type

## PORTABLE FIRE PROTECTION SYSTEMS

A portable fire extinguishing system consists of using portable fire extinguishers.

Fire extinguishers are the first line of defense against fire. Portable fire extinguishers can save life and property by putting out or containing fires within the capability of the extinguisher. Since a majority of the fires have small beginnings, prompt use of a portable fire extinguisher can prevent the spread of fires. But, it is important to use the right type of extinguisher for the particular fire, for the extinguisher to be effective. Also, sometimes the type of extinguisher used may prove to be dangerous than helpful, under the circumstances.

Portable fire fighting equipment are only first aid fire fighting equipment.

Fire extinguishers are not designed to fight a large or spreading fire. Even against small fires, they are useful only under the right conditions. They are intended for use only on incipient fires and their value may be negligible if the fire is not extinguished or brought under control in the early stages.

The disadvantage of portable fire extinguishers are that they provide only a limited amount of extinguishing agent, which may be from a few seconds to a few minutes at the most. Also, extinguishers are not universal, with suitability for all classes of fires. But the advantage lies in the fact that they can be deployed and used on the fire very quickly, thereby acting on the fire in its infancy stage itself and hence have been very effective in fighting fires.

Portable fire fighting equipment can be used only during the incipient stage and the smouldering stage of a fire. Once the fire reaches the flame and growth stage, portable fire extinguishers are ineffective and not advisable to be used. Portable fire extinguishers are only first aid fire fighting systems.

## WATER AND SAND BUCKETS

Simple and very handy, the water and sand buckets can be very effective in putting out fires.

Water and sand buckets are used to put out small fires. The sand is thrown at the seat or base of the fire to put it off.

**Fig. 2.7:** Water and sand buckets.

### Sand

Some burning materials, like metals, which cannot be extinguished by the use of water, may be dealt with by means of dry earth, dry sand, powdered graphite, powdered talc, soda ash or limestone, all of which act as smothering agents. Dry sand may also be used to prevent burning liquids, such as paints and oils, from flowing down drains, basements, etc. Sand should never be used for extinguishing fires in machinery such as electric motors.

Sand can also be kept bagged and in suitable enclosing boxes.

For fire buckets to be useful, fire buckets should not be used for any purpose other than for which they are intended.

Fire buckets should be kept filled at all times. To prevent breeding of insects and to comply with rules of local bodies, the water in fire buckets should be refilled with clean water every week. If necessary, disinfectants such as phenyl in small quantities can be added to the water in the buckets.

It is preferable to cover the fire buckets with a lid to prevent evaporation and contamination.

It may be necessary to keep spare buckets in stock.

| Acceptable replacement | Buckets of water | | Water-type extinguisher 9 litres (2 gallons) |
|---|---|---|---|
| | *For one bucket* | *For three buckets* | *Extinguishers* |
| Dry sand | 1 bucket | 3 buckets | |
| $CO_2$ extinguisher (IS 2878) | 3.0 kg | 9 kg (in not less than 2 extinguishers | 9.0 kg |
| Dry powder extinguisher (IS 2171) | 2.0 kg | 5.0 kg (in 1 or more extinguishers) | 5.0 kg |
| Foam extinguishers | 9.0 litres | 9.0 litres | 9.0 litres |

Acceptable replacements for water buckets and water type extinguishers in occupancies where class B fires are anticipated

## Water Spray

A water spray effectively extinguishes class A fires, but cannot be used on class B and class C fires. For areas where ordinary combustibles are present, such as schools, barns, etc. extinguishers should be placed near to where they may be used and can be prominently seen. But not too close to prevent access to it during an emergency.

Water should never be used on an oil-based fire. For example, flames will shoot from a burning kitchen pan, if water is added to tackle the blaze.

The types of extinguishers are (based on extinguishant content):

* Water extinguishers
* Foam extinguishers
* Powder extinguishers
* Gas extinguishers
* Vapourizing liquid extinguishers.

## EXPELLING METHODS

Extinguishers use different mechanisms for expelling the contents. Extinguishers are normally operated with the help of gas pressure in the upper part of the container, which forces the extinguishant out through the discharge nozzle. The pressure required is generated by different methods as below.

*Chemical reaction:* Two or more chemicals react inside the container to produce an expellant gas when the extinguisher is activated by the operating mechanism.

*Gas cartridge:* The pressure is produced by means of compressed or liquefied gas released from a gas cartridge fitted into the extinguisher.

*Stored pressure:* The expellant gas is stored with the extinguishant in the body of the extinguisher and thereby the container is permanently pressurized. In the case of gas extinguishers, however (e.g. $CO_2$ extinguisher) the expellant gas itself is also the exting-uishant.

## WATER EXTINGUISHERS·

Common water fire extinguishers hold up to 9 liters (6 litres capacity is also available) of water and have a reach (throw) of up to six meters. Water may be stored under pressure in the cylinder or the water may be pressurized with the aid of a gas cartridge which releases gas upon opening. They can be nitrogen pressurized or by $CO_2$ cartridge. For cold climates, the water can have added anti-freezing solutions.

In fires that involve energized electrical equipment, the electrical non-conductivity of the extinguishing media is of importance. (When electrical equipment is de-energized, fire extinguishers for class A or B fires can be used safely.)

Water is limited to use on class A fires. A portable pressurized water extinguisher can achieve a rating of 2A (American rating system).

But a water extinguisher utilizing high pressure of about 7–8 bar can achieve a rating of A and C. For a class B rating for water, water in the form of a fine mist is required. Water and air are stored together under pressure and released simultaneously to produce a fine liquid mist, capable of an ABC rating.

### Water Extinguisher—Old Type

This contains water mixed with sodium bicarbonate with a bottle of sulphuric acid kept at the top.

### Soda Acid Type Extinguisher

#### 9-Liter Soda Acid Water Type Extinguisher

These extinguishers consist essentially of a cylindrical or conical container; which is normally fitted with a soda solution to an indicatory mark. This solution is ejected on to the fire by pressure of a gas generated by the chemical action set up by allowing a charge of acid to react with carbonate or bicarbonate which has been previously dissolved in the water. The gas thus generated exerts downward force on the surface of the liquid and drives it out of the extinguisher through a nozzle. Sodium bicarbonate and sulphuric acid are normally used. When the solution of sodium bicarbonate and sulphuric acid react, $CO_2$ is formed and this $CO_2$ expels the water through the nozzle.

$$2NaHCO_3 \; + \; H_2SO_4 \longrightarrow Na_2SO_4 \; + \; 2H_2O \; + \; 2CO_2$$

| sodium bi- + | sulphuric | sodium | + water + | carbon |
| carbonate | acid | sulphate | | dioxide |

This is one of the most common types and is cylindrical in shape with a nozzle at the top. Inside the body, a discharge tube which runs to the bottom is fitted with a strainer to prevent small particles entering into the tube and blocking the nozzle. At the top, there is a screwed cap in the form of a brass casting with a knob and plunger in the center. Safety holes are provided in the brass head, which is screwed to the main body, to facilitate any excess gas formed to escape. On unscrewing the head, a cage is seen positioned in the cylinder in which a glass bottle of sulphuric acid is stored. The body of the extinguisher contains water with sodium bicarbonate. Striking the plunger breaks the glass bottle and allows the acid to chemically react.

### Use

Soda acid extinguishers are usually used to extinguish small class 'A' fires. While using a soda acid extinguisher, the fire should be approached from as close as possible before activating the fire extinguisher. The jet should be directed on to the burning material in the heart of the fire, regardless of flame and smoke. There are two types of operation normally used–upright and turnover. Care should be taken while operating in either case. If it is operated wrongly, the gas will escape and the extinguisher will become ineffective. When the fire is extinguished, the extinguisher may be turned over in both the methods of operation, to stop the discharge and to avoid causing damage to materials due to excessive water.

### Working Principle

When the knob is struck, the pointer breaks the glass bottle. The pieces of glass remain in the cage. Sulphuric acid comes out from the holes of the cage. It gets mixed with sodium bicarbonate, producing carbon dioxide and water. The reaction creates pressure in the cylinder and this pressure presses the water downwards and the water then comes out through the dip tube to the nozzle.

### Recharging

Recharging of the extinguisher may be done in the following order:
1. Before opening the head cap, make sure that the vent holes or any other venting devices are checked. Then keeping the head away from trajectory of the cap, open the cap slowly.
2. Remove the cage or bottle holder.
3. Wash out the extinguisher properly and remove any foreign materials.
4. Dissolve the bicarbonate of soda in a clean bucket using lukewarm water. Pour solution into container and pour water if required, till the level reaches the mark.
5. Examine nozzle, vent hole, discharge hose, etc.
6. If a plunger is used, see that it works freely. Grease very lightly.
7. Insert the new bottle of sulphuric acid in the cage, making sure that it is positioned properly. Replace the cage in the extinguisher. The sulphuric acid bottle contains 57 gm.
8. Service the head, apply lubricants if required and replace the operating head.
9. The proportion of charges should not be altered under any circumstances.
10. 454 gm of sodium bicarbonate is mixed in 9 liters of warm water to make the solution.

When turned upside down to activate, the acid spills and reacting with sodium bicarbonate solution produces $CO_2$ which then pressurizes the water in the cylinder to force the water out at a pressure.

But now, stored pressure or gas cartridges are used for pressurizing the water. The body cylinder of the extinguisher is filled with pressurized air to about 10 bar using a compressor. The extinguisher can be provided with a pressure gauge on the cylinder to indicate the pressure in the cylinder. The pressure is released by a spring loaded valve on top of the central dip pipe. When the handle provided on the cylinder is activated, by squeezing or pressing, the spring loaded valve is released and allows the discharge of the contents of the cylinder.

These are extinguishers made of corrosion resistant stainless steel cylinder, filled with water under pressure. Normally of 10 liters capacity with a discharge range of 10–15 m and operating pressure of 7 bar. The water is rechargeable.

- Standard capacity –10 litres
- Throw (discharge range) –10 to 15 meters.

*Proportion of Extinguishant Discharged*

An extinguisher is usually designed so that after being fully charged and operated in its normal working position, the proportion of contents discharged should not be less than:

- Water and foam  – 95%
- Powder  – 85%
- Halon  – 85%
- $CO_2$ gas  – 75%

*Pressurized Water Fire Extinguisher*

These are for protection of wood, paper, fabric and other ordinary combustibles. They consist of a corrosion resistant steel cylinder which is usually wall mounted. It is filled with water at a pressure of 7 bar.

Standard capacity  – 2.5 gallons
  (10 litres)

Throw  – 30–45 ft
(discharge range)  (10–15 meters)

The water can be refilled and recharged.

Fig. 2.8

| Fire extinguisher size and placement for class A hazard | | | |
|---|---|---|---|
| | Light hazard (low occupancy) | Ordinary hazard (moderate occupancy) | Extra-hazard(high occupancy) |
| Minimum rated extinguisher | 2 A | 2 A | 4 A |
| Maximum floor area per unit of A | 300 sq m | 150 sq m | 100 sq m |
| Maximum floor area for extinguisher | 1000 sq m | 1000 sq m | 1000 sq m |
| Maximum travel distance to extinguisher | 25 mts | 25 mts | 25 mts |

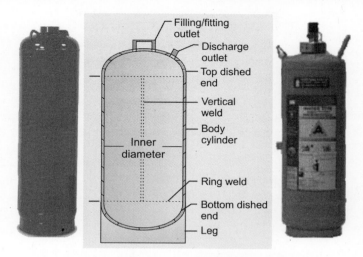

**Fig. 2.9:** Pressure vessel for portable extinguishers.

## Hydrostatic Test

The hydrostatic test, i.e. the application of hydraulic pressure to determine the ability of the vessel to hold pressure, is the method used. While it is possible to test-pressurize the container with a gas (e.g. with air)—the hazardousness of such a procedure negates its use. The energy stored by a gas under pressure is considerably greater than that for a liquid, since liquids are practically incompressible. The failure of a vessel under hydrostatic test will not burst with as great explosive force as would a similar vessel under test with a gas at an equal pressure. In most cases, the hydrostatic test is conducted at one and a half times the design working pressure of the vessel (or up to 3 times the normal working pressure). In some cases, where a test is performed to determine the ultimate bursting pressure, the vessel will be tested to five times or more, of the design working pressure.

A number of precautions are indicated for hydrostatic testing. First is the need for assuring that the air in the container is removed. In the case of irregularly shaped vessels,

---

### Delay in operation of an extinguisher

The time duration between the activation of the release mechanism and the commencement of discharge should not be more than 4 seconds.

---

### Powders of the metals

Hafnium, plutonium, thorium, titanium, uranium and zirconium are highly combustible and can cause dust explosions.

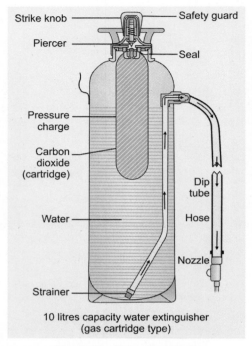

Strike knob — Safety guard

Piercer —

— Seal

Pressure charge

Carbon dioxide (cartridge)

Dip tube

Water — Hose

Nozzle

Strainer —

10 litres capacity water extinguisher
(gas cartridge type)

**Fig. 2.10:** Strike knob type.

this may be best accomplished by rotating the container to bring the air to the top and allowing it to bleed off. Second is the use of rupture discs, particularly if the test pressure is high. Third is the need for a protected test area. This is governed by the container type, its material of construction, the pressure to be applied, and the nearness of people and other equipment that might be damaged. The best possible test enclosure is a pit. In some cases, a steel box or test cell is constructed to contain a possible failure. Steel walls in such construction should be built with steel that will not brittle—fracture when struck by a high velocity projectile from a test failure. A safety consideration is the limiting of the liquid volume in the test vessel. This will reduce the amount of energy stored when the test pressure is applied. A successful means for accomplishing this is the insertion of solid pieces of material in the container to take up as much of its volume as possible.

Water reduces the temperature of the combustible material until the material is at too low a temperature to burn. But, when a fire is extinguished by spraying water on the fire, less than about 8% of the water is generally effective in extinguishing the fire, due to loss of water, such as by run-off or evaporation of the water.

Recent innovations have introduced mixing the water with water-absorbing resins or super absorbent polymers which, being insoluble in water and forming a gel, can absorb up to 20 times their weight of water. They help in increasing the effective utilization of water.

*Fire buckets should be kept filled*

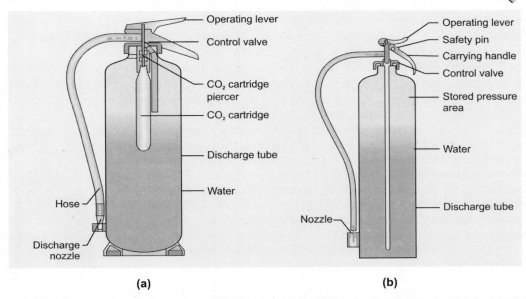

Fig. 2.11: (a) Water extinguisher (gas-cartridge type) (b) Water-stored pressure type extinguisher.

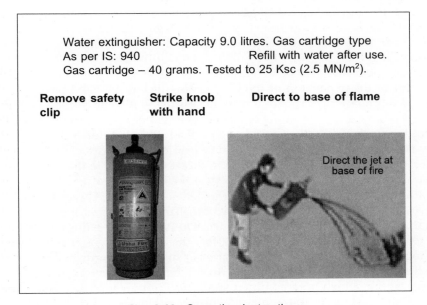

Fig. 2.12: Operating instructions.

After discharge, water-extinguishers must be washed out with fresh water using at least 2 changes. Refill with clean water and attach a new gas-cartridge.

## CARBON DIOXIDE FIRE EXTINGUISHERS

Carbon dioxide is stored under pressure as a liquid, and this liquid flashes to carbon dioxide gas after ejection, as it is no longer under pressure.

$CO_2$ is a clean extinguishing agent. Hence, carbon dioxide extinguishers are intended for use in data-processing centers, laboratories, communication rooms, food storage and processing plants, where contamination and clean up times are important. $CO_2$ does not contaminate food, valuable materials or clothing and leaves no residue. $CO_2$ gas reduces temperature and oxygen flow quickly, to smother flames and suffocate fire. $CO_2$ deprives the fire of oxygen by sweeping and dilution effects. $CO_2$ fire extinguishers are made of seamless carbon steel to withstand high storage pressure. $CO_2$ gas is electrically non-conductive and therefore safe for fighting fires in electrically powered equipment. $CO_2$ is a colourless gas which is heavier than air. $CO_2$ can also be stored in aluminium cylinders and designed for wall mounting. $CO_2$ can be used as an alternative clean agent for the prohibited Halon 1211. Can be used on electrical fires also as $CO_2$ is a non-conductive agent.

### Construction

Cylinders are manufactured to IS: 7285 and brass forge valve to IS: 3224. The complete extinguisher is manufactured to IS specification IS: 2878, latest revision and $CO_2$ extinguishers are normally available in capacity of 2 kg, 3 kg, 4.5 kg, and 6.5 kg. The $CO_2$ gas charge used should conform to IS: 307 and filled with liquefied $CO_2$ gas with a filling ratio of less than or equal to 0.667 (filling ratio is the ratio of the charge volume to the cylinder volume).

**Fig. 2.13**

On account of the expansion of the discharging gas and its ability to freeze, careful design of the discharge mechanism is very necessary. A discharge tube is fitted into the cylinder so that liquid $CO_2$ from the bottom of the cylinder is released through the valve in the cylinder head. This valve should open quickly and provide a clear passage for the $CO_2$. The expansion may commence in the flexible hose, if fitted, but it mostly takes place in the discharge horn, which is used to direct the gas on to the seat of the fire. The main purpose of the discharge horn is to stop the entrainment of air with the $CO_2$ by reducing the velocity of the gas. Without this horn, the jet of $CO_2$ gas and air would act like a blow torch and increase the intensity of the fire. Never hold the extinguisher from the discharge metal pipe to avoid cold burn. Do not lift the extinguisher more than 2 meter as $CO_2$ is heavier than air, it may affect the user. The gas makes a considerable noise during discharge. If a fireman has not used such an extinguisher before, the sound may take him off his guard and cause him to discharge the extinguisher in the form of a very dense vapour which, in a confined space will impair visibility considerably. $CO_2$ extinguisher should be sent to the manufacture or other specialist firms for recharging.

*Although a colour code for extinguishers is specified, in India, at present, all fire extinguishers are generally painted red.*

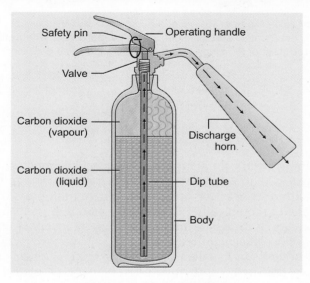

**Fig. 2.14:** CO$_2$ extinguisher.

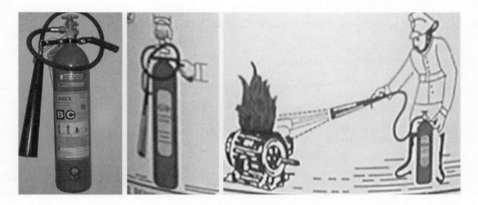

**Fig. 2.15**

## Operating Instructions

Carbon dioxide fire extinguishers contain an extinguishing agent that will not support life when used in sufficient concentration to extinguish a fire. The use of this type of fire extinguisher in an unventilated space can dilute the oxygen supply. Prolonged occupancy of such spaces can result in loss of consciousness due to oxygen deficiency.

## TAC Recommendation for Electrical Equipment

For electrical equipment, it would be necessary to provide extinguishers as under:

> Carbon dioxide extinguisher: as per IS: 2898; 4.50 kg capacity. Hydraulically tested to 250 bar. Inspect monthly and if weight is less by more than 10%, refill. Keep discharge fittings clean. Keep in shade and cool place.
>
> | Keep upright and hold in hand firmly | Remove safety pin and turn anticlockwise | Discharge directly at the base of the flame |
> | --- | --- | --- |

For rooms containing electrical transformers, switchgears, motors and/or of electrical apparatus only, not less than 2 kg. Dry powder or carbon dioxide type extinguishers shall be provided within 15 m of the apparatus. Where motors and/or other electrical equipment are installed in rooms other than those containing such equipment only, one 5 kg dry powder or carbon dioxide extinguisher shall be installed within 15 m of such equipment in addition to other requirements.

Where electrical motors are installed on platforms, one 2 kg dry powder or carbon dioxide type extinguisher shall be provided on or below each platform. In the case of a long platform with a number of motors, one extinguisher shall be accepted as adequate for every three motors on the common platform.

## Extinguisher Labelling

All extinguishers should be clearly marked with the following details:

- Type of fire extinguisher
- The extinguishing medium (extinguishant)
- Discharge type–stored pressure or gas cartridge
- Nominal charge of the extinguishant (kg or litres)
- The class of fire for which the extinguisher can be used
- Operating instructions
- Working pressure and test pressure.

## FOAM FIRE EXTINGUISHER

Portable foam type extinguishers are best suited to put out class 'B' fires involving flammable liquids like oil, solvents, petroleum products, etc. The foam expelled by activation of the extinguisher, forms a blanket over the surface of the liquid on fire and cuts out contact of the burning liquid with air and thus extinguishing the fire.

There are basically two ways of producing foam from portable extinguisher:

a. By chemical reaction
b. By self-aspiration of a foam solution through a foam branch pipe (mechanical foam).

### 9-Liter Chemical Foam Type Extinguisher

The chemical foam extinguisher consists of a cylindrical inner container and a large cylindrical outer container. The inner container is usually made of polythene and filled with 680 gm aluminum sulphate mixed in water and the outer container with 680 gm of sodium bicarbonate mixed in water. A wheel cap fitted with 'T' handle connected with diaphragm works as a blank seat for inner container so that the two solutions cannot be mixed accidentally. The two solutions, when mixed, produce aluminum hydroxide, sodium sulphate and carbon dioxide. The pressure created by the formation of carbon dioxide gas forces the foam out.

$$Al_2(SO_4)_3 + 6NaHCO_3 = 2Al(OH)_3 + 3Na_2SO_4 + 6 CO_2$$

A 6-meter jet can be maintained for a period of 30 seconds and discharge of expelled foam must be completed in 90 seconds. Foam solutions are electrically conductive and hence not recommended for use on fire involving live electrical/electronics equipment. Though foam used as a spray is less conductive than straight stream, because foam is cohesive and contains materials that allow water to conduct electricity, a spray foam is more conductive than a water fog.

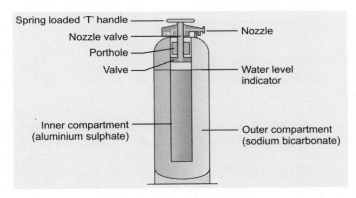

Fig. 2.16

## Method of Operation

Pull the plunger to open the valve and rest it on the notch provided. Block the nozzle with finger and turn the extinguisher upside down twice, at the same time shaking the extinguisher to ensure mixing of the two solutions.

When the pressure is felt on the finger, point the nozzle towards the fire, keeping extinguisher upside down. Direct the jet at the inner edge of the container or at an adjoining vertical surface above the level of the burning liquid. If this is not possible, direct the jet in an arc with gentle sweeping movement allowing the foam to float on the surface of the liquid.

The foam jet should not be directed into the liquid, as there is a possibility of splashing the burning liquid on to the surrounding and spreading the fire.

*Accidents are caused by unsafe acts, unsafe conditions or a combination of both*

## Mechanical Foam Extinguisher

The foam is produced by entrainment of air in the foam solution. The air entrainment can be done by various means such as using a venturi mixer or using a fan to blow air into a spray of the foam solution.

The foam solution (i.e. 3.6 liters water and 0.4 liters foam compound) is contained in the main shell of the extinguisher and carbon dioxide gas (120 gm) is held at high pressure in a sealed cartridge. When the extinguisher is operated, the cartridge seal is broken, allowing gas to escape to the main shell and push out the foam solution from the extinguisher.

Foam used in portable fire extinguishers is of two types:

- Fluoroprotein foam–which can be used for class B fires.
- AFFF (Aqueous film forming foam) which is intended for both class A and class B fires. A concentrated solution of AFFF contains perfluorosurfactants, which with addition of water, causes a reduction of the surface tension. Generally about 1 to 3% of AFFF added to water is sufficient.

## Precautions in Operation

The extinguisher should be operated by holding it upright and piercing the gas container by applying pressure on the knob of the cartridge's piercing mechanism or by operating the wheel valve of the gas container. Carbon dioxide gas is released from the gas cartridge into the container and the foam solution is expelled. The foam is forced out the delivery hose and aspirated by specially designed branch. Never block the port hole of the discharge of the specially designed branch (nozzle), allow air to be sucked in and allow the foam to expand. Where a liquid in a container is on fire, direct the jet at the far inner edge of the container or at an adjoining vertical surface above the level of the burning liquid. A jet from a foam extinguisher should have a length of at least 6 meters for a duration of about 90 seconds.

## Care and Maintenance

At least once in a week check the nozzle outlet and vent hole on the threaded portion for clogging. Check that the plunger can be raised to the fully exerted position and it is to be cleaned, if necessary.

Once in three months, dismantle the components, check for any damage, clean and grease them as required, stir the solution in the container with clean and dry stick.

Annually, operate 50% of the extinguisher and check it by projecting a jet to a distance of not less than 6 meter for a minimum period of 30 seconds.

Clean the extinguisher thoroughly and examine its inside for any corrosion. Subject rusty or corroded extinguishers to a pressure test, even if they are not due for it and their performance otherwise is satisfactory.

Carry out the pressure test every two years on each extinguisher by means of a hydraulic test pump. The test pressure is up to 17.5 bar and is applied for 2½ minutes.

| Selection of extinguishers | | | | | | | |
|---|---|---|---|---|---|---|---|
| *Extinguisher* | *Type* | *Method of operation* | *Class of fire applicable* | | | | |
| | | | A | B | C | D | E |
| Water | Soda acid | Inverted | Y | Y | Y | X | X |
| | Water $CO_2$ | Upright | YY | X | X | X | X |
| | Stored pressure | Upright | YYY | X | X | X | X |
| Foam | Chemical foam | Inverted | Y | Y | X | X | X |
| | AFFF $CO_2$ cartridge type | Upright | YY | YYY | X | X | X |
| | AFFF stored pressure type | Upright | YY | YYY | X | X | X |
| Powder | Stored pressure and gas cartridge type | | | | | | |
| | Sod bicarbonate | Upright | Y | YYY | YY | X | Y |
| | Pot bicarbonate | Upright | Y | YYY | YYY | X | Y |
| | Mon amm phosphate | Upright | Y | Y | YY | X | Y |
| | TEC | Upright | Y | Y | YY | Y | Y |
| Gas | $CO_2$ | Upright | YY | Y | YY | X | YYY |
| | Halon | Upright | YY | Y | YY | X | YYY |

X–Not suitable; Y–average; YY–good; YYY–very good.
Class E–electrical energized fire

Note: Soda acid type and chemical foam types are now obsolete.

## Operating Instructions

AFFF mechanical foam extinguisher:
Capacity –9.0 litres.
Tested to 25 Ksc.

For class A and B fires
As per IS: 10204.

After discharge, the extinguisher must be washed out carefully with fresh water with at least 2 changes.
Recharge at least after every 2 years.

Fire involving liquids with a specific gravity lower than that of water is very difficult for extinguishing with the use of water. When water is applied to the burning surface of such a liquid, it lowers the temperature momentarily and then sinks below the surface of such a liquid. Foam, which is relatively insoluble in most liquids and because of its light weight, floats on the surface of the liquid. It also forms a blanket capable of covering the

| Foam extinguisher | | | |
|---|---|---|---|
| Specification | ISI | IS:10204 | IS: 13386 |
| Capacity | Litres | 9 | 50 |
| Diameter | Mm | 175 | 300 |
| Thickness | Mm | 1.6 | 3.15 |
| Hydraulic test pressure | Bar | 30 | 30 |
| Discharge range | Mtrs | 6 | 10 |
| Discharge time | Secs | 25 to 60 | 60 to 180 |
| Min discharge quantity | % | 90 | 95 |
| Height | Mm | 600 | 1185 |
| Empty weight | kg | 6.1 | 37.8 |
| Full weight | kg | 15.1 | 87.8 |
| Temperature range | Deg C | 0 to 60 | 0 to 60 |
| Fire rating | | NA | NA |

surface of the burning liquid and so extinguishes the fire. It also serves as a radiant heat barrier, which is important in extinguishing oil and petrol fires.

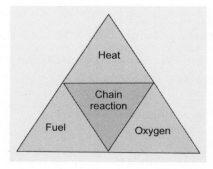

Fig. 2.17

## HALON EXTINGUISHER

Halogenated agent-type fire extinguishers contain agents whose vapour has a low toxicity. However, their decomposition products can be hazardous. When using these fire extinguishers in unventilated places, such as small rooms, closets, motor vehicles, or other confined spaces, operators and others should avoid breathing the gases produced by thermal decomposition of the agent.

Portable halon extinguishers generally range from 500 gm to 7 kg and are invariably of the stored pressures. They are normally pressurized to about 10 bar with dry nitrogen to ensure efficient discharge. Most of the extinguishers have their discharge controlled by a lever but a few designs have a striker.

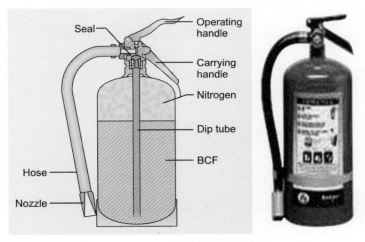

**Fig. 2.18**

Several manufacturers incorporate some sort of pressure indicator and a method of showing whether the extinguisher has been used. By their design, halon extinguishers, once used, need to be recharged either in workshops of the manufacturers or other specialists. Some smaller types have an expendable body, the operating mechanism capable of removal and replacement by another cartridge. Some others are completely disposable.

### Parts of a Fire Extinguisher

*Cylinder:* This is the body of the extinguisher. It is pressurized and holds some combination of extinguishing agent and expellant gas.

*Handle:* A grip for carrying or holding the extinguisher. Lifting an extinguisher by the handle will not cause the unit to discharge.

*Trigger:* A short lever mounted above the handle at the top of the extinguisher. The unit will discharge when the trigger is squeezed.

*Nozzle:* This is at the top of the extinguisher where the extinguishing agent is expelled and often a hose is attached to the nozzle.

*Pressure gauge:* The effective range of an extinguisher and its ability to expel its entire agent, both decrease as the pressure drops. Recharge if the pressure drops below normal operating level.

*Locking mechanism:* A locking mechanism is provided to prevent accidental discharge. The mechanism must be removed or released for the extinguisher to work.

## POWDER EXTINGUISHERS

ABC dry chemical powder extinguisher (stored pressure type)–made to IS: 13849 (Figs 2.19 and 2.20).

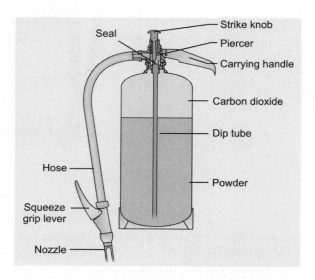

Fig. 2.19: Dry powder extinguisher–stored pressure type.

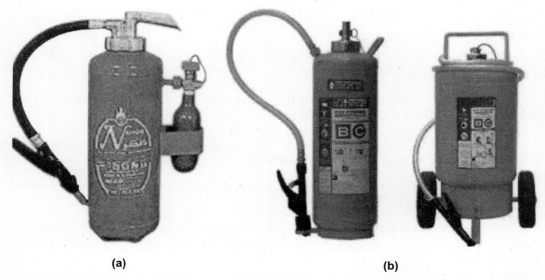

Fig. 2.20: (a) Dry powder extinguisher fires with external gas cartridge; (b) Dry chemical extinguisher for B and C.

## Expellant Gas

Carbon dioxide used in expellant gas cartridges should meet the following specifications:

1. The vapour phase should not be less than 99.5% carbon dioxide.
2. The  water content of the liquid phase should not be more than 0.01% by weight [–34.4° C dew point].
3. The oil content should not be more than 10 ppm by weight.

Nitrogen used as an expellant gas should be standard industrial grade with a dew point of –52.2° C or lower.

| | | |
|---|---|---|
| **Using a dry powder extinguisher** | | |
| Remove safety clip | Strike hand on knob and press grip, if any. | Press nozzle and direct the jet towards the near edge of the fire with a rapid sweeping motion |

Fig. 2.21

**Operating instructions:** B C dry powder: 10 kg capacity
Inspection—Unscrew cartridge from cap periodically (at least once/ year) and check by weight. If weight stamped on the extinguisher is reduced by 10%, the extinguisher needs recharging.
Do not use water for cleaning the extinguisher.
Testing: Cylinders are tested to 30 bar as per IS 2171.

*Invert or shake a dry powder or dry chemical extinguisher, monthly, to loosen the powder*

| Dry powder fire extinguisher | | | |
|---|---|---|---|
| *Extinguisher* | *unit* | | |
| Specifications | | IS: 2171 | IS: 2171 |
| Capacity | kg | 5 | 10 |
| Diameter | Mm | 150 | 175 |
| Thickness | Mm | 1.6 | 2 |
| Height | Mm | | |
| Hydraulic test pressure | Bar | 30 | 30 |
| Discharge time | Secs | 15–20 | 23–30 |
| Empty weight | kg | 5.4 | 10.4 |
| Full weight | kg | 10.4 | 20.4 |
| Storage temperature range | ° C | 0–60 | 0–60 |
| Fire rating | | NA | NA |

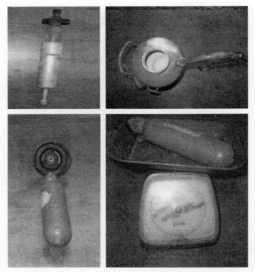

**Fig. 2.22:** Cartridge with container; powder filled cylinder; weight check of cartridge.

## Typical composition of a BC dry chemical powder

Sodium bicarbonate –97%; magnesium stearate –1.5%; dry calcium phosphate –1%; magnesium carbonate –5%.

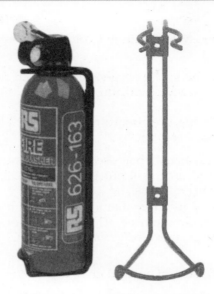

Fig. 2.23: A small 0.6 kg powder extinguisher for vehicles with stand for fixing.

Fig. 2.24: Container for placing the cartridge showing the rubber cap at the end and the discharge hole after removing the rubber cap.

The container ensures that the gas is discharged from the bottom of the powder.

The types of hazards and equipment that can be protected using dry chemical extinguishing systems include the following:

1. Flammable or combustible liquids
2. Flammable or combustible gases
3. Combustible solids including plastics, which melt when involved in a fire
4. Electrical hazards such as oil-filled transformers or circuit breakers
5. Textile operations subject to flash surface fires
6. Ordinary combustibles such as wood, paper, or cloth
7. Restaurant and commercial hoods, ducts, and associated cooking appliance hazards such as deep-fat fryers.

## Limitations

Dry chemical extinguishing systems are not considered satisfactory protection for the following:

1. Chemicals containing their own oxygen supply, such as cellulose nitrate
2. Combustible metals such as sodium, potassium, magnesium, titanium, and zirconium
3. Deep-seated or burrowing fires in ordinary combustibles where the dry chemical cannot reach the point of combustion

Dry chemical fire extinguishers, when used in a small unventilated area, can reduce visibility for a period of up to several minutes. Dry chemical, discharged in an area, can also clog filters in air-cleaning systems.

A dry chemical fire extinguisher containing ammonium compounds should not be used on oxidizers that contain chlorine. The reaction between the oxidizer and the ammonium salts can produce an explosive compound ($NCl_3$).

> Halon extinguishers should not be used on fires involving oxidizers, since they can react with the oxidizer.

Multipurpose dry chemical shall not be used on machinery such as carding equipment in textile operations and delicate electrical equipment. Before dry chemical extinguishing equipment is considered for use in protecting electronic equipment or delicate electrical relays, the effect of residual deposits of dry chemical on the performance on electronic equipment shall be evaluated.

The following factors shall be considered in determining the amount of dry chemical required:

1. Minimum quantity of dry chemical
2. Minimum flow rate of dry chemical
3. Nozzle placement limitations including spacing, distribution, and obstructions
4. High ventilation rates, if applicable
5. Prevailing wind conditions, if applicable

### Precautions to be Taken While Recharging Powder Extinguishers

Certain precautions are to be ensured during the cylinder recharging or maintenance work. We have to make sure that any powder remaining in the discharge tube, hose and nozzle is fully cleared. The interior of the cylinder should be thoroughly dried. When the new powder container is opened for refilling, the powder should be transferred immediately into the extinguisher and the appliance sealed. It is recommended that the person who handles it may wear protective facemask. Mixing of different types of powder has to be avoided. Some powders may react with each other, which may not be observed immediately, but at a later stage. Only one type of powder extinguishers should be opened at one time to avoid mixing. The charging is to be conducted in a clean and dry room.

| ABC dry powder extinguisher | | | |
|---|---|---|---|
| *Extinguisher* | | | |
| Specification | | IS: 13849 | IS: 13849 |
| Capacity | kg | 5 | 10 |
| Diameter | mm | 78 | 175 |
| Thickness | mm | 1.6 | 2.0 |
| Height | mm | 280 | 605 |
| Hydraulic test pressure | Bar | 30 | 30 |
| Min discharge quantity | % | 85 | 85 |
| Discharge range | mtr | 1.5 | 4 to 6 |
| Empty weight | kg | 0.5 | 6.2 |
| Full weight | kg | 1.5 | 16.2 |
| Temperature range | ° C | 0–60 | 0–60 |

## Safety Requirements

Where total flooding and local application systems are used and there is a possibility that personnel could be exposed to a dry chemical discharge, suitable safeguards shall be provided to ensure prompt evacuation of such locations.

Safety procedures shall provide a means for prompt rescue of any trapped personnel.

Safety points to be considered shall include, but not be limited to, the following:

1. Personnel training
2. Warning signs
3. Pre-discharge alarms
4. Discharge alarms
5. Respiratory protection

A dry chemical extinguishing agent is a finely divided powdered material that has been specially treated to be water repellent and capable of being fluidized and free-flowing so that it can be discharged through hose lines and piping when under expellant gas pressure.

Dry chemical, when discharged, will drift from the immediate discharge area and settle on surrounding surfaces. Prompt clean-up will minimize possible staining or corrosion of certain materials that can take place in the presence of moisture.

Monoammonium phosphate and potassium chloride are slightly acidic and, in the presence of moisture, can corrode metals such as steel, cast iron, and aluminum. Potassium bicarbonate, sodium bicarbonate, and urea-based potassium bicarbonate are slightly basic and, in the presence of moisture, can corrode metals such as aluminum, aluminum brass, aluminum bronze, and titanium. Such corrosion will vary from a dull or tarnished finish to mild surface corrosion. Corrosion should not be of concern when accompanied by prompt clean-up. For the most part, these dry chemical agents can be readily cleaned up by wiping, vacuuming, or washing the exposed materials. A monoammonium phosphate-based agent will need some scraping and washing if the exposed surfaces were hot when the agent was applied. Upon exposure to temperatures in excess of 121° C or relative humidity in excess of 50%, deposits will be formed that can be corrosive, conductive, and difficult to remove.

### Foam-compatible Dry Chemicals

It may sometimes necessary to apply powder extinguishing agent as well as foam. This dual application can provide the effect of both the agents. To achieve this combined action, the powder and the foam need to be applied together, in the same stream. The powder should be consistent for mixing with water and the foam solution. When foam-dry chemical systems are used or proposed for the protection of a hazard, the manufacturer of the powder and the foam should be consulted as to the compatibility of the agents. Mixtures of certain dry chemicals can generate dangerous pressures with the foam solution.

### USING A FIRE EXTINGUISHER

It is advisable to resort to the use of fire extinguishers only when it is safe to do so considering the size and position of the fire. There should be clear unrestricted access to the fire and a quick, safe retreat should be possible at all times.

It is important to have knowledge of:

- Which is the best type of extinguisher to use for a particular type of fire?
- How is it operated?
- How is it used efficiently on a fire?

Many fire extinguishers deliver their entire quantity of extinguishing material in 8 to 10 seconds (although some take 30 seconds or longer to discharge). The agent needs to be applied correctly at the outset since there is seldom time for experimentation. In many fire extinguishers, the discharge can be started or stopped by a valve. When using some fire extinguishers on flammable liquid fires, the fire can flare up momentarily when the agent is initially applied.

Portable fire extinguishers can save life and property by putting out or containing fires within the capability of the extinguisher. But, it is important to use the right type of extinguisher for the particular fire, for the extinguisher to be effective. Also, sometimes the type of extinguisher used may prove to be dangerous than helpful, in the circumstances.

### Precautions in Selection and Use of Portable Fire Extinguishers

Physical conditions that affect selection are following.

*Gross weight:* In the selection of a fire extinguisher, the physical ability of the user should be contemplated. When the hazard exceeds the capability of a hand portable fire extinguisher, wheeled fire extinguishers or fixed systems should be considered.

*Corrosion:* In some fire extinguisher installations, there exists a possibility of exposing the fire extinguisher to a corrosive atmosphere. Where this is the case, consideration should be given to providing the fire extinguishers so exposed with proper protection or providing fire extinguishers that have been found suitable for use in these conditions.

*Agent reaction:* The possibility of adverse reactions, contamination, or other effects of an extinguishing agent on either manufacturing processes or on equipment, or both, should be considered in the selection of a fire extinguisher.

*Wheeled units:* Where wheeled fire extinguishers are used, consideration should be given to the mobility of the fire extinguisher within the area in which it will be used. For outdoor locations, the use of proper rubber-tired or wide-rimmed wheel designs should be considered according to terrain. For indoor locations, the size of doorways and passages should be large enough to permit ready passage of the fire extinguisher.

*Wind and draft:* If the hazard is subject to winds or draft, the use of fire extinguishers and agents having sufficient range to overcome these conditions should be considered.

For confined spaces, prominent caution labels on the fire extinguisher, warning signs at entry points, provision for remote application, extra-long-range fire extinguisher nozzles, special ventilation, provision of breathing apparatus and other personal protective equipment, and adequate training of personnel are among the measures that should be considered.

Most fires produce toxic decomposition products of combustion, and some materials, upon burning, can produce highly toxic gases. Fires can also consume available oxygen or produce dangerously high exposure to convected or radiated heat. All of these can affect the degree to which a fire can be safely approached with fire extinguishers.

In the fire extinguishers manufactured in the country, different designs are available without any standardization. Although standards and specifications have been developed by BIS (Bureau of Indian Standards), the standards are not completely adopted by the extinguisher manufacturers. They can differ in the design of the handles, pressurizing system or delivery method. This can sometimes confuse users. Correct labelling and identification is necessary. The users should be trained in the selection and use of portable fire extinguishers.

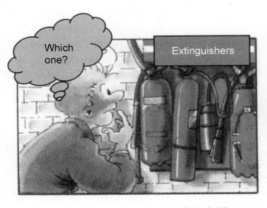

Fig. 2.25

## MAINTENANCE OF FIRE EXTINGUISHERS

Fire extinguishers are mechanical devices. They need care and maintenance at periodic intervals to ensure that they are ready to operate properly and safely. Parts or internal chemicals can deteriorate in time and need replacement. They are pressure vessels, in most cases, and so need to be treated with respect and handled with care.

Maintenance procedures should include a thorough examination of the basic elements of a fire extinguisher such as:

* Mechanical parts of all fire extinguishers
* Extinguishing agent of cartridge- or cylinder-operated dry chemical, stored-pressure, or other type of fire extinguishers
* Expelling means of all fire extinguishers

Periodically check the fire extinguisher gauge to determine, if there is pressure in the extinguisher. If the gauge indicates empty or needs charging, replace or recharge the extinguisher immediately. To test non-gauge extinguishers, push the plunger indicator (usually green or black) down. If it does not come back up, the extinguisher has no pressure to expel its contents.

*Conversion of fire extinguisher types:* Fire extinguishers should not be converted from one type to another, and should not be converted to use a different type of extinguishing agent. Fire extinguishers shall not be used for any other purpose than that of a fire extinguisher.

*Leak test:* After recharging, a leak test should be performed on stored-pressure and self-expelling types of fire extinguishers.

In a powder type of extinguisher, do not pull the pin and expel the contents to test. If a portion of the powder extinguisher is used, it has to be refilled or replaced. Invert and

**Extinguisher maintenance chart**

Date of refilling _____

Serviced by _____ HP test on _____

Due date for refilling _____

Type (e.g: $CO_2$. 4.5 kg)   Extinguisher no._____

Remarks _____

_____

shake a dry-powder or dry-chemical extinguisher monthly to loosen the powder. If the powder gets jam-packed, the extinguisher may be ineffective.

When a fire extinguisher is refilled, a shoot-off of the charge can be done to see how far it throws and how long the charge lasts.

Bags and boxes help in mounting on walls. Boxes protect the extinguisher in areas with heavy activity or traffic or rough usage locations. They help to keep the extinguishers clean and ready to use. The covers are flame resistant.

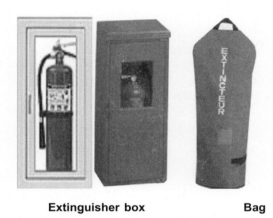

**Extinguisher box**           **Bag**

Fig. 2.26: Extinguisher box and bag.

## Foam

Fire-fighting foam is a stable aggregation of small bubbles of density lower than oil or water that exhibits a tenacity for covering horizontal surfaces. Air foam is made by mixing air into a water solution, containing a foam concentrate, by means of suitably designed equipment.

## Carbon dioxide extinguishers

Carbon dioxide will not extinguish fires where the following materials are actively involved in the combustion process:

1. Chemicals containing their own oxygen supply, such as cellulose nitrate
2. Reactive metals such as sodium, potassium, magnesium, titanium, and zirconium
3. Metal hydrides

| Parameters | BC dry chemical powder | Carbon dioxide | Water | Foam AFFF | ABC powder |
|---|---|---|---|---|---|
| Applicable | B and C | B and C | Only A | Only B | A, B, C, electrical fires |
| Contents | Sodium/potassium bicarbonate may generate corrosion if exposed to atmosphere | Liquefied carbon dioxide gas may choke if used in closed compartment unless ventilated | Water may cause electrocution or fire may spread if used on wrong fire | Sodium bicarbonate + foam stabilizer/ aluminum sulphate | Monoammonium phosphate + stabilizer/ aluminum sulphate |
| Tendency of contents stored | Granulation | Has to be kept in a cool place | Corrosion | Tends to settle | Free flowing no caking |
| Method of usage | Knob has to be banged | Difficult to operate chances of cold bun | Has to be banged | Has to be banged | Thumb press operation |
| Extinguisher design | Cylindrical | Cylindrical | Conical/ cylindrical | Cylindrical | Simple shape, easy to use |
| Net wt Gross wt Ratio | 1 kg/3 kg (3 times) 2 kg/4.6 kg (2.3 times) 5 kg/11.3 kg (2.3 times) | 2 kg/6 kg (3 times) 4.4 kg/18.4 kg (4 times) 6.8 kg/25.5 kg (3.7 times) | 9 ltrs/12.4 kg (1.4 times) | 9 ltrs/14.4 kg (1.6 times) | 1 kg/2.1 kg (2.1 times) 2 kg/3.6 kg (1.8 times) 5 kg/7.5 kg (1.5 times) |
| Pressurization method | Gas cartridge | Stored pressure | Spot pressure | Spot pressure | Stored pressure |
| Discharge time | 1.2 kg (8.30 seconds) Depending on model | 1.2 kg (14–30 seconds) Follow in all directions | 1.2 kg (60–120 seconds) Fire can spread again | 1.2 kg (85–90 seconds) Fire can spread again | 1.2 kg (6–12 seconds) 5 –10 kg (15 –25 seconds) |

*(Contd.)*

| Parameters | BC dry chemical powder | Carbon dioxide | Water | Foam AFFF | ABC powder |
|---|---|---|---|---|---|
| Discharge activation and control | To be banged on the plunger to free $CO_2$ which throws powder. Can be controlled | Valve to be opened - can be shut as required but cooling of valves make it difficult | Banged hard on floor upright or inverted- depending on model Once activated cannot be stopped | Banged hard on floor upright Inverted- depending on model Once activated cannot be stopped | Aim at the base, press the gun and leave Once activated cannot be stopped |
| Maintenance | Only in some models to be inspected, checked, refilled annually whether used/not used | Annual inspection and servicing varies with model | Inspected, refilled annually whether used/ not used | Inspected, refilled annually whether used/ not used | Nil-restricted to observation of pressure gauge |
| Pieces in kg | 1 kg 2 kg 5 kg 10 kg | 2 kg 4.5 kg 9 kg | 9 litres | 4.5 litres 9 litres | 1 kg 2 kg 5 kg |
| British standards fire rating | 8B 21B 34B 55B | 21B 34B 55B | 13A | 13A 13B | 8A 13A 21A 21B 34B 55B |
| Shelf life | 2 years | 2 years | 2 years | 2 years | 5 years |

### Multipurpose dry chemical

Ammonium phosphate based extinguishing agent that is effective on fires involving both ordinary combustibles, such as wood or paper, and fires involving flammable liquids.

Monoammonium phosphate and potassium chloride are slightly acidic and, in the presence of moisture, can corrode metals such as steel, cast iron, and aluminum.

Potassium bicarbonate, sodium bicarbonate, and urea-based potassium bicarbonate are slightly basic and, in the presence of moisture, can corrode metals such as aluminum, aluminum brass, aluminum bronze, and titanium.

## HOW TO USE A FIRE EXTINGUISHER

To actuate a fire extinguisher, one or more of the following steps are required:

*Position for operation:* The intended position for operation is usually marked on the fire extinguisher. When the position of operation is obvious (such as when one hand holds the fire extinguisher and the other hand holds the nozzle), this information can be omitted.

*Removal of restraining or locking devices:* Many fire extinguishers have an operation safeguard or locking device that prevents accidental actuation. The most common device is a lock pin or ring pin that needs to be withdrawn before operation. Other forms of such devices are clips, cams, levers, or hose or nozzle restrainers. Sometimes there may be some type of tamper-indicators such as wires or lead-seals which will break by itself upon handling.

*Start of discharge:* This requires one or more of several actions such as turning or squeezing a valve handle or lever, pushing a lever, or pumping. These can cause a gas to be generated, release a gas from a separate container, open a normally closed valve, or create a pressure within the container.

*Agent application:* This act involves direction of the stream of extinguishing agent on to the fire.

Even though extinguishers come in a number of shapes and sizes, they all operate in a similar manner. Here's an easy acronym for fire extinguisher use: **PASS** (Pull, aim, squeeze, and sweep).

Fig. 2.27: How to use a fire extinguisher.

Pull the pin at the top of the extinguisher that keeps the handle from being accidentally pressed.

Aim the nozzle toward the base of the fire.

**Kerosene**

Flash point generally between 40–65° C. Explosive limits of 0.7 to 5.0%. Kerosene consists mostly of C9 through C17 hydrocarbons. Kerosene is the most common 'incidental' accelerant, as it is used in numerous household products ranging from charcoal lighter fluid to lamp oil to paint thinner to insecticide carriers. It is also used as jet fuel. K-1 kerosene has a low sulfur content required for use in portable space heaters.

Stand approximately 8 feet away from the fire and squeeze the handle to discharge the extinguisher. If you release the handle, the discharge will stop.

Sweep the nozzle back and forth at the base of the fire. After the fire appears to be out, watch it carefully since it may re-ignite!

Position yourself so that the extinguisher can be aimed at the base of the fire. Get as near to the fire as possible. When in position, start discharging the extinguisher with a sweeping, back and forth motion across the base of the fire.

## Training in Use of Fire Extinguishers and Fire Protection Facilities

Training can ensure correct use of portable fire extinguishers, which is necessary for effective fire fighting.

Portable fire extinguishers are appliances to be used principally by the occupants of a fire-endangered building or area, who are familiar with the location and operation of the extinguisher through education or training. Portable fire extinguishers are primarily of value for immediate use on small fires. They have a limited quantity of extinguishing material and, therefore, need to be used properly so that this material is not wasted.

Fig. 2.28

Fire extinguishment can be tested on extinguishing preplanned fires of determined size and description as follows:

Class A rating:   Wood
Class B rating:   5 cm depth n-heptane or diesel fires in square pans
Class C rating:   No fire test.
Class D rating:   Special tests on specific combustible metal fires

Fire extinguishers can represent an important segment of any overall fire protection programme. However, their successful functioning depends upon the following conditions having been met:

1. The fire extinguisher is properly located and in working order.
2. The fire extinguisher is of the proper type for a fire that can occur.
3. The fire is discovered while still small enough for the fire extinguisher to be effective.
4. The fire is discovered by a person ready, willing, and able to use the fire extinguisher.

## INSTALLING A FIRE EXTINGUISHER

### Distribution of Fire Extinguishers

Location of a fire extinguisher–where should it be installed? Portable fire extinguishers are most effectively utilized when they are readily available in sufficient number and with adequate extinguishing capacity for use by persons familiar with their operation. In fire emergencies where fire extinguishers are relied upon, someone usually has to travel from the fire in order to obtain the device, and then return to the fire before beginning extinguishing operations. This takes time, with the number of seconds or minutes governed mainly by the travel distance involved in securing the fire extinguisher and placing it in operation. Sometimes fire extinguishers are purposely kept nearby (as in welding operations); however, since a fire outbreak usually cannot be prejudged as to location, fire extinguishers are more often strategically positioned throughout areas.

Where fire hazards exist, portable, hand-operated fire extinguishers shall be located where fires will not block their access. While locating portable fire fighting equipment such as extinguishers, we have to consider the nature of the risk to be covered. The equipment should be placed in conspicuous positions and should be readily accessible for immediate use. It is important to keep in mind that portable fire extinguishers are intended only for use on incipient fires and their utility is negligible if the fire is not extinguished or brought under control in the early stages.

| Fire extinguisher size and placement for class A hazard | | | |
|---|---|---|---|
| | Light hazard (low occupancy) | Ordinary hazard (moderate occupancy) | Extra hazard (high occupancy) |
| Minimum rated extinguisher | 2 A | 2 A | 4 A |
| Maximum floor area per unit of A | 300 sq m | 150 sq m | 100 sq m |
| Maximum floor area for extinguisher | 1000 sq m | 1000 sq m | 1000 sq m |
| Maximum travel distance to extinguisher | 25 mts | 25 mts | 25 mts |

The most convenient location is–wall mounted, preferably near the exit.

## FIRE BLANKETS

Blanketing is another method by which fire can be extinguished. Covering the fire with a blanket, especially of persons whose clothing is on fire, is blanketing. The person should be laid down and covered or rolled in a rug, coat, jacket, woolen blanket etc. Oil in pans and small utensils can be smothered with a fire-resistant blanket or similar materials. Fire blankets are available in two specifications, i.e. light duty and heavy duty blankets. Light duty fire blankets are used for extinguishing small fires in containers of cooking oils or fires in clothing worn by people.

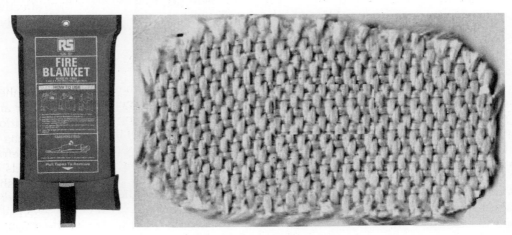

**Fig. 2.29:** Fire blanket.

Heavy-duty blankets are used for industrial applications, to resist penetration by molten metal while cutting or welding and protect against radiant heat. It may also be used for light duty applications.

Fire blankets are highly efficient fire fighting units, which can extinguish a fire in its early stages by smothering. The blanket can withstand very high temperatures. Fire blankets can be used on small fires in schools, hospitals, residences.

The blankets are generally made from glass fiber material (420 gm/sq.m). Some blankets are impregnated on both sides with either neoprene rubber or silicone rubber, which is self-extinguishing. These blankets can also be used for protecting personnel. Blankets are available in standard sizes of 120 cm × 120 cm or 180 cm × 120 cm.

Fire blankets can also be used (apart from using it to stop a fire) to protect people and also to protect equipment. Water proof covers may also be used to cover equipment during fire fighting.

## FIRE SERVICES

Fire protection services may be provided by public agencies or large establishments may have individual fire emergency service to cater to its particular fire safety needs. The fire service should be capable of providing the required services at immediate notice. Being prepared at all times is extremely important.

The efficiency of the fire services depends upon the following factors:

- Efficient manpower (personnel)
- Efficient appliances (fire engines, etc.)
- Efficient equipment (ladders, BA sets, etc.)

Fire service appliances and equipment should be properly designed to enhance the efficiency of the fire services.

An optimum design of the fire service appliances:

- Enhances the efficiency of the fire service and the degree of fire protection provided.
- Minimizes wear and tear of the fire appliances and equipment and idle time of the appliance, saves financial loss of the organization.
- Adds to the progress and development of the fire service.

### Manpower Planning for Fire Fighting

Fire fighting operation is a team work. Fire fighting operations have to be orchestrated and coordinated between the various team members, outside agencies, public and the affected parties. This coordination responsibility should be borne and handled by the captain or leader of the team. It is important that other members of the team participating in the fire fighting operation have predetermined roles and responsibilities. Individual members should be trained in their respective roles. Cross-training can also be imparted for exigencies. The role and responsibilities of the leader is very important. The leader should be an experienced person who should be able to direct the team and command respect from the team members. It is quite necessary that, working in emergencies, the team members should have faith and repose confidence in the team leader or captain.

For the fire services to provide efficient and reliable fire fighting operations, they should have at-hand a thorough knowledge of the premises at the scene of fire. These include:

- The general layout of the premises/building
- The location of the entry, exits and access points, drains, water sources
- Details of the fire protection system installed in the premises
- Details of the materials in stock and the equipment in the building
- Effect of the fire, smoke and the extinguishant on the materials and equipment in the fire area
- Presence of people inside the fire scene–where they are, how many and their details.

### Communication during Fire Fighting and Rescue Operations

Communication plays a crucial function in the coordination not only amongst the team members, but also between the operation teams and the other agencies, with the public and the people inside the fire incident premises. Today with technological advancement, wireless communication is available which can be very useful and efficient. Battery operated loud address systems can convey messages over a vast area. Full utilization of such means helps increase the capability of the operations.

### Types of Appliances Used by the Fire Services

#### Regular

- Trailer fire pump
- Water tender

- Water tanker
- Turn table ladder
- Hydraulic platform
- Miscellaneous equipment.

*Special*

- Hose laying lorry
- Foam and DCP tender
- Crash tender/RIV (rapid intervention vehicle)
- Emergency tender/rescue tender
- Mobile control van
- Breathing apparatus van
- Breakdown van
- Canteen van and other minor appliances.

Fig. 2.30: Fire engines (fire tenders).

(a)                                          (b)

Fig. 2.31: (a) One of the earliest, fire engines in America; (b) A modern fire engine.

**Fig. 2.32:** Water tender type of fire engine.

### Fire Engine Design

*Design Considerations*

- Geographical conditions.
- Road and traffic conditions.
- Capacity of the water tank.
- Type of ladders and other equipment to be mounted.
- Position of the pump mounting.
- Matching of the engine speed with that of pump through power take off.
- Required power of the fire engine.
- Required capacity of the chassis.
- Load distribution criteria.

### Typical Features of a Fire Engine

*Features of 'x' Type Standard Fire Engine (as per specification IS-6067–1972)*

- Engine
- Fuel system
- Water tank
- Hose reel
- Pump
- Primer
- Body work and
- Other accessories.

Terms pertaining to the fire engine:

- BHP: Power of the engine required to run the appliance as well as the pump. It depends on total weight of the vehicle and required acceleration.
- Torque: The rated revolution of the engine at which maximum power is transmitted through the shaft.
- GVW: Gross vehicle weight, the chassis capacity is calculated in this term.
- Pay load: Refers to the total weight of a driver's cabin, body, pump, accessories, equipment, lockers, ladders, etc. (3.5 tonnes) and the weight of the empty water tank including its shunting. (1.5 tonnes) and water contents.
- Pay load: Gross vehicle weight—tare weight = pay load.
- Tare weight: The chassis weight is referred as tare weight.

- Front axle load: Pay load permissible on the front axle, i.e. the load of cabin, engine, etc.
- Rear axle load: The pay load permissible on the rear axle. The rear axle load is nearly double that of front axle load and it covers load of water tank, pump, ladder and other accessories, etc.
- Equitable load distribution: The distribution of load equally gives longer life to the chassis.
- Power take off: Arrangement of gears by which the power is transmitted from engine to run the pump is called power take off. (The power take off shaft alignment from gear box to the pump should be horizontal, if possible.)
- 15 to 25% power is lost in transmission. It varies with the type of power take off.

In India, all fire engines are diesel driven, having a maximum bhp at 2800 rpm (revolutions per minute), but the maximum torque is at 800 to 2000 rpm (Torque means the rated revolution of the engine at which maximum power is transmitted through the shaft.) The chassis capacity is calculated by the term gross vehicle weight (GVW).

Permissible pay load is gross vehicle weight minus tare weight for the standard chassis (tare weight means weight of the chassis).

| Weight calculation of the fire engine for Tata chassis LPT 1210E | |
|---|---|
| • GVW | 12 tonnes |
| • Tare (chassis weight) | 3.5 tonnes |
| • Pay load capacity | 8.5 tonnes |
| • Driver's cabin, body, accessories equipment, lockers, ladder, etc. | 3.544 tonnes |
| • Weight of the empty water tank as per BIS specification (3 mm thick MS sheet) with baffles | 1.5 tonnes |
| • Maximum allowable water tank capacity | 3000 ltrs |

## Load Distribution Criteria for Fire Engine

Permissible pay load on rear axle is nearly double than that on the front axle. The front axle normally takes the load of the engine and the cabin. The mounting of pump immediately behind the driver's cabin is preferred to utilize the balance capacity of the front axle load. The maximum load, i.e. water tank, body, accessories, etc. is normally embossed on the rear axle. (The best equilibrium of load distribution extends the life of the chassis.) The vehicle is a 4 × 2 drive in case of a water tender.

## Selection of the Engine for the Water Tender

- Either petrol or diesel fuel type shall be capable of developing break horse power not less than 56 kW at maximum rpm.
- Shall have five or more cylinders.
- Shall be provided with a device for quick starting in addition to electrically operated starter of adequate power.
- Shall be provided with a well-designed hand starting drive.
- Shall be capable of driving the fully laden appliance at speed from starting up without any preliminary running period.

- Shall have thermostatic control device to control operating temperature of the engine cooling water.
- Shall be provided with a cooling system to permit being continuously stationary without over heating. It shall also permit the continuous running of the pump without over heating the engine. (This can be achieved by indirect open circuit type cooling system.)
- Shall be provided with suitable gauges and lubricating systems.

## Water Tank

- Shall have required capacity.
- Shall be of mild steel with 3.5 mm thickness and elliptical in shape.
- Shall be mounted immediately behind the cabin to allow the full contents to flow to the pump.
- Shall be suitably baffled.
- Shall be shaped and mounted to bring the center of gravity as low as possible in the chassis.
- Shall be fitted with a cleaning hole of 285 cm diameter.
- Shall be fitted with over flow and filling pipe.
- The overflow pipe shall not less be than 15 mm in diameter and be fitted with a delivery hose connection.

## Hose Reel

- One hose reel shall be provided on the water tender.
- Hose reel shall be mounted at easily accessible place for use from either side of the appliance.
- The hose reel shall not be less than 60 mt in length of 12 mm bore.
- The hose reel shall be fitted with a cut-off nozzle.

## Fire Pump

- Fire pump shall be centrifugal type to afford easy access to the impeller.
- Fire pump shall be preferably single stage design and as light as possible.
- The impeller length shall be renewable and be manufactured from high quality bronze.
- A drain plug shall be provided at the bottom of the casing.
- The suction inlets shall be of 100 mm diameter connected with internal strainer and a blank cap.
- The strainer shall be readily removable but retain firmly in position.
- The pump shall be mounted at front, rear or mid-ship as per specific requirement.

## Primer

- Primer shall be capable of lifting water through at least 7.2 m at a rate of not less than 30 cm per second.
- The primer shall be preferably fully automatic.
- The selection of primer shall be as specified.

## Acceptance Tests

The appliance shall be subject to the following test before putting into commission. This test may be made at the manufacturer's works or else where as agreed between the purchaser and the manufacturer.

- Road test—it relates to the acceleration and performance test
- Stability test
- Pump test
- Primer test

## Salient Features of the Water Tender Type-x

- It has a larger capacity water tank (6000 litres or more)
- It is provided with a centrifugal pump capable of delivering not less than 3200 litres of water per minute at 7 bar.

## Hose Laying Lorry

- Meant for carrying a large number of hoses to compensate the need at a serious fire.
- The appliance is constructed on a bigger chassis.
- The body is built up of mild steel of 18–20 gauge sheet on an iron angle frame.

Fig. 2.33: Hose laying lorry.

It also carries hose ramp used to protect hose lines from being damaged by traffic.

Fig. 2.34: Other types of vehicles used by the fire services.

## Ambulance

### Ambulance Equipment

Ambulance should have the following equipment:

Fig. 2.35: Ambulance.

*General:* A wheeled stretcher with folding and adjusting devices; Head of the stretcher must be capable of being tilted upward.

*Others:*
- Fixed suction unit with equipment
- Fixed oxygen supply with equipment
- Pillow with case
- Sheets
- Blankets
- Towels
- Emesis bag
- Bed pan
- Urinal
- Glass

*Safety equipment:*
- Flares with life of 30 minutes
- Flood lights
- Flash lights
- Fire extinguisher dry powder type
- Insulated gauntlets

### Emergency Care Equipment

*Resuscitation:*
- Portable suction unit
- Portable oxygen unit
- Bag-valve mask, hand operated artificial ventilation units
- Airways
- Mouth gags

- Tracheostomy adaptors
- Short spine board
- IV Fluids with administration unit
- BP monitor
- Cugg
- Stethoscope

*Immobilization:*
- Long and short padded boards
- Wire ladder splints
- Triangular bandage
- Long and short spine boards

*Dressings:*
- Gauze pads — 10 cm × 19 cm
- Universal dressing — 25 cm × 90 cm
- Roll of aluminium foils
- Soft roller bandages 15 cm × 5 m
- Adhesive tape in 7.5 cm roll
- Safety pins
- Bandage sheets
- Burn sheet

*Poisoning:*
- Syrup of Ipecac
- Activated charcoal

*Pre-packeted in doses:*
- Snake bite kit
- Drinking water

Emergency medicines — as per requirement.

**Fig. 2.36:** Breathing apparatus van.

Fig. 2.37: DCP/foam tender.

### Rescue Tender (emergency tender)

The emergency tender is very useful at a fire scene and for other emergencies also. The emergency tender carries:

- Breathing apparatus
- Special equipment for illumination.
- Other equipment required for breaking, cutting, rescue work and fire fighting.

The appliance shall be as compact as possible.

Fig. 2.38: Rescue tender (emergency tender).

### Breakdown Van

When a fire appliance develops a minor breakdown, it is not possible, neither desirable, to remove the appliance to a workshop from the scene of incidents and therefore it is necessary to rectify such defect at the place of incident itself. The breakdown van is useful to carryout such repairs. The breakdown van usually contains all emergency tools and gears required to carryout a running repair to vehicles and the pump system.

### Communication and Control

#### The Mobile Control Van

- It provides a mobile control point, to have effective fire ground control and communication with the main control at the scene of incident.
- It is required preferably on a serious fire or other serious incident where the number of fire pumps and the number of fire personnel is large, as well as when the area of incident is larger.

Fig. 2.39: Breakdown van.

- It should be fully mobile and should accommodate equipment for communication, such as VHF-set and telephone equipment, as well as other required articles, (i.e. clock, walky-talky, register book, etc.)

## Hydraulic Platform

The hydraulic platform is a mobile truck mounted ladder which is operated with engine power.

- It consists of two or three booms hinged together.
- To lower the booms, pivot in vertical plane on each other and on the fulcrum frame.
- The fulcrum frame is mounted on a turn table on the center line of chassis, over the rear axle.
- At the head of the top boom, the platform is hinged and kept parallel to the chassis by gravity or using a hydraulic ram.
- Hydraulic motors are driven by the power take off from the rear engine which affects all movements including that of jacks.
- The hydraulic oil is stored in a tank of subframe.
- The length, width and height, turning circle, acceleration, breaking, road holding, etc. vary according to type.

Fig. 2.40: Hydraulic platform.

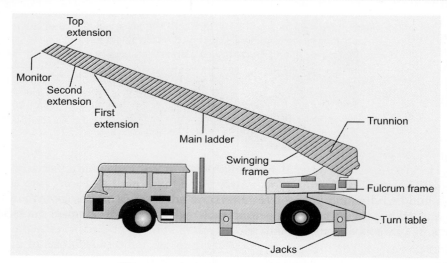

Fig. 2.41: Parts of a hydraulic platform.

## Parts of a Hydraulic Platform

### Devices and fittings on hydraulic platform

- Controlling devices
- Booms and turn table
- Jack and plumbing devices
- Cage
- The safety devices
- Monitor
- Communication devices
- Others

## Turn Table Ladder

- It is a mechanical device comprising ladder to reach heights above 18 meters.
- It enables quicker movements of the ladder in various angles, extensions and directions without moving the chassis.
- Its operation demands specialized training.
- Length of the ladder varies from 30 to 45 meters.

### Devices and Fittings on Turn Table Ladder

- The ladder assembly
- Fulcrum frame
- Turn table
- Swinging and elevating frame
- Pumping gear
- Ladder
- Automatic ladder pawl
- Monitor and brackets

- Platform
- Communication devices
- Axle locking devices
- Safety devices
- Controlling devices
- Other equipment

## Parts of a Mobile Fire Fighting Equipment

A typical fire engine has the following equipment on board (details given for a medium-sized fire engine):

- Foam compound—2700 liters
- Water—4000 litres.
- Pump capacity—3000 lpm
- Delivery hose—2 ½ in–1000 ft.
- Hand control and other type—6 nos branch pipe.
- Hose fittings of different types.
- Dry chemical type extinguisher – 65 kg capacity each – 2 nos.
- $CO_2$ gas cylinder – 22.5 kg capacity each. – 6 nos.
- Foam branch pipe – 100 lpm – 6 nos.
- Foam monitor – 1 no.
- Ladder 30 ft – 1 no.
- Fire entry suit – 2 nos.
- B-A set and other protective equipment – 1 set.
- Cutting tools – 1 set each.
- First aid box.

## First Aid Box Contents (Suggested)

a. Six small size sterilized dressings.
b. Three medium size sterilized dressings.
c. Three large size sterilized dressings.
d. Three large size sterilized burn dressings.
e. One (60 ml) bottle of centrimide solution (1%) or a suitable antiseptic solution.
f. One (60 ml) bottle of mercury-chrome solution (2%) in water.
g. One (30 ml) bottle containing sal-volatile having the dose and mode of administration indicated on the label.
h. One pair of scissors.
i. One roll of adhesive plaster (2 cm × 1 m).
j. Six pieces of sterilized eye pads in separate sealed packets.
k. A bottle containing 100 tablets (each of 325 mg) of aspirin or any other analgesic.
l. Polythene wash bottle (1/2 liter, i.e. 500 cc) for washing eyes.
m. A snake-bite lancet.
n. One (30 ml) bottle containing potassium permanganate crystals.
o. One tourniquet.
p. A supply of suitable splints.

## WATER RELAYS

### Need for Water Relays

There can be a situation, in which a fire may be very remote from the supply of water. In such a situation, it may be very difficult to get the required pressure from a single pump for carrying the water to the fire. Intermediate pumps are therefore brought in at needed places to get the required pressure for carrying water to the fire. This system of carrying water to the scene of fire is known as water relay. Standards, such as NFPA, require that the first-responding unit must start flowing water within five minutes of arrival. The minimum flow rate as per standards is 250 gpm. Large-diameter hose relays are a more efficient way to move water for distances up to about 1000 meters, while tanker shuttles are more efficient for distances more than 2000 meters. For distances in between, either method can work well.

### Truck Relay

With the plenty availability of truck mounted water tanks, supplying the required water to a scene of fire by such tanker trucks would be most suitable and convenient option today. With the availability of water tanker trucks, the easiest and most feasible water supply arrangement today is to establish a relay of water tankers from the water source to the fire scene. The water tankers also have pumps mounted on them which can be used to fill the tankers with water and also transfer the water to fire tenders. In exigencies, arrangements can be made to directly pump the water from the tankers to the fire, in which case delivery hoses have to be connected to the outlets of the pump mounted on the tankers. These water tankers have a higher capacity of water storage and the water quantity can be up to 16000 litres. The water tankers can run between the water source and the fire scene and thereby establish a series and continuous supply of water for the fire fighting operations.

### Types of Water Relay

There are two types of water relay system.

    a. Open circuit relay system
    b. Closed circuit relay system

#### Open Circuit Relay System

In this type of relay, a number of intermediate portable dams and pumps are kept between the supply pump (the first pump or lifting pump) and the fire. The water is lifted from the source of water by the supply pump and is discharged through the deliveries to the second dam (or sump), where a second pump, kept near it, lifts the water and carries it through the deliveries into the third dam, and so on, until the water is carried to the fire.

    This relay has certain disadvantages:

    a. Time is wasted in laying suction hose at each intermediate pump and in their pumping.
    b. Loss of manpower—additional manpower is required at every dam.
    c. Portable dams are required for collecting water.

## Closed Circuit Relay System

In this type of relay, intermediate pumps are arranged to be kept between the supply pump and the fire. Water is carried from one pump to another through the deliveries and collecting heads. This relay system is very much preferred to the open circuit relay system.

*Advantages:*

a. It saves time and work.
b. It enables an economy in equipment.

### Relay from a Water Hydrant System

The most common available source of water is the fire hydrant. The primary objective of the fire fighting for a large fire then is to get the largest amount of available water from these hydrants on to the fire in the minimum amount of time. Water may be required at various points in the fire scene. A four-way valve can be used to draw water from the hydrant to an intermediate tank and then pump the water from the tank to various other points.

### Arrangement of Water Relay System

While operating a water relay, the following points should be kept in mind:

- The largest pump should be placed near the water source to act as the supply pump.
- Non-percolating hoses should be preferred.
- Hoses of the largest diameter should be employed.
- Two or more lines should be employed.
- Proper ramping at road junctions must be made.

The water pressure, quantity of discharge and distance between the pumps should be as follows.

| Quantity of water and distance to be pumped between pumps | Pressure at supply pump | Size of non-percolating hoses |
|---|---|---|
| **1500 LPM** | 7 bar | |
| 100 m | | Single 70 mm |
| 400 m | | Twin 70 mm |
| 850 m | | Single 90 mm |
| 1400 m | | Twin 90 mm |
| **2200 LPM** | 7 bar | |
| 200 m | | Single 90 mm |
| 600 m | | Twin 90 mm |

### Capacity of Hoses

An important consideration while planning the water relay is the friction loss in the hoses. The loss of head due to friction is enhanced as the square of the velocity of the water in the hose. It is very important that the velocity should be as low as possible. In order to maximize the volume of water delivered to the pumps and to reduce the number of pumps to the minimum required, hose of the largest diameter should be used. If possible, two or more lines of hose should be provided between pumps. If twin lines are used instead of a single line, it will reduce the velocity to half and the loss of head due to friction is thereby reduced to one quarter. Using multiple lines, with a given pump pressure, it is quite possible to deliver over four times the distance obtainable with a single line of a hose of the same diameter.

The nominal working pressure of hose used by the fire service is 7 bar. Pressures in excess of 7 bar should not be normally used unless the short-term situation demands it. Pressures in excess of 8.5 bar should never be used continuously in a relay.

If a relay does not produce sufficient water within the capabilities of the pumps in use and the pressure is already more than the recommended 7 bar, the output can only be improved significantly by increasing the number of lines or the size of the hose. The output can also be increased by cutting short the distance between pumps by using more booster pumps.

## Practical Considerations

The efficiency of a water relay depends on the following important factors:
- a. All pumping units and other equipment must be in good working condition, so that the maximum output can be delivered regardless of their position in the relay.
- b. The layout of the relay has to be organized in such a way that all equipment can work at maximum capacity.
- c. The skill of the operators.

## Breakdowns

When a booster pump breaks down there is no need of shutting down the relay until the defective pump is ready to be replaced. There may be a drop in the output. But the relay will continue to function. The throttles of other booster pumps must be adjusted as required by the changed conditions. It is recommended to have a spare pump ready, with crew, to set into the relay.

## Communication

For the efficient operation of a water relay, it is important to maintain good communications along the route so that changes in conditions, orders to shut down, etc. can be implemented quickly. The type of communication adopted depends on conditions, availability of manpower, etc. It is the responsibility of the relay officer to employ an appropriate system.

Operators of the intermediate pumps must control the relay by observing their gauges. Their compound gauges should have a reading of zero. The throttle has to be adjusted to give the reading of zero. If it shows zero, it means they deliver the same amount of water, which is received from the supply pump.

The quantity of water delivered at this pressure is considered the nominal output of the pump. At full throttle, operating at lower pressure than 7 bar normally yields only a small increase in output according to the pump characteristics. The operation at higher pressure leads to a reduction of output which, in case of some pumps, may be quite severe. So, the operation at 7 bar is recommended.

## FIXED FIRE PROTECTION SYSTEMS

### (Water Hydrant System Standpipe System)

A water hydrant system consists of a high pressure water piping distributed all around a facility with a high pressure pumping system (usually powered by a separate generator)—for the purpose of general fire protection.

In areas subject to freezing temperatures, only a portion of the hydrant is above ground. A valve rod extends from the valve itself, up through a seal at the top of the hydrant. This design is known as a "dry barrel" hydrant. In a dry barrel hydrant, the barrel, or vertical body of the hydrant, is normally dry. A drain valve underground opens when the water valve is completely closed. This allows all water to drain from the hydrant body to prevent the hydrant from freezing.

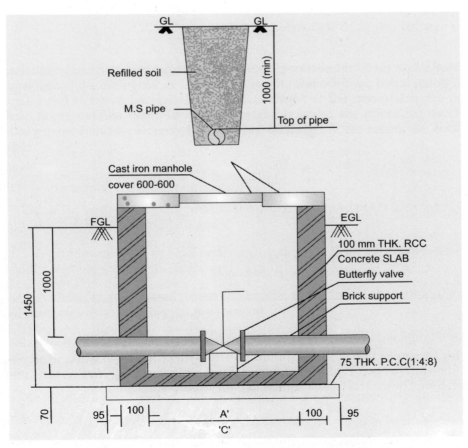

**Fig. 2.42:** Hydrant pipe laying—underground pipe and valve chamber.

## Water Supply System

The objective of a fire protection water supply system is to supply water for fire fighting throughout a facility. The system consists of water storage tanks, pumps, underground and overhead piping network and associated control valves.

The capacity of a hydrant service as recommended by TAC, considering the occupancy hazard and size of installation.

| Nature of risk | Number of hydrants | Pump capacity M³/hour | Delivery pressure bar |
|---|---|---|---|
| 1. Light hazard | Not exceeding 20 | 96 | 5.6 |
| | 20 to 55 | 137 | 7 |
| | 55 to 100 | 171 | 7 |
| | More than 100 | 171 +171 for every additional 125 hydrants | 7/8.8 |
| 2. Ordinary hazard | Not exceeding 20 | 137 | 7 |
| | 20 to 55 | 171 | 7 |
| | 55 to 100 | 273 | 7 |
| | More than 100 +273 for every additional 125 hydrants | 273 | 7/8.8 |
| 3. High hazard-A | Not exceeding 20 | 171 | 7 |
| | 20 to 55 | 272 | 7/8.8 |
| | 55 to 100 | 410 | 7/8.8 |
| | More than 100 + 410 for every additional 150 hydrants | 410 | 7/8.8/10.5 |
| 4. High hazard-B | Not exceeding 20 | Two of 171 | 7 |
| | 30 to 55 | Two of 273 | 7/8.8 |
| | 55 to 100 | Two of 410 | 7/8.8 |
| | More than 100 + one of 410 for every additional 200 hydrants | Two of 410 | 8.8/10.5 |

## Fire Water Storage

Water for fire protection systems should preferably be from fresh water sources such as river, tube-well or lake. Where fresh water is not easily available, fire water supply may be seawater or other acceptable source like recycled water.

## Fire Water Requirement

The regulatory requirements for fire water storage and pumping capacity differ from region to region. The storage capacity is to be based on the area of the premises or facility to be provided with fire protection and the facilities requiring special fire protection. But a typical guideline can be: The fire water storage requirement is calculated from the flow requirement. The flow requirement can be calculated from:

Flow requirement = ((A + B + C)/20) + D litres per minute, where,

A. Building area—total area in sq meters of all floors of the buildings in the premises.
B. Storage area—total area in sq meters of all the floors and open spaces where combustible materials are stored.
C. Elevated floor area—total area in sq meters of all floors over 15 meters above ground.
D. Area requiring special protection—such as for storage (outdoors) of flammable gases and liquids such as LPG, fuel oils, etc.

Only the numerical value of the area is to be taken for the above calculation. In areas where the fire risk involved does not require use of water, such areas under B, C or D may, for the purpose of calculation, be halved.

## Example

For a facility having a total building area of 10000 sq. m—(A); 500 sq. m of combustible material storage area–(B); 1000 sq. m of elevated floor area—(C); and a special area of 500 sq. m–(D); the flow requirement is:

$$= ((10000 + 500 + 1000)/20) + 500$$
$$= 575 + 500 = 1075 \text{ lpm.}$$

Considering a 90 minute flow requirement.

The storage capacity = 1075 × 90 = 96750 litres or 1.0 lakh litres.

If the site or the factory under consideration is situated at not more than 3 km from an established city or town fire service, the pumping capacity based on the amount of water arrived at by the formula above may be reduced by 25%.

Where the areas under B, C or D are protected by permanent/fixed individual automatic fire-fighting, installations for the purpose of calculation, can be halved.

The storage capacity should be sufficient to maintain the above flow requirement for a minimum period of 90 to 120 minutes.

## Overhead and Ground Level Storage

Overhead storage tanks may be provided if the fire water requirement is less than 50000 litres. It is accepted practice to provide ground level storage for fire water, if the requirement of fire water is more than 50000 litres. When the water is stored in ground level tanks, suitable pumps and associated piping has to be installed.

| Water storage recommended by TAC based on type of risk | |
|---|---|
| *Nature of risk* | *Capacity reserved for static storage exclusively for fire hydrant service* |
| 1. Light hazard | Not less than 1 hour's aggregate pumping capacity with a minimum of 135000 litres. |
| 2. Ordinary hazard | Not less than 2 hour's aggregate pumping capacity. |
| 3. High hazard (A) | Not less than 3 hour's aggregate pumping capacity. |
| 4. High hazard (B) | Not less than 4 hour's aggregate pumping capacity. |

The capacity of the reservoir for ordinary and high hazard class occupancies may be reduced by the quantum of inflow (of one hour in case of ordinary hazard, 90 minutes in case of high hazard (A) and two hours in cases of high hazard (B) occupancies), from a reliable source (other than town's main), but in no case should the reservoir capacity be less than 70% of that mentioned above.

Areas which are provided separate fire protection systems, such as gaseous fire protection, need not be considered in the above.

## Storage Design

Water for hydrant service should be stored in easily accessible surface/above ground reservoirs or tanks of steel or concrete masonry. Reservoirs of and over 250000 litres capacity should be in two interconnected compartments to facilitate cleaning and repairs.

The storage should be capable of delivering the flow calculated as above for at least 90–100 minutes.

As a general requirement, storage tanks of a capacity of 4,50,000 liters is normally provided in industrial establishments. If the requirements exceed, the storage tanks can be distributed around the facility.

| Hydrogen |
|---|
| Hydrogen is the simplest element. Atomic number is 1. Hydrogen gas has a specific gravity of 0.0694 (air = 1), so it is much lighter than air. Hydrogen is highly flammable, forming water upon combustion. Explosive limits are 4 to 75%. |

| BIS standards pertaining to hydrant system design | |
|---|---|
| *Part/detail* | *Specification* |
| Pump | IS: 1520 |
| Motor | IS: 325 |
| Diesel engine | IS: 5514 |

*(Contd.)*

| BIS standards pertaining to hydrant system design | |
|---|---|
| *Part/detail* | *Specification* |
| MS pipe | IS: 1239 (heavy grade) up to 150 NB; IS: 3589; 6.35 mm thick (200 NB to 350 NB) |
| MS pipe fittings | ASTM–A –105 or ASTM –A 234 Gr WPB |
| Hydrant valve (gunmetal) | IS: 5290 type A |
| Water monitor | IS: 8442 |
| Butterfly valve | BS: 5515 |
| Non-rising spindle sluice valve | IS: 780/IS: 2906 |
| Non-return valve | IS: 5312 |
| Hose box | M.S 3 mm sheet, fabricated, epoxy painted. |
| Hose | IS: 8432 |
| Branch pipe (gun metal) | IS: 903 |
| Cable | IS: 1554 (part -1) |
| Gate valve | IS: 778 |

## FIRE WATER PUMPS

A fire pump is an integral component of a total fire protection system. A fire protection system at a facility may include automatic sprinkler systems, standpipes, hose stations, and/or fire hydrants.

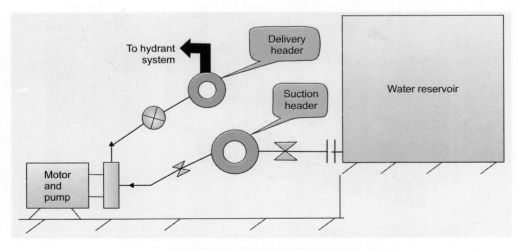

**Fig. 2.43:** Fire water pump house schematic.

**Fig. 2.44:** Fire water pump house showing two main pumps and a jockey pump.

## Engine-driven Pumps

To incorporate a high reliability in the operational availability of fire water systems, along with motor-driven pumps, engine-driven pumps are also installed in parallel to provide the pumping in case of non-availability of external power supply. It should be possible to start engine driven pumps at any time. Engine-driven pumps can be operated on variable speed basis by regulating the speed of the engine.

(a)                                    (b)

**Fig. 2.45:** (a) Engine-driven pump (Mather and Platt), (b) Vertical submersible pump.

## Working Principle of a Centrifugal Pump

The purpose of a fire pump is to provide or enhance the water supply pressure from public mains, suction tanks, gravity/elevated tanks, lakes, and other bodies of water.

Fire water supplying pumps should conform to standards such as BIS and pass performance tests to ensure that the equipment has been thoroughly examined and appropriately tested for fire pump installations.

A responsive fire protection system requires a good and reliable back up of pumps. Fire water pumps should be capable of delivering at a minimum pressure of 7 bar.

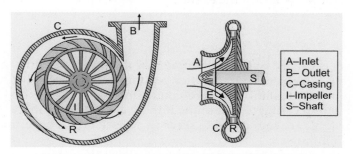

A–Inlet
B–Outlet
C–Casing
I–Impeller
S–Shaft

**Fig. 2.46:** Working principle of a centrifugal pump.

| Typical fire pump specification | | |
|---|---|---|
| *Description* | *Main pump* | *Jockey pump* |
| Make | Kirloskar/Mather Platt | Kirloskar/Mather Platt |
| Capacity (m³ per hour) | 273 | 25 |
| Motor kW | 125 | 18.5 |
| Design duty (head) (metres) | 88 | 88 |
| Type of pump | End suction; top centrifugal discharge | End suction; top discharge centrifugal |
| Speed (rpm) | 2900 | 2900 |
| Drive | Electrical motor | Electrical motor |
| Motor kW/volts | 125/415 | 18.5/415 |
| Motor type | TEFC; Class "F" | TEFC; Class "F". |

## Types and Characteristics

The two most common types of fire pumps are the horizontal-shaft centrifugal pump and the vertical-shaft turbine pump. The horizontal shaft centrifugal pump is one in which the pressure is developed by impelling water outward from a center of rotation. The impeller is mounted on a horizontal shaft. The vertical shaft centrifugal pump is similar, but the impeller is mounted on a vertical shaft.

These pumps can be driven by electric motor, internal combustion engines, or steam turbine. Fire pumps are available with rated capacities from 25 to 5,000 gallons per minute (gpm). Pressure ratings range from 3 to 28 bar for horizontal pumps and 2 to 35 bar for vertical-shaft turbine pumps.

Horizontal-shaft centrifugal fire pumps are required to be installed to operate under positive suction head. If the water supply is such that suction lift cannot be avoided, a vertical-shaft turbine fire pump should be installed. Fire pumps are designed to provide their rated capacity with a built-in safety factor (150% of rated capacity at 65% of rated pressure) to provide a cushion in the event that there is a greater than expected demand at the time of a fire.

### Testing of Pumps

Testing of pumps is carried out to determine that the pump performs as specified and designed. The tests which are carried out on a pump are discussed below.

*Hydrostatic test:* Hydrostatic tests are conducted by applying hydraulic pressure internally to the pump casing to determine if there are any leaks in the casing joints, to find out if the castings are sound and to serve as a check on the cover plate bolting. The test is done for 15 minutes.

*Running test:* Running test is carried out to establish the performance characteristics of the pump. To determine the pump performance, the head rise and the power input to the pump are usually measured as a function of the flow-rate through the pump at a constant speed.

**Fig. 2.47:** Oil rig on fire.

### Auxiliary Equipment

Fire pump accessories have an important bearing on the complete functioning of a fire pump. An understanding of the following auxiliary equipment is worthy of attention.

A motor controller is a critical component to ensure the successful operation of the pump. The controller includes timers, disconnecting means, circuit breakers, and similar devices.

The power supply to the fire pump should be positioned upstream from the facility's main electrical disconnect. In the event of a fire, this will allow the fire pump to continue to run, even though the power to the facility has been disconnected (e.g. standard operating procedures for the fire department may be to disconnect the main power supply to minimize the danger of electrical shock to fire department personnel). All electrical wiring should be in accordance with NFPA 70, National Electrical Code.

Circulation relief valves are used to prevent the pump from overheating. Their function is to open at slightly above the rated pressure when there is little or no discharge, so that sufficient water is discharged. These valves are not needed on diesel pumps where cooling water is taken from the pump discharge.

Relief valves are required on the discharge line when the operation of the pump can result in excess pressure.

Jockey pumps maintain pressure in the underground, compensate for leakage, and reduce the number of times the fire pump starts. Jockey pumps are not needed on all fire pump installations; however, they are usually found where there is an extensive underground piping system. They are equipped with gauges, control valves, and a check valve.

Hose valves are used in testing pumps. 2.5" valves are attached to a manifold placed outside the pump room to avoid any damage to the pump, driver, and controller.

## Flushing and Hydrostatic Tests

New fire pump installations require the suction piping to be flushed at designated flow rates or at the hydraulically calculated demand rate of the system, whichever is greater. The designated flow rates are dependent upon the size of the suction pipe.

| Pipe size (inches) | 4 (100 mm) | 5 (125 mm) | 6 (150 mm) | 8 | 10 | 12 |
|---|---|---|---|---|---|---|
| Flow (GPM) | 390 | 620 | 880 | 1560 | 2440 | 3520 |

Suction and discharge piping should be hydrostatically tested at not less than 14 bar pressure, or at 3.5 bar in excess of the maximum pressure to be maintained in the system, whichever is greater. The pressure should be maintained for two hours.

## Field Acceptance Tests

Field acceptance tests are required when a new fire pump has been installed. This test procedure ensures that the pump, driver, controller, and auxiliary equipment has been properly installed and is operating as per the manufacturer's specifications. Acceptance tests are generally carried out in the presence of the pump manufacturer's representative.

## Inspection, Maintenance, and Testing

The owner or a representative (e.g. management, company) is responsible for the maintenance of the fire pump. Fire pumps should be inspected, maintained, and tested as per the manufacturer's specifications.

A comprehensive maintenance programme is generally broken down into three components—inspection, maintenance, and testing.

## Inspection

A visual examination of the fire pump is performed to verify that it appears to be in operating condition and is free of physical damage. Examples include:
- Ambient temperature in the pump room is minimum 10° C (20° C if diesel engines are used) (important in cold environments)
- Pump suction, discharge, and bypass valves are open.
- Controller pilot light (power on) is illuminated.

## Maintenance

Maintenance is the work that is performed on the fire pump to make repairs or to keep it operable. Examples include:
- Lubricate pump bearings
- Clean pump room louvers
- Clean strainer and filter in diesel fuel system

## Operational Testing

A procedure used to determine the status of the fire pump and auxiliary equipment by conducting periodic physical checks. Examples include:
- Conduct a weekly churn test (run pump without water flowing)
- Conduct an annual full-flow performance test
- Operate alarm, supervisory, and trouble signals

## PUMP HOUSE

Pump rooms and power facilities should be as free as possible from exposure to fire, explosion, flood, and windstorm damage. Light, heat, ventilation, and floor drainage should be provided for pump rooms. Pumps should be located in fire resistive or non-combustible buildings. A dry location above grade is recommended. Pump rooms should be large enough to accommodate personnel as well as all equipment and devices for inspection, maintenance, and testing.

### A Typical Pump House for a Large Industrial Establishment

*Main pump:* An electric motor-driven pump with a capacity of 410 m³/hr at 88 m WC (water column) with a TEFC electric motor of 160 kW.

*Standby pump:* An identical pump as the main pump, but driven by a diesel engine which acts as a 100% standby in the event of failure of the main pump.

*Jockey pump:* A 10.8 m³/hr at 88 m WC electric motor-driven pump with a 15 kW 2900 rpm TEFC motor acting as the jockey pump to keep the water network system under pressure round the clock to ensure automatic operation of the system.

### Types of Centrifugal Pumps

There are generally centrifugal pumps of the following types:
- Horizontal end suction type
- Horizontal split-case type
- Vertical shaft, turbine type.

All the above three types can be used for fire protection systems. The inlet or suction is generally gravity fed or flooded.

---

### Fire-stopping techniques

Fire-stopping techniques are used to minimize ignition potential and fire-spread. These techniques include the following:

a. Avoidance of mass concentration of combustible materials near potential heat or ignition sources
b. Spatial separation and configuration to minimize or eliminate flame propagation paths
c. Thermal damping by judicious placement of fire-resistant heat-sink masses
d. Flashover barriers
e. Sealed packaging, such as inerted compartments and fire-resistant encapsulation
f. Automatic fire detection and suppression such as infrared thermography, fire detectors, and fixed suppression systems

---

Examples of where flammable liquid and dust can be present at the same time are at a coal-handling facility, where there is methane gas and coal dust, and in an automotive paint spray shop, where flammable paint and powdered metal flecks are sprayed.

### Operation of the Pumps

Fire pump operation procedure for (Fig. 2.49) the installation is as follows:

- System pressure is maintained at 7 bar using jockey pump.
- Jockey pump starts on/off automatically depending upon the system pressure.
- Main pump 2 will start automatically, if the system pressure is reduced to 6 bar.
- Main pump 1 will start automatically, if the system pressure is reduced to 5 bar.
- Diesel pump will start automatically, if the system pressure is reduced to 4 bar.

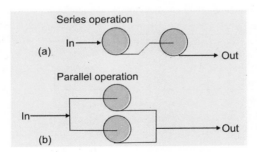

**Fig. 2.48:** Pumps connected in series and parallel.

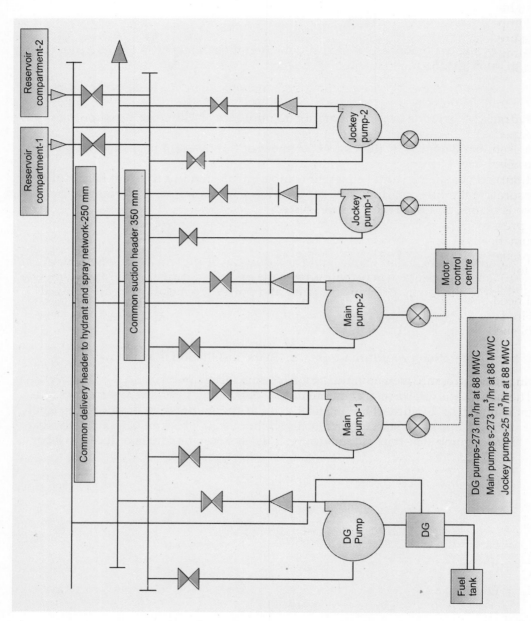

**Fig. 2.49:** Schematic of a typical fire—water pump-house.

## Multistaging of Pumps

Centrifugal pumps can be combined to achieve the duty requirements—higher pressure or higher flow. The pumps can be combined either in series or in parallel.

Pumps in series—the objective of running pumps in series is to increase the head or pressure.

Pumps in parallel—when pumps are connected in parallel, the effect is an increase in the total flow or discharge.

## FIREWATER PIPING

A hydrant system consists of a network of water piping. The fire hydrant system is designed to provide water at any given time.

A useful and efficient hydrant system is the backbone of any fire protection installation. Whenever water is required for fire fighting, the hose (available in a hose box near the hydrant) is to be connected to the hydrant valve and the valve is to be opened. The water pressure in the piping network is generally maintained above 8.8 bar. When the hydrant valve is opened to draw water, the network pressure drops below 8.8 bar. The drop in the pressure activates a pressure switch and actuates the hydraulic pumps to start automatically. The pumps are divided into–main pump with standby and jockey pump. At predetermined pressure levels, the main pumps start automatically in order to make up for the pressure loss. The jockey pump is designed to stop automatically as soon as the main pump switches on. The main and the standby pumps do not stop automatically and need to be stopped manually.

Fire service pipe lines should be sized as per the codes of "Fire Protection Manual" published by the Tariff Advisory Committee (TAC) as well as relevant Indian standards.

### Fixed Hydrant System Piping

Piping material–IS: 1239(heavy)/IS 3589 (6.35 mm thick) is used for fire-water piping.

The network is analyzed for the pressure loss and velocity at each and every nodal point using methods such as the Hardy-cross method of network analysis.

### Hazen William's Formula for Pressure Loss in Pipes (and Hoses)

$$P_f = \frac{10.9 \times 10^5 \times Q^{1.85} \times L}{D^{4.87} \times C^{1.85}}$$

Where,

$P_f$ = Loss of water pressure in the  pipe (metres of water)
$Q$  = Flow inside the pipe in lit/min.
$D$  = Inside diameter of pipe in m.
$C$  = Hazen William's coefficient = 120
$L$  = Equivalent length of pipe in m.

### Hose Boxes

Hose boxes containing water delivery hoses and branch pipes are installed near all water hydrant locations. They are generally kept in a key box. In case of a fire, the glass of the box is broken and the hose box can be opened with the help of a key placed inside. One end of the delivery pipe is connected to the fire hydrant and to the other end the branch pipe is to be connected, so that the water from the fire hydrant can be directed to the seat of the fire.

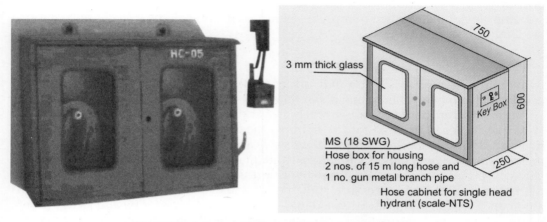

Fig. 2.50: A hose box along side a manual call point.

Fig. 2.51: A hydrant valve and representation of a hose reel.

The power source for an electric motor-driven fire pump must be reliable and have adequate capacity to carry the locked-rotor currents of the fire pump motor and accessory equipment. These two main requirements ensure that the fire pump will operate in the event of a fire without being accidentally disconnected and that the fire pump will continue to operate until the fire is extinguished, the fire pump is purposely shut down, or the pump itself is destroyed.

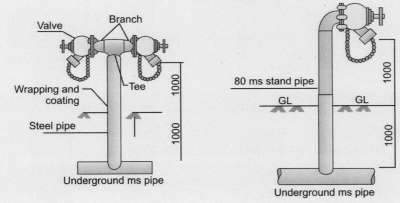

**Fig. 2.52:** Hydrants of various types.

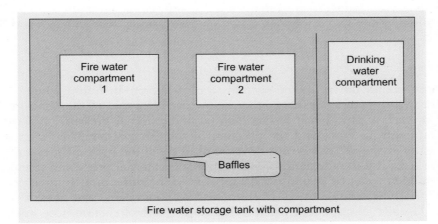

Fire water storage tank with compartment

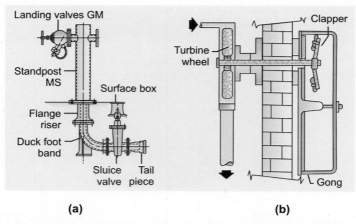

**(a)**  **(b)**

**Fig. 2.53:** (a) Underground hydrant pipe; (b) Water flow alarm.

| Class and specification of pipes for hydrants | | |
|---|---|---|
| *Type of pipe* | *Class of pipe* | *Specification* |
| i.   Horizontally cast iron pipes | B | IS: 7181 |
| ii.  Vertically cast iron pipes | A | IS: 1537 |
| iii. Centrifugally cast  (spun) iron pipes | A | IS: 1536 |

## HOSE REEL AND HOSE FITTINGS

### Hose

Hose is one of the most basic essential requirements of fire fighting equipment. It is necessary to convey water either from open water supplies or pressure supplies. Mainly, hoses are divided into two categories, based on their uses as suction hoses and delivery hoses.

### Suction Hose

Suction hose is laid down on the suction side of the pump (inlet) where the water passing through it is at a pressure either below or above that of the atmosphere. It is designed in such a way that it resists internal and external pressure. It should have sufficient strength to withstand the pressure of the external air when a vacuum has formed inside. It should also be strong enough to resist the hydrant pressure, usually encountered. As it will withstand both internal and external pressures, it can be used either from open water or pressure fed supplies. At the same time, it must possess the maximum flexibility and lightness. Suction hose is an inevitable part of every pumping appliance. Usually an appliance has to carry about 10 m of suction hose in either 3 m or 2.5 m length depending on the storage capacity. The diameter of the hose depends on the capacity of the pump and three standard sizes such as 75 mm, 100 mm and 140 mm are generally used.

*Partially embedded suction hoses:* This is usually made of tough rubber lining embedded fully as a spiral, from tempered galvanized steel wire. This embedding is so arranged that it provides a full water-way and a relatively smooth internal surface. The wall of the hose is prepared from several layers of canvas and rubber lining in such a way that turns of each one lies midway between two turns of the other. The complete wall is consolidated by vulcanizing.

*Fully embedded (smooth bore) suction hoses:* This has a thick internal rubber lining embedded fully with a spiral wire. The wall is built up in the normal way with piece of fabric and rubber as for the partially embedded type. Suction hose should be constructed to withstand a pressure of 10.5 bar.

*Delivery hose:* Delivery hose is laid down from the delivery side of the pump (outlet) to the place of requirement and the water passing through it is always at a pressure greater than that of the atmosphere. Delivery hose are divided into two categories.

   i. Percolating Hose (unlined hose)
   ii. Non-percolating Hose (Lined hose)

*Percolating hose:* It is mainly used for forest fire fighting applications. The seepage of water through the hose provides protection to the hose against damage by glowing embers falling on to it or the hose being laid on hot ground. Percolating hose is made from a jacket, which contains a proportion of cotton yarns to provide wetting and wicking properties, lined with a very thin lining, which will allow water to seep through without jetting. This keeps the outer jacket wet.

*Non-percolating hose:* In fire services, non-percolating hoses are generally used for delivering water. These consist of a reinforced jacket, made from polyester or nylon yarns. This type of hose has an inner lining of vulcanized rubber fixed to the jacket by an adhesive. The coated lining is inserted into the jacket and steamed with supper heated steam. The use of non-percolating hoses is recommended as friction losses will be much lesser than that of percolating hoses.

Lined hoses are mainly divided into 3 types:

*Type 1: Lined hose without external jacket treatment:* Such hose absorbs liquid into the reinforcement jacket and requires drying after use.

*Type 2: Coated lined hose:* This has a thin elastomeric outer coating which reduces liquid absorption into the jacket and may slightly improve abrasion resistance.

*Type 3: Covered lined hose:* In this case, a thicker elastomeric cover is applied preventing liquid absorption but also adding substantial improvements to abrasion and heat resistance.

(a)                   (b)

**Fig. 2.54:** (a) High pressure coated hose, (b) Flexible fire hose.

## Selection and Care of Hoses

### Characteristics of Hoses

a. *Flexibility:* Hose must be flexible enough to be handled easily and without kinking.
b. *Durability:* The durability and wearing capacities of hose should be as high as possible. Materials used, particularly in the wrap, should have high resistance to abrasion and be able to withstand the rough usage.
c. *Resistance to rot:* Natural fibers such as flax and cotton are liable to be affected by mildew rot. When these materials are used in the construction of the hoses, they must be given rot-proofing treatments.
d. *Friction loss:* A rough internal surface helps to increase the resistance to the flow of water through the hose. The surface should therefore be as smooth as possible for reducing the loss of pressure through friction.
e. *Weight:* The weight of a hose is important not only for handing, but also for stowing and storing.
f. *Hose pressure and acceptance test:* Fire hose has to withstand very high internal pressure.

During fire fighting, this pressure may be as high as 10 bar and shock pressure may be even higher. So it is usually so constructed as to withstand at least twice the pressure to which it is likely to be subjected. Each length of hose is to be tested for a pressure of 20 bar. Manufacturers usually carry out the acceptance tests and certify the hoses.

### Materials Used to Make Hoses

a. *Flax:* Flax has great strength and durability with high absorbent qualities. But when wet, the weight of lined hose gets increased. It must be given a rot proofing treatment also.

b. *Cotton:* Cotton is strong and has high resistance to abrasion. It also has high absorbent qualities and requires rot proofing treatment.

c. *Nylon:* Nylon is a synthetic rot resistant fiber, light in weight and of great strength with low degree of absorbency. Nylon stretches well enough under pressure and its use in the hose is normally limited to the weft.

d. *Terylene:* Terylene is a rot resisting fibre, light in weight and has great strength with low absorbent qualities. It resists stretching to a great extent than synthetic fibers and can be used for warp as well as for weft.

## Construction of Hoses

Jacket for non-percolating hoses is woven on circular looms. It is either of plain or twill weave. Plain weave is generally used. The type of material used and method employed for making jackets of non-percolating hoses varies with each manufacture. Rubber or plastic can be used in combinations with any of the jackets.

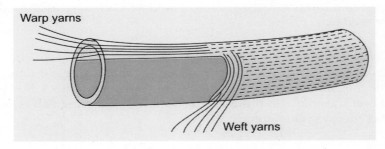

**Fig. 2.55**

a. Plain view jacket
b. Twill jacket

The various weaving instructions are defined as follows:

a. Warp: The yarn running length wise down a hose.
b. Weft: The yarn running circumferentially around the hose.

A jacket is the layer in which the weft passes over and under each warp alternately. A jacket can also be with the weft yarn passing over one warp and two consecutive warps (or vice versa).

## Hose Reel Hose

It consists of a rubber tube or lining strengthened by wrapping with layers or plies of rubberized woven fabric applied spirally. An outer cover of abrasion resistant rubber is applied and wall is vulcanized together. Hose-reel hose is a non-percolating delivery hose of small internal diameter, normally, 19 mm, carried on a rotating drum. The hose must be flexible, but at the same time should not flatten unduly when coiled on. It is used for taking a hose line quickly inside the building using the available water supply carried on the appliance such as water tender.

### Damage to a Hose

It is possible to avoid most of the damages caused to the hoses. It can be avoided by strictly carrying out the hose drill as instructed. Every opportunity must be made use of to impress on personnel, the necessity for care in the use of hose. The causes of damages in hoses are given below.

### Abrasion

Dragging of hose is an important point to be considered. Abrasion is likely to happen more rapidly when the hose is wet. It should never be pulled around a corner where it might scrape against wall. The vibration of a pump will certainly cause severe chaffing at the point where the hose touches the ground leaving the pump delivery.

### Mildew

Mildew is a fungus which can damage the hose very severely. Hoses not dried enough or are in a damp condition cause the growth of mildew. Lack of ventilation also expedites the growth of mildew.

### Shock

Rolls of hoses thrown roughly to the ground causes shock. If water flows too rapidly, without giving time for the hose to take the strain, also causes a shock. Sudden closure of hand-control branches can also bring about shock. While closing valves, special care is to be taken to prevent water hammer.

### Rubber Acid

Rubber lined hoses stored without being properly drained, will form water which will turn to acid in pockets. When the water evaporates through slow drying, the acid concentration increases. This accumulated acid contaminates the hose jackets.

### Chemicals

Hoses are liable to damages when they come into contact with acid, alkalis, oils, grease, paints, etc.

### Methods of Storing Hoses

#### The Roll

In this method, the hose is laid out flat on the surface and roll is made from the female coupling end. The coupling is doubled down on the hose. Hose is then rolled up until the male coupling is reached. A hose strap is finally passed through the centre of the roll and secured behind the male coupling.

#### Dutch Roll

The hose must be laid down flat on the surface and the female coupling has to be drawn back along the hose towards the other end so that the female coupling rests on the top of the hose and about 1 meter short on the male coupling. After the upper layer has been

**Fig. 2.56:** The roll.

arranged exactly over the lower layer, the hose is rolled up from the bite so that the coupling comes together on the out side of the roll.

**Fig. 2.57:** Dutch roll.

### Flaking

In this method, the hose is doubled back and forward on itself and subsequently secured at the centre of the folds.

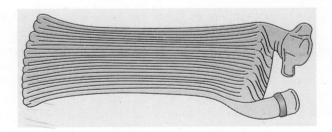

**Fig. 2.58:** Flaking method of storing hoses.

### Figure of Eight

This is a method of flaking avoiding sharp folds. But this type of storing requires more space than the other types of storing.

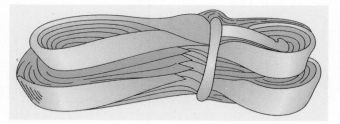

Fig. 2.59: Figure of eight flaking.

The following are the general hints for prolonging the life of hoses:

- New hoses must never be allowed to remain in the case in which it is received. Bends have to be removed and the coil loosened.
- Hoses must be stored in cool, dry and well-ventilated places.
- When storing hoses, short bends must be avoided as far as possible. If it is necessary to store hoses in folds, the bend must be removed at intervals to prevent the fabric from taking a permanent set at the bend, particularly hoses with rubber lining.
- Those, which are used for a longer period of time, should not be left to remain on the appliance. They have to be removed and placed on racks or towers.
- Water should be passed periodically through rubber-lined hoses to keep the lining in proper condition. They should be properly drained and dried in towers or in a warm room.
- Rubber gets spoiled more rapidly at high temperatures. Therefore, rubber-lined hoses should not be exposed to hot air or direct sunlight.
- Hoses should never be bent or rested at too acute an angle especially under water pressure. This would cause severe strain on the hose and may lead to breakage.
- Hoses, which are laid down across road, must be ramped to prevent damage by the passing vehicles. If hoses are contaminated with acids or alkalis, they should be thoroughly washed with water immediately.
- Hoses should be drained by under running before storage.
- For storing hoses on appliances, rolled or coiled hose is preferable to that made by other roll methods. The coiled hoses (rolls) are less acute and less liable to damage.
- When rolling hoses, the female coupling should be rolled first and the first coil should be slightly loose. Care should be taken when the hoses are stored on appliances to avoid possibility of hoses chaffing against lockers or with the other contents of the locker, due to vibration.
- Appliance locker doors should be opened at frequent intervals to allow circulation of air.

## Hose Drying Cabinet

These are used for drying wet hoses. The interior of the cabinet is made warm and ventilated by means of heaters and fans. The temperature is usually fixed at approximately 40° C and arrangements are usually made that the temperature shall not rise above the determined point. When hose is thoroughly dried, it should be soft and flexible.

## Hose Drying Plant

Modern hose drying plants generally consist of bridles having a number of drying chambers. Hoses are coupled to manifolds through which hot air is blown. In addition, hot air is circulated by a fan for drying outer sides of the hoses. An exhaust fan in each chamber maintains constant circulation of air.

## Repairing of Hoses

Immediately after use, hoses should be cleaned and tested carefully. They should to be examined for damages. The damages can be repaired by the following procedures.

### Hose Clamps and Bandages

Holes in the hose can be temporarily covered by means of clamps and bandages while in use. This will prevent the hose from bursting and also will save wastage of water. Several types of clamps are available to be used as bandages.

### Darning

Hoses can be permanently repaired by darning. The operation is like darning of cloths to cloth.

### Patching

Leakage of hoses can be repaired satisfactorily by patching, if the hose is in good condition. At present, this method is practiced widely. The patch is applied in a similar manner to the repairing of a punctured inner tube of a tyre.

## Hose Fittings

Water is the most widely used extinguishing medium in majority of fires. Since water is very cheap and easily available, fire services all over the world use it extensively. Water for fire fighting must be taken from a source, conveyed on to the scene of fire and delivered in the desired shape or mode. The method of application varies depending upon the type of fire such as jet, spray, fog, etc. We may not get water on the scene of fire where it is required. Fire fighting pump draws water from the available source, imparts pressure and delivers it to the scene of fire. Suction hose draws water up to the pump arid delivery hoses help to carry it to the required place. To join hoses, fittings, which are used in conjunction with hoses, are known as "hose fittings". The different types of hose fittings, which are commonly used in fire services, are:

1. Couplings
2. Branches and nozzles
3. Collecting heads and suction hose fittings
4. Breechings
5. Adaptors
6. Miscellaneous hose fittings

## Couplings

Fittings used for joining two lengths of hose together or to an apparatus or vice-versa, are called couplings. They are fitted to either end of hose lengths used for suction or delivery purpose. The following couplings are normally used.

Interlocking couplings
  (i) Instantaneous male/female
 (ii) Sure lock coupling

Hose reel couplings
  (i) Hermaphrodite coupling
 (ii) Macdonald coupling

Screw type couplings
  (i) Round-threaded
 (ii) V-threaded

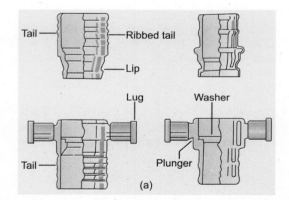

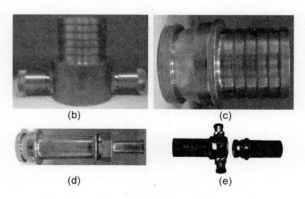

**Fig. 2.60:** Couplings.

All couplings are always a pair, one of which is named male and the other female. They differ in shape and function except the hermaphrodite coupling where each of the pair is similar in shape and function. Instantaneous couplings are used for delivery hoses and pump outlets whereas screw type couplings are used on suction hose and pump inlets.

## Water Monitors

A water monitor is a permanently mounted water nozzle, often elevated to prevent obstructions so that the monitor can direct a high pressure stream of water over a wide area. Water monitors are provided in emergency areas and inaccessible areas where it is impractical to lay hose lines.

A water monitor is a piece of equipment designed for mounting on hydrant stand pipes or built trolleys. Water monitors can be operated by a single person. It can be fixed at the desired elevation and can be rotated in any desired direction. The water monitors are installed at strategic locations such as LPG storage yard and can be used to direct water from the hydrant system on to the required area from a distance. The monitors can give a horizontal throw of about 55 meters and vertical 30 meters.

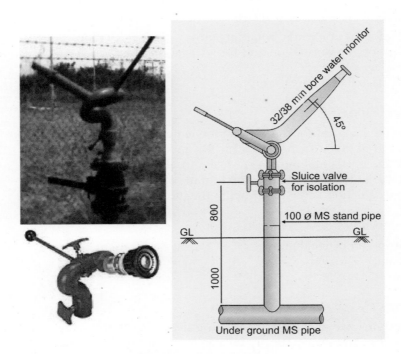

**Fig. 2.61:** Water monitor.

Water monitors can also be used with fog nozzles or with a foam branch pipe using instantaneous type (quick-fit) coupling fittings.

**Fig. 2.62a:** Water and foam application monitor for a flammable liquid storage facility (Williams hazard control).

**Fig. 2.62b:** Mock drill in a LPG bottling plant of Hindustan Petroleum at Mysore.

Water monitors are provided with:

- Drain cock
- Reducer
- Swivel joint for horizontal rotation and for vertical rotation.
- Lock for horizontal rotation and for vertical rotation.
- A nozzle
- Handle, base flange, grease nipple.

### Water Monitor Operation

- Release the lock-nut for the vertical movement of the monitor nozzle.
- Adjust the jet towards the fire vertically and lock it.
- Release the horizontal movement lock nut, if required.
- Adjust the jet towards the fire horizontally and lock it.
- Open the isolation valve provided in the monitor, gradually.

### Monitor Movement

Water monitors can be rotated by:
- Horizontal –0° to 360°.
- Vertical –45° to 80°.

### SPRINKLER SYSTEMS

Sprinkler systems are automatic fire extinguishing systems which can detect a fire, announce the fire and automatically take action to spray the extinguishant to suppress the fire.

A sprinkler system consists of a network of water carrying pipes, fitted with spray nozzles, installed around a control space so as to discharge water at a designed flow rate and pressure to control or extinguish a fire. The system is actuated by suitable signal mechanisms or by manual operation.

A sprinkler is the most important part of an autosprinkler fire extinguishing system. It can detect and extinguish fire automatically. Sprinklers with initiation temperature ranging from 57 to 182° C can be provided according to the nature of fire risk. Latest design concealed sprinklers can be provided to keep up the places like hotels, supermarkets, offices, etc. In case of fire, the sprinkler's cover automatically drops off when the surrounding temperature is 10 to 15° C lower than nominal action temperature of the sprinkler head. As the heat increases, the exposed sprinkler sprays water/chemical to extinguish the fire.

Sprinklers are made mandatory in many regions of the world.

The whole philosophy of sprinklers is based upon the premise of applying the right amount of water (as little as possible) in the right place (the seat of the fire) at the right time (as quickly as possible).

This system is an active fire protection measure, consisting of a water supply, providing adequate pressure and flow rate to a water distribution piping system, on to which fire sprinklers are connected.

1. Control mode sprinklers are the standard design. These stop a fire from spreading by dumping water directly on the fire when it starts, lowering its core temperature to the point where the fire can no longer sustain its heat. This design also pre-wets flammable material adjacent to the fire.
2. Suppression sprinklers are specially suited to work quickly and handle fast-growing and challenging fires. Instead of pre-wetting the area as the control mode design does, the suppression sprinklers release a deluge of water directly on the core of the fire and thereby lowering the temperature quickly and efficiently. This design is often preferred in buildings containing highly flammable materials, as they quickly stop an already-severe fire from growing.

Low-pressure sprinklers are designed to handle fire protection needs in tall buildings where water pressure may be reduced in the upper floors. This design is often used in skyscrapers and tall tenement buildings. Using this design can be more cost-effective than other designs. It will also reduce pipe size and reduce the need for a fire-water pump.

These sprinklers use only about a hundredth of the water used by the fire brigade to extinguish a fire.

The main extinguishing effect of water is based on its heat absorbing ability. The cooling effect interrupts heat development and prevents the evaporation of combustibles by wetting combustible materials. Water extinguishes fires by cooling and prevents ignition of unburned materials.

- Sprinklers provide an ideal fire protection.
- Sprinklers are reliable without complications.
- Sprinklers are economic and environment friendly.
- When sprinklers are used correctly, provides extremely high degree of safety.

- Sprinklers operate only when necessary.
- Sprinklers offer structural fire protection.
- Sprinkler systems can be automatic and independent of other systems.

Types of automatic sprinkler systems:

- Alternate
- Dry pipe
- Pre-action
- Wet pipe

## Water Storage and Supply for a Medium Velocity Water Spray (MVWS) System

Sprinkler systems need a reliable source of water supply of ample capacity. A water spray system is under pressure at all times so that water will discharge immediately when the MVWS system operates. An automatic alarm valve triggers a warning signal when water flows through the spray piping.

| Design considerations for an MVWS system for a double bullet LPG system of 75 cu.m each (total 75 MT) | |
|---|---|
| 1. Bullet dimension | : 4 m dia × 10.7 m length of shell |
| 2. Dish-end 2 nos | : 4 m diameter |
| 3. Surface area with product piping | : 429.4 m$^2$ |
| 4. Design density of spray | : 10.2 lpm/m$^2$ |
| 5. Water demand | : 4379.88 lpm |
| 6. Deluge valve selected | : Diameter 150 mm (max output 14100 lpm) |
| 7. No. of spray rows | : 6 |
| 8. No. of sprayers per row | : 8 |
| 9. Distance between spray nozzles | : 2 m |
| 10. No. of spray nozzles. | : 110 |
| 11. K-factor for nozzle selected | : K 35 |
| 12. Minimum pressure assumed at the remotest nozzle | : 1.4 bar |
| 13. Discharge flow rate assumed at the remotest nozzle with min pressure of 1.4 bar | : 41.41 lpm |
| 14. Proposed water flow rate | : 5601.95 lpm |
| 15. Max pressure considered in the network | : 2.61 bar (less than 3.5 bar) |
| 16. Max velocity considered in the network | : 3.79 m/sec (less than 5 m/sec) |

The passages in the water spray nozzles are small and can easily get clogged by foreign material. Hence it is necessary to provide and install strainers in the pipe lines.

Over one-third of all sprinkler system failures can be attributed to inadvertently closed water supply.

Piping for sprinkler systems can be:

- Wet pipe system.
- Dry pipe system.

## Wet Pipe System

In a wet pipe system, the sprinkler pipes are always filled with water under pressure.

*Features:*

- All parts of the system piping up to the sprinkler heads are always filled with water under pressure.
- When heat actuates the sprinkler system, water is sprayed over the area below.

*Disadvantages of the wet pipe system:*

- Shutting off and draining a wet pipe system is hazardous in that it removes the protection when it may be necessary.
- Once the fusible bulb melts, water flow continues until the main water supply is manually closed.

## Dry Pipe System

Dry pipe sprinkler systems are installed in areas where there is a risk of freezing of the water in the sprinkler system. The sprinkler pipes are not filled with water, but down stream of the dry-pipe alarm valve, the pipes are filled with compressed air.

## Pre-action Dry Pipe System

Pre-action dry pipe system is a combination of a fire detection and sprinkler system. Such systems are installed where any unintentional or mistaken accidental release of water needs to be prevented. This can happen due to reasons like mechanical damage to the sprinkler system. Confirmatory signal measures are incorporated in the system to achieve high credibility.

MVWS systems have an automatic fire/heat detection and actuation system. Detection is by a quartzoid bulb detector. A standard quartzoid bulb detector has a liquid filled glass bulb with a temperature rating of 79°C. The basic principle of water spraying is to give a complete surface wetting with a pre-selected water spray density. The design of the system takes into consideration—the nozzle types, sizes, spacing and amount of water required.

Fire sprinklers can be automatic or of open orifice design. Automatic fire sprinklers operate at a predetermined temperature, utilizing a fusible element, a portion of which melts, or a frangible glass bulb containing liquid which breaks, allowing the plug in the orifice to be pushed out of the orifice by the water pressure in the fire sprinkler piping, resulting in water flow from the orifice. The water stream impacts a deflector, which produces a specific spray pattern designed in support of the goals of the sprinkler type (i.e. control or suppression). Modern sprinkler heads are designed to direct the spray downwards. Spray nozzles are available to provide spray in various directions and patterns. The majority of automatic fire sprinklers operate individually in a fire. The entire sprinkler system does not activate, unless the system is of a special deluge type.

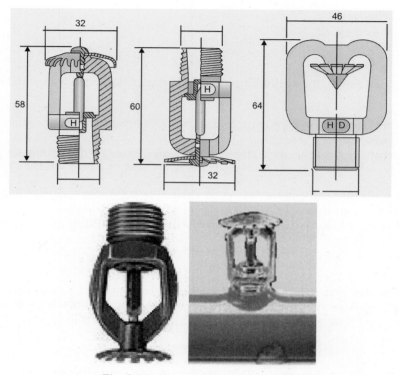

**Fig. 2.63:** Types of sprinkler heads.

Open orifice sprinklers are only used in water spray systems or deluge sprinkler systems. They are identical to the automatic sprinkler on which they are based, with the heat sensitive operating element removed.

Automatic fire sprinklers utilizing frangible bulbs follow a standardized colour coding convention indicating their operating temperature. Activation temperatures correspond to the type of hazard against which the sprinkler system protects. Residential occupancies are provided with a special type of fast response sprinkler with the unique goal of life safety.

| Maximum ceiling temperature | Temperature rating | Temperature classification | Colour code (with fusible link) | Glass bulb colour |
|---|---|---|---|---|
| 38° C | 57–77° C | Ordinary | Uncoloured or black | Orange (57° C) or red (68° C) |
| 66° C | 79–107° C | Intermediate | White | Yellow (79° C) or green (93° C) |

*(Contd.)*

| Maximum ceiling temperature | Temperature rating | Temperature classification | Colour code (with fusible link) | Glass bulb colour |
|---|---|---|---|---|
| 107° C | 121–149° C | High | Blue | Blue |
| 149° C | 163–191° C | Extra high | Red | Purple |
| 191° C | 204–246° C | Very extra high | Green | Black |
| 246° C | 260–302° C | Ultra-high | Orange | Black |
| 329° C | 343° C | Ultra-high | Orange | Black |

Adopted from NFPA–13. The table indicates the maximum ceiling temperature, nominal operating temperature of the sprinkler, colour of the bulb or link and the temperature classification.

(a)

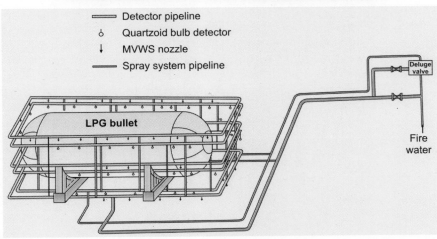

Detector pipeline
ᵒ Quartzoid bulb detector
↓ MVWS nozzle
Spray system pipeline

LPG bullet

Deluge valve

Fire water

(b)

**Fig. 2.64:** (a) Sprinkler system array around an LPG bullet; (b) Bullet-sprinkler schematic.

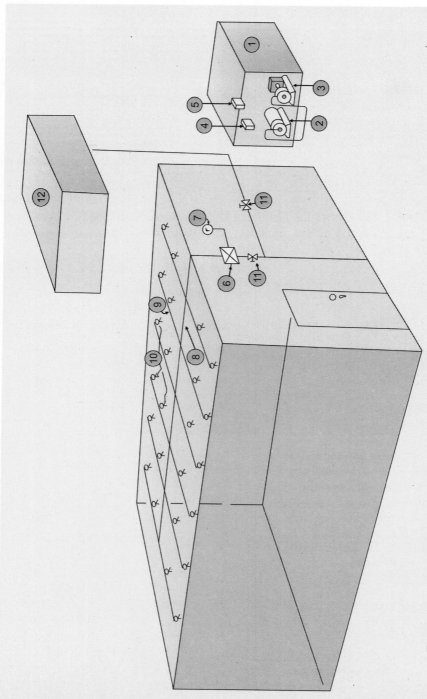

**Fig. 2.65:** Sprinkler system—schematic view.

1. Pump room, 2. Motor-driven pump, 3. Engine-driven pump, 4. Motor control panel, 5. Engine control panel, 6. Installation control valve, 7. Water motor gong, 8. Distribution pipe, 9. Branch pipe, 10. sprinklers 11. sprinklers 11. Sluice valve, 12. Gravity tank

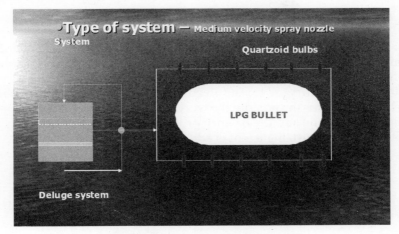

**Fig. 2.66:** Fire fighting system in LPG.

## The Deluge System for a Medium Velocity Water Spray System (MVWS)

The deluge system controls the water supply to the MVWS system and is actuated by an automatic detection system.

The deluge valve is a quick release, hydraulically operated, diaphragm actuated type valve. It has three chambers, isolated from each other by the diaphragm operated clapper and seat seal. In the seat position, water pressure is transmitted through an external by-pass check valve and restriction orifice from the system supply side to the top chamber, so that the supply pressure in the top chamber acts across the diaphragm operated clapper holding the seat against the inlet supply pressure because of its two to one differential design. Upon fire detection, the top chamber is vented to atmosphere through the outlet port via the opened actuation devices. The top chamber pressure cannot be replenished through the restricted inlet port and thus it reaches less than half the supply pressure instantaneously and the upward force to the supply pressure lifts the clapper, allowing water to enter the system pipe and the alarm devices.

Other components in the deluge valve system are the main drain valve, priming connection, drip check valve, emergency release valve and pressure gauges.

To actuate the deluge valve electrically, a solenoid valve is provided to drain the water from the top chamber of the deluge valve. A pressure switch is provided to activate an alarm, to shut down the desired equipment and to give deluge valve tripped indication. In addition, the pressure switch can also monitor low air pressure and the fire condition.

### Vapour pressure

Vapour pressure is the pressure exerted by the vapours of the liquid at any given temperature at which the vapour and liquid are in equilibrium in a closed vessel. Lower the flash point, higher is the vapour pressure at any given temperature.

### Gaseous fuels

**Hydrogen:** The simplest element. Atomic number 1. hydrogen gas has a specific gravity of 0.0694 (air = 1), so it is much lighter than air. Hydrogen is highly flammable, forming water upon combustion. Explosive limits are 4 to 75%.

**Natural gas:** A mixture of low-molecular weight hydrocarbons obtained in petroleum bearing regions throughout the world. Natural gas consists of approximately 85% methane, 10% ethane and the balance propanes, butanes and nitrogen. Since it is nearly odourless, an odourizing agent is added to most natural gas prior to final sale.

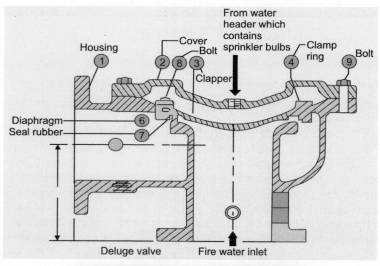

**Fig. 2.67:** Deluge valve 1800 lpm.

### Resetting of the Deluge Valve

- Close the upstream side stop valve provided below the deluge valve.
- Open both the drain valves and close them when the flow of water has stopped.
- Inspect and replace if required, or close the portion of the detection system that has been subjected to fire condition.
- In case of hydraulic release system, allow the water gauge to build to full service line pressure and when the water gauge on the hydraulic release side no longer rises, open the upstream side valve provided below the deluge valve. No water should flow to the system. This can be checked by depressing the drip check valve knob.

Do not close priming valve, down stream and upstream valves while the system is in service.

The releasing device must be maintained in the open position when actuated to prevent the deluge valve from re-closing.

### High Velocity Water Spray System (HVWS)

A high velocity water spray where the water is atomized is generally employed for electrical transformers. Transformers are always filled with oil as a means of cooling the windings. The oil, if it gets overheated can leak or even explode from the transformer and can then catch fire also. The system consists of high velocity water spray in the form of a conical spray consisting of water droplets travelling at very high velocity. The principle of action of this system is: Droplets of water traveling at high velocity bombard the surface of the oil to form an emulsion of oil and water that does not support combustion. The system, when forming an emulsion, intersperses cold water with the oil, thus cooling it and reducing the rate of vapourization and in addition prevents further escape of flammable vapour. It also provides a smothering effect, as while the water droplets are passing through the flame zone, some of the water is transformed into steam and this dilutes the oxygen feeding and creates a smothering effect.

A pipeline containing water is located on top and below the transformer. When the oil in the core of the transformer gets heated up above a certain temperature, quartzoid bulb detectors provided in the pipeline surrounding the transformers burst, thereby dropping the pressure in the system. This activates a deluge valve which in turn activates the high pressure pump.

For external/outdoor/open fire hazards, high velocity water spray system is the most economical and reliable fire protection system with almost 95% reliability in eliminating fire and smoke.

### FOAM SYSTEM

### Types of Foam System

- Low expansion system
- Medium expansion system
- High expansion system

Fig. 2.68: Foam application and off-shore fire.

### Low Expansion System

A low expansion system is typically used in portable fire extinguishers. Portable fire extinguishers can mix only a limited volume of air with the water-foam solution. Portable fire extinguishers have a small eductor at the end of the discharge pipe for mixing air.

### Medium Expansion System

Medium expansion foam system use a foam maker to provide the air required to achieve the expansion of the foam. The foam maker can be held by hand or fixed to a pipeline.

This type is mainly used for liquid hydrocarbon storage fire fighting.

The deluge valve or water hydrant, inline foam inductor and foam maker are important parts of this system.

### Inline Foam Inductor

Inline foam inductors are used to induct foam liquid in water stream to make available, proportioned solution of liquid concentrate and water to foam.

(a)

(b)

Fig. 2.69: (a) Fuel oil tank farm, (b) Inline foam maker with foam tank.

Fig. 2.70: A large fuel tank on fire.

Large fuel tanks pose a challenge to fire fighting. The cost of a fire protection (foam fire protection with other systems) is high.

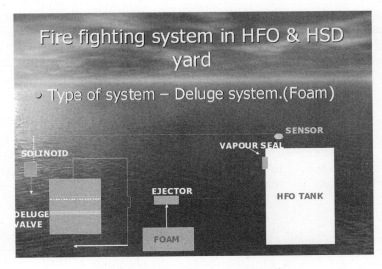

Fig. 2.71: Foam system schematic.

*Deluge System–Sequence of Operation*

- Sensor actuates at 79° C and opens the solenoid valve.
- Vapour seal will break
- High pressure water above diaphragm will be released and diaphragm opens and allows the main supply.
- Due to ejection principle, water sucks the foam.
- Water along with foam enters the tank at the top.

## Control and Exposure Protection

Control of fire to allow controlled burning of combustible materials where extinguishment is not possible and exposure protection to reduce heat transfer from an exposure fire can be accomplished by water spray or foam, or both, from these special systems such as MVWS and HVWS or foam systems. The degree of accomplishment is related strongly to the fixed discharge densities provided by the system design.

Foam of any type is not an effective extinguishing agent on fires involving liquefied or compressed gases (e.g. butane, butadiene, propane), on materials that will react violently with water (e.g. metallic sodium) or that which produce hazardous materials by reacting with water, or on fires involving electrical equipment where the electrical non-conductivity of the extinguishing agent is of primary importance.

### Foam Concentrates

Foam concentrate or concentrate is a concentrated liquid foaming agent as received from the manufacturer.

### Protein Foam Concentrate

Protein-foam concentrate consists primarily of products from a protein hydrolysate, plus stabilizing additives and inhibitors to protect against freezing, to prevent corrosion of equipment and containers, to resist bacterial decomposition, to control viscosity, and to otherwise ensure readiness for use under emergency conditions. They are diluted with water to form 3 to 6% solutions depending on the type. These concentrates are compatible with certain dry chemicals.

### Fluoroprotein Foam Concentrate

Fluoroprotein foam concentrate is very similar to protein-foam concentrate but has a synthetic fluorinated surfactant additive. In addition to an air-excluding foam blanket, they also can deposit a vapourization-preventing film on the surface of a liquid fuel. It is diluted with water to form 3 to 6% solutions depending on the type. This concentrate is compatible with certain dry chemicals.

### Alcohol-resistant Foam Concentrate

This concentrate is used for fighting fires on water-soluble materials and other fuels destructive to regular, AFFF, or FFFP foams, as well as for fires involving hydrocarbons. There are three general types. One is based on water-soluble natural polymers, such as protein or fluoroprotein concentrates, and also contains alcohol-insoluble materials that precipitate as an insoluble barrier in the bubble structure. The second type is based on synthetic concentrates and contains a gelling agent that surrounds the foam bubbles and forms a protective raft on the surface of water-soluble fuels; these foams can also have film-forming characteristics on hydrocarbon fuels. The third type is based on both water-soluble natural polymers, such as fluoroprotein, and contains a gelling agent that protects the foam from water-soluble fuels. This foam can also have film-forming and fluoroprotein characteristics on hydrocarbon fuels. Alcohol-resistant foam concentrates are generally used in concentrations of 3 to 10% solutions, depending on the nature of the hazard to be protected and the type of concentrate.

## Aqueous Film-forming Foam (AFFF) Concentrate

This concentrate is based on fluorinated surfactants plus foam stabilizers and usually diluted with water to a 1%, 3%, or 6% solution. The foam formed acts as a barrier both to exclude air or oxygen and to develop an aqueous film on the fuel surface capable of suppressing the evolution of fuel vapours. The foam produced with AFFF concentrate is dry chemical compatible and thus is suitable for combined use with dry chemicals.

## Film-forming Fluoroprotein (FFFP) Foam Concentrate

This concentrate uses fluorinated surfactants to produce a fluid aqueous film for suppressing hydrocarbon fuel vapours. This type of foam utilizes a protein base plus stabilizing additives and inhibitors to protect against freezing, corrosion, and bacterial decomposition, and it also resists fuel pickup. The foam is usually diluted with water to a 3% or 6% solution and is dry chemical compatible.

> **Foam solution:** A homogeneous mixture of water and foam concentrate in the proper proportions and is commonly called as solution.

## Action of Foam

Foam being lighter than all liquid fuels, it floats on their surfaces.
Foam blanket helps knockdown and extinguishes fire in the following ways:

- By excluding air (oxygen) from the fuel surface.
- By separating the flames from the fuel surface.
- By restricting the release of flammable vapour from the surface of the fuel.
- By forming a radiant heat barrier which can help to reduce heat feedback from flames to the fuel and hence reduce the production of flammable vapour.
- By cooling the fuel surface and any metal surfaces as the foam solution drains out of the foam blanket. This process also produces steam, which dilutes the oxygen around the fire.

**Minimum application rates of foam solution for the production of low expansion foam for use on liquid hydrocarbon fuel (class B) fires**

| Foam type | Tanks D < 45 m | D >=45 m D < 81 m lmp/m² | D >81 m | Spill/bund tanks minutes | Fuel flash point > 40° C minutes | Fuel flash point < 40° C minutes |
|---|---|---|---|---|---|---|
| Protein foam | 6.5 | NR | NR | 15 | NR | NR |
| Fluoroprotein foam | 5 | 8 | 9 | 15 | 45 | 60 |
| AFFF | 4 | 6.5 | 7.3 | 15 | 45 | 60 |
| FFFP | 4 | 6.5 | 7.3 | 15 | 45 | 60 |
| AFFF-AR | 4 | 6.5 | 7.3 | 15 | 45 | 60 |
| FFFP-AR | 4 | 6.5 | 7.3 | 15 | 45 | 60 |

< — less than, > — more than; D–diameter of tank; NR–not recommended

Spill containment can be accomplished by any of the following:

- Non-combustible, liquid-tight raised sills, curbs, or ramps of suitable height at exterior openings
- Sloped floors
- Open-grate trenches or floor drains that are connected to a properly designed drainage system.

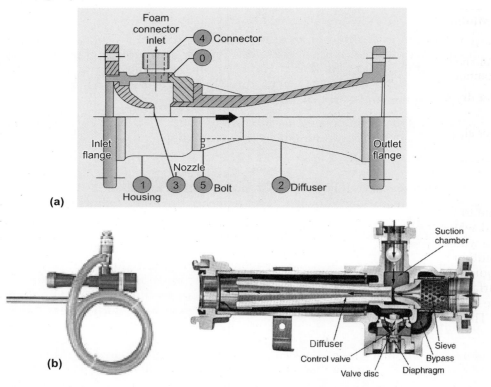

**Fig. 2.72:** (a) Inline foam eductor (simple type); (b) Inline eductor (inductor).

### Water-soluble flammable liquid fires (polar solvents)

AFFF- and FFFP-types of fire extinguishers shall not be used for the protection of water-soluble flammable liquids, such as alcohols, acetone, esters, ketones, and so forth.

- The full form of CNG is compressed natural gas. Basically, it is methane gas and there are many composition of the gas. However, one composition of CNG is given as under:
    - $CH_4$ - 93.7%
    - $C_2H_6$ - 3.2%
    - $C_3H_6$ - 0.3%

## Foam Maker

Foam makers are used to generate air foam by aspirating air into a stream of foam solution. The premixed foam solution used by foam maker to generate air foam is made available through inline foam inductor. A combination of foam maker and inline foam inductor are designed for predetermined solution rate and pressure required at the base of the foam maker.

| Minimum rate of discharge from foam hose streams protecting vertical tanks containing liquid hydrocarbons | |
|---|---|
| For chemical foam systems with stored solutions (A and B solutions) | 0.8 gpm of A solution and 0.8 gpm of B solution per sq m of liquid surface area |
| For dry powder foam generator systems | 1.6 gpm of water (including stabilizer) to generate foam for each sq m of liquid surface area |
| For air foam systems | 1.6 gpm of water to the foam maker per sq m of liquid surface area. |

### Function

Due to fire around the tank, temperature will rise, at 60° C, sprinklers tube will burst and water inside that tube will be released. This actuates the deluge valve to allow the fire water in the line. When water flows through inline foam inductor, it sucks the foam solution from the tank and makes foam in the foam maker. This foam is applied on the liquid surface. Foam forms a layer on the liquid and hence cuts off contact of the liquid surface from the atmospheric air. By smothering effect, due to non-availability of oxygen, the fire will be extinguished.

The capacity of foam application system is based on the liquid surface area of the tank.

---

### Some flammable liquids

**Diesel fuel:** Diesel fuel consists mostly of hydrocarbons ranging from C10 to C24. Diesel fuel has a flash point of 48° C to 72° C and explosive limits of 0.7 to 5%.

**Ethanol:** Ethanol is ethyl alcohol. Grain alcohol. Flammable, water soluble alcohol. Flash point of 13° C. Explosive limits of 3.3 to 19%.

**Toluene:** Toluene is methylbenzene. An aromatic compound having the formula $C_6H_5CH_3$. A major component of gasoline. Toluene has a flash point of 5° C and explosive limits of 1.2 to 7%.

**Turpentine:** Contains

1. Gum. The pitch obtained from living pine trees. A sticky viscous liquid.
2. Oil. A volatile liquid obtained by steam distillation of gum turpentine, consisting mainly of pinene and diterpene. Turpentine is frequently identified in debris samples containing burned wood.

**Fig. 2.73:** Mobile foam generator.

## Foam Systems and Combinations

### Foam-water Sprinkler System (FWSS)

FWSS is a special system that is pipe-connected to a source of foam concentrate and to a water supply. The system is equipped with appropriate discharge devices for extinguishing agent discharge and for distribution over the area to be protected. The piping system is connected to the water supply through a control valve that usually is actuated by operation of automatic detection equipment that is installed in the same areas as the sprinklers. When this valve opens, water flows into the piping system, foam concentrate is injected into the water, and the resulting foam solution discharging through the discharge devices generates and distributes foam. Upon exhaustion of the foam concentrate supply, water discharge follows and continues until shut off manually. Systems can be used for discharge of water first, followed by discharge of foam for a specified period, and then followed by water until manually shut off. Existing deluge sprinkler systems that have been converted to the use of aqueous film-forming foam or film-forming fluoroprotein foam are classified as foam-water sprinkler systems.

*Foam-water deluge system:* A foam-water sprinkler system employing open discharge devices, which are attached to a piping system that is connected to a water supply through a valve that is opened by the operation of a detection system, which is installed in the same areas as the discharge devices. When this valve opens, water flows into the piping system and discharges from all discharge devices attached thereto.

*Foam-water dry pipe system:* A sprinkler system employing automatic sprinklers or nozzles that are attached to a piping system that contains air or nitrogen under pressure, the release of which (as from the opening of a sprinkler) permits the water pressure to open a valve known as a dry pipe valve. The water then flows into the piping system and out the opened sprinklers.

*Foam-water pre-action system:* A sprinkler system employing automatic sprinklers or nozzles attached to a piping system containing air that might or might not be under pressure, with a supplemental detection system installed in the same area as the sprinklers. Actuation of the detection system opens a valve that permits water to flow into the sprinkler piping system and to be discharged from any sprinklers that have activated.

## Foam system definitions

**Pre-primed system:** A wet pipe system containing foam solution.

**Proportioning:** Proportioning is the continuous introduction of foam concentrate at the recommended ratio into the water stream to form foam solution.

**Rate:** The total flow of solution per unit of time, which is expressed in L/min (gpm).

## Design considerations for a foam system for a diesel storage tank

1. Tank type: fixed roof tank.
2. Tank diameter: 13 mts (ID).
3. Liquid surface of each tank: 132.7 m$^2$
4. Foam application rate: 5 lpm/m$^2$.
5. Quantity of foam solution required/min with an application rate of 5 lpm/m$^2$: 5 × 132.7 = 663.5 lpm.
6. Size of inline inductor selected (one per tank): 80 mm NB (450–900 lpm)
7. Deluge valve selected (one per tank): 80 mm NB (max output 1100 lpm)
8. Size of foam maker selected (one per tank): 80 mm NB (450–900 lpm)
9. Foam solution required for 55 min of application: 55 × 663.5 = 36492.5 litres.
10. Foam concentrate: 3%.
11. Foam concentrate required: 0.03 × 36492.5 = 1094.8 litres
12. Foam tank capacity: 1200 litres.
13. Pipe size considering velocity of 5 m/sec (foam tank to inductor): 80 mm NB.

## GASEOUS FIRE FIGHTING SYSTEMS

For a gaseous system, it is important that the control space should be able to maintain a sufficient extinguishing agent concentration for the required holding time. Room integrity tests should be done during the system commissioning to check the above.

Gaseous systems should not reduce the oxygen level below 12% if the room cannot be evacuated in less than 60 seconds and if below 10%, the room should be evacuated in less than 30 seconds and should not be normally occupied. The $CO_2$ level in the space should not be more than 5%.

**Fig. 2.74:** Foam spray unit at top of a fuel tank.

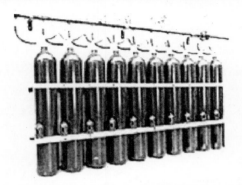

**Fig. 2.75:** Bank of cylinders and a spray system for gaseous system (halon).

### Halon 1301 System

System design requirements for halon 1301 total flooding fire extinguishing system:

- The system has fully charged cylinders of halon 1301 gas in a ready and operable condition.
- The control area is sealed (wall openings) to ensure adequate concentrations of the agent.
- Halon gaseous systems normally have an interlocking tie-in between the halon control panel and air conditioners, overhead doors, roof ventilation units and ventilation dampers to prevent agent loss.
- Cross-zoned detectors (usually of ionization type) are provided to sense fire.

#### Alarm Sequence

- The first detector to be activated sounds the pre-alarm, shuts off any outside air conditioning, closes the ventilation.
- The second detector activates the final warning alarm in both the protected zone and at a central panel and initiates a 30 second time delay sequence prior to the agent's automatic release.
- A manual, dead-man type abort switch is provided at the exits from the protected zone.
- Halon concentration: the systems are designed to achieve total flooding for 10 minutes or longer holding minimum concentration of 5.0% in all locations of the protected area. The total discharge is to be achieved in 10 seconds.
- Halon storage containers: One or any interconnected combination of single cylinders pressurized with dry nitrogen to 25 to 40 bar at 21° C as applicable.

The cylinders are distinctively marked with contents specification and pressurization gauge.

### Restrictions on Halon Systems

Halon 1301 was one of the most generally effective fire protection agents. But because of the damage halon causes to the ozone layer, similar to the action of the CFCs, production and use of halon has been curtailed since 1992 and the halon mandatory decommission comes into effect from 1st Jan 2004. The choice for gaseous alternatives for halon has narrowed down to:

- Use of $CO_2$ systems.
- Use of inert gas systems.
- FM 200.
- NAF SIII.

Any substitute should be highly reliable and an effective clean agent and should not have any long-term environmental problems.

### Argon Gas System (Inert Gas System)

Argon is a noble gas, chemically very inert, 1.38 times the density of air. It is present in the earth's atmosphere at 0.93% by volume. Other uses of argon have been in inert gas–shielded welding, as a carrier gas for gas chromatography and as filler gas in the lamp industry.

Argon has been chosen as an effective alternative to halon.
- Argon is a genuine green agent.
- Argon exhibits excellent mixing upon discharge, as it is heavier than air and provides security against low level fires.
- Argon does not pose any hazard to personnel in the affected space.
- Argon can be used as a single gas, without any mixing with any other gas.
- There is no breakdown of the argon extinguishing agent.
- Argon does not cause fogging of the room atmosphere upon discharge.
- Argon has minimal short-term cooling, so avoiding condensation on equipment
- In an argon system, long pipe runs from cylinder to nozzle is possible.

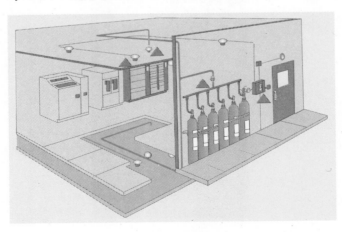

Fig. 2.76

All electric motors and lighting fittings, and switches should be flame-proof and dust-proof in hazardous areas as defined as per IS: 5572 (Part 1)-1978.

All electrical installations should be in accordance with IS: 1646-1982 and there shall be separate source of supply to main and ancillary connections.

## $CO_2$ System

Carbon dioxide ($CO_2$) is a clean agent. It can be used to protect a wider range of risks than any other clean agent, and at a lower cost.

- Effective clean agent
- Used in a variety of ways.
- Highly reliable.

$CO_2$ (carbon dioxide) gas used in the system has a high ratio of expansion, which facilitates rapid discharge and allows three-dimensional penetration of the entire hazard area quickly. $CO_2$ extinguishes a fire by reducing the oxygen content of the protected area below the point where it can support combustion. Due to the extreme density of the $CO_2$, it quickly and effectively permeates the protected hazard area and suppresses the fire. Rapid expansion of the 10–15 cm of $CO_2$ snow to gas reduces the ambient temperature in the protected hazard area, which aids in the extinguishing process and retards re-ignition. $CO_2$ is electrically non-conductive and does not normally damage sensitive electronic equipment. $CO_2$ has no residual clean-up associated with its use as a fire-suppressing agent. When it is properly ventilated, the gas escapes to atmosphere after the fire has been extinguished.

Fig. 2.77: A bank of $CO_2$ cylinders.

## Advantage of $CO_2$ System

Carbon dioxide gas has properties and advantages no other extinguishing agent offers. It is fast in action, effectively stops difficult fires, is non-damaging to the equipment and non-conducting.

## Flooding Systems Using $CO_2$

The types of fires that can be extinguished by total flooding methods are divided into two categories:

1. Surface fires involving flammable liquids, gases, and solids
2. Deep-seated fires involving solid subject to smouldering

Surface fires are the most common hazard particularly adaptable to extinguishment by total flooding systems. They are subject to prompt extinguishment when carbon dioxide is quickly introduced into the enclosure in sufficient quantity to overcome leakage and provide an extinguishing concentration for the particular materials involved.

For deep-seated fires, the required extinguishing concentration needs to be maintained for a sufficient period of time to allow the smouldering to be extinguished and the material to cool to a point at which reignition will not occur when the inert atmosphere is dissipated. In any event, it is necessary to inspect the hazard immediately thereafter to make certain that extinguishment is complete and to remove any material involved in the fire.

Practically, all hazards that contain materials that produce surface fires can contain varying amounts of materials that could produce deep-seated fires. Proper selection of the type of fire that the system should be designed to extinguish is important and, in many cases, will require sound judgment after careful consideration of all the various factors involved.

| Volume of space $m^3$ | Flooding factors for $CO_2$ | | Calculated quantity(kg) (not less than) |
|---|---|---|---|
| | Volume factor $CO_2$ | | |
| | $m^3/kg\ CO_2$ | $kg\ CO_2/m^3$ | |
| Up to 3.96 | 0.86 | 1.15 | – |
| 3.97–14.15 | 0.93 | 1.07 | 4.5 |
| 14.16–45.28 | 0.99 | 1.01 | 15.1 |
| 45.29–127.35 | 1.11 | 0.90 | 45.4 |
| 127.36–141.5 | 1.25 | 0.80 | 113.5 |
| over 141.5 | 1.38 | 0.77 | 113.5 |

## Application of $CO_2$ System

Transformers • Power generators • Ships holds • Machinery spaces • Electrical equipment-alternators.
• Chemical equipment-flammable liquids • Chemical works and stores • Switch rooms • Electric movers • Solvent stores • Dip tanks.

## FM 200 System

FM 200 is an environmentally acceptable total flood extinguishing system that is safe for the personnel and accepted for use for both occupied and unoccupied spaces. The system has been developed specially for risks that require a fast acting, non-damaging extinguishant.

- Rapid extinguishing in 10 seconds.
- Safe for people
- Alternative for halon

FM 200 offers zero ozone depletion, can be used in low concentrations, and has low toxicity properties.

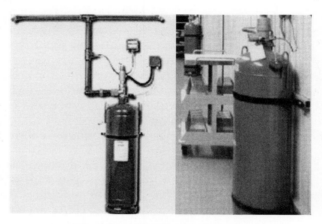

Fig. 2.78: FM 200 cylinder line.

FM 200 is the worlds most accepted protection system to facilities containing valuable corporate assets from people and processes to equipment and software. FM 200 discharges a gas within 10 seconds after receiving an actuation signal and reaching into all areas of the protected facility. FM 200 suppresses fire by "flooding" the protected area with a 7% concentrate of FM 200. Instead of inerting the entire area, FM 200 actively attacks the fire itself, soaking up heat like a sponge soaks up liquid and breaking down the fire's molecular structure and leaving no residue to damage sensitive equipment and does not require costly cleanup. This rapid action suppresses a fire in its incipient stages before it has an opportunity to grow or cause significant damage and enables to get back to normal conditions.

FM 200 is non-toxic when used properly. FM 200 is the most viable alternative to halon 1301. FM 200 causes no breathing problems for people and won't obscure vision in an emergency situation, has zero ozone depletion potential (ODP), has a low atmospheric lifetime (31 to 42 years) and hence there are no usage restrictions.

## *FM 200 Application*

Telecommunications • Switch gear rooms • Automotive battery rooms • Data processing computer rooms • Medical high density/high value areas • Military delicate electrical equipment • Petroleum industrial control rooms • Transportation • Flammable liquid storage.

| Combustible solids |
|---|
| **Magnesium:** A silvery metal used in some metal incendiaries. The dust is highly explosive. Ignition point of 340° C.<br>  **Sulfur:** A non-metallic yellow element. A constituent of black powder, sulfur burns readily when in powdered form. |

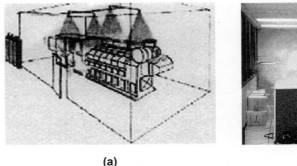
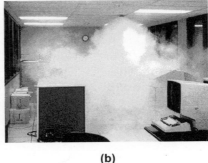

(a)                                                                                     (b)

**Fig. 2.79:** (a) Spray arrangement for an equipment from a bank of cylinders; (b) FM 200 spray.

| Definition of flammable and combustible liquids | |
|---|---|
| *Agency* | *Classification by flash point ° C* |
| OSHA | Flammable < 37.8<br>Combustible >= 37.8 |
| NFPA | Flammable class I < 37.8<br>Combustible class II > = 37.8 to < 60<br>class IIIA >= 60 to < 93 class IIIB > = 93 |

## FIRE PROTECTION ENGINEERING

### Reliability of Fire Protection Systems

Absolute reliability is essential as such systems are required to operate suddenly and sometimes in severe conditions. They should be fool-proof and simple to operate and use, without complications. Hence skilled fire protection engineering design is vital.

Fire protection systems operate on the basis of detection of fire at an early stage.

Fire protection systems are unique, since the majority of their service life is spent in a static, no-flow condition. However, when required in an emergency, their operation is critical. Hence design reliability of fire protection systems is very important.

Many codes have been developed to assure that fire protection systems are readily available when needed.

In India, apart from the various codes and standards, the TAC also enforces the design of the fire protection system.

Fire protection system design is covered under many regulations and standards, such as:

- NFPA standards
- Factories Act and factory mutual regulations.
- Regulations from insurance companies
- BIS specifications and standards
- TAC (tariffs advisory committee) specifications
- Standards like ASME, ASTM, AWWA, ANSI, etc.

## FIRE FIGHTING

The two basic aspects of fire fighting systems are—extinguishing the fire and reducing the loss due to fire. These two aspects are entirely dependent on the efficiency of the fire fighting systems installed and the awareness about the usage and principles of fire protection by the operating personnel. It is therefore imperative that the fire protection systems available be maintained in a state of peak performance at all times.

**Fig. 2.80:** Fireman fighting fire.

### Tips to fighting fire

The best way to fight fires from flammable liquids and gases is to stop the flow of the fuel, whenever that is possible.

Do not use water on burning metals and chemicals. They may react with water.

Water must never be used when the fires involves electrical equipment which are energized.

## Importance of Wind Direction

When fighting fires, it should always be carried out from the windward side, i.e. along the direction of the wind. Wind cocks or wind flags are installed near strategic locations to indicate the wind direction to the fire or emergency handling personnel.

During the design of hazardous installations, the wind direction is to be considered in selecting the sites for storage of flammable materials.

### Wind Rose

A wind rose is a diagram in which statistical information concerning the direction and speed of the wind at a particular location may be conveniently summarized. In the standard wind rose, a line segment is drawn in each of perhaps eight compass directions from a common origin. The length of a particular segment is proportional to the frequency with which winds blow from that direction. Parts of a given segment are given various patterns, indicating frequencies of occurrence of various classes of wind speed from the given direction.

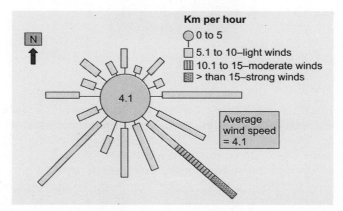

**Fig. 2.81:** Wind rose diagram for a site near Chennai showing the wind direction and velocity for a 24-hour period over a year.

Based on the above wind data, the fuel oil and LPG storage at the site was established at the north-west location.

**Fig. 2.82:** Sign showing prohibition of water usage.

### Personal Protection Equipment (PPE) for Fire Fighting

During fire fighting, PPE is essential since the situation is an emergency and the hazard has already developed. All fire-fighting personnel should be equipped with relevant PPE for their personal safety. It is also required for carrying out the fire-fighting operations effectively.

**Fig. 2.83:** Fire protection suit used during fire fighting.

### BREATHING APPARATUS

Poisonous atmosphere is one among the many hazards which create hurdles in the way of fire fighters to extinguish fire or to make a search of premises to find injured or trapped occupants. This has led to the development of breathing apparatus which enables the wearer to work safely in irrespirable atmospheres.

The human body needs a constant supply of oxygen in order to survive. Oxygen is a constituent of air and the body normally gets this from the atmosphere. When at rest, the body's requirements are comparatively low and air is inhaled and exhaled by the lungs by breathing movements at the rate of about 15 to 18 times per minutes. When more energy is exerted, either through work or nervous excitement, the breathing rate is higher. It is probable and may be as much as 30 times a minute.

### Respiratory Cycle

The body must draw air into the lungs, hold for a sufficient time for oxygen required to be absorbed, and then breathe out. This process is known as respiration and consists of two spontaneous actions, inhalation (breathing in) and exhalation (breathing out). The inhalation is done through a muscular effort, which raises ribs and lowers diaphragm, thus enlarging the chest cavity and creating partial vacuum, which causes air to enter. Exhalation normally needs no effort, because when breath is released, the ribs fall and diaphragm rises automatically. This contracting of the chest cavity and forcing the air out is called the respiratory cycle. The parts of the respiratory system are:

- Nose
- Throat

- Wind pipe
- Air pipe
- Lungs
- Brain

| Breathing under varying degree of effort | | |
|---|---|---|
| | Breathing rates | |
| Degree of exertion | Air breathed litres/min | Oxygen consumed litres/min |
| Rest in bed | 7.7 | 0.24 |
| Rest in standing | 10.7 | 0.33 |
| Walking at 3 km/hr | 17.5 | 0.73 |
| Walking at 5 km/hr | 25.8 | 1.11 |
| Walking at 8 km/hr | 60.9 | 2.54 |

An increase in the amount of oxygen needed may also be due to the obstruction of the air passage, emotional excitement, etc. It also makes some people faint on hearing bad news, the shock of which causes a sudden demand for more oxygen than is immediately available. Since nervous excitement leads to the increase of the breathing rate automatically, it is highly important that fire fighters are emotionally stable so that the sudden and abnormal demands on any breathing apparatus they may be wearing are reduced to a minimum.

## Oxygen Deficiency

Oxygen deficiency results in breathing difficulty and therefore we should know the safety level of oxygen at different percentages.

| Oxygen level | Condition |
|---|---|
| 23% and above | Enriched |
| 21% | Normal |
| 19.5% | Minimum safe |
| 15.19% | Workability reduced |
| 12.1% | Increase in respiration |
| 8 to 10% | Fainting |
| 6 to 8% | Death after 5 minutes |
| 4 to 6% | Coma within one min |

## Types of Breathing Apparatus

The types of breathing apparatus used in assisted breathing are:

CABA–Compressed air breathing apparatus
ELSA –Emergency life support (saving) apparatus
FBA–Forced breathing apparatus

CABA (also called as self-contained breathing apparatus—SCBA)

The SCBA provides complete respiratory protection to the wearer who breathes from the supply carried on his back. It is self-contained, absolutely dependable and a compact unit for negotiating in a confined space.

SCBA are standard accessories in all fire engines.

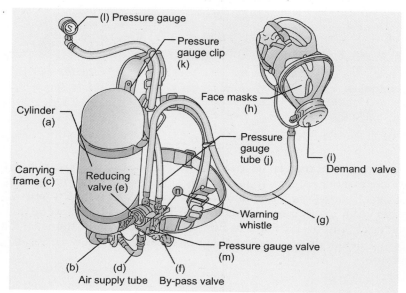

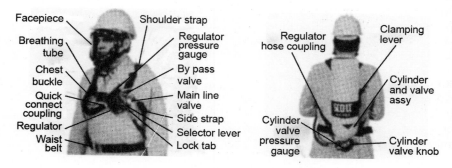

Fig. 2.84: SCBA (scott make).

## Uses

Apart from fire-fighting operations, the SCBA can be used under the following situations.

1. Toxic atmosphere
2. Rescue operations
3. Underground work
4. Working in confined spaces

## Body Parts

### 1. Cylinder

It consists of a single cylinder with a capacity of 1800 litres of air at 200-bar pressure when fully charged. Cylinders are made of alloy steel and are attached to a carrying plate or frame and secured to the wearer's back with a harness.

### 2. Harness

It includes shoulder straps, waist straps, etc. The harness is so designed that it will enable the wearer to lower the apparatus quickly and easily without others help.

### 3. Pressure Gauge

Breathing apparatus must be fitted with a pressure gauge calibrated in bars to indicate the volume of the air in the cylinder. The gauge is placed in the front side of the wearer so as to be able to read the pressure easily.

### 4. Demand Valve (first breath demand valve)

The provision of demand valve guarantees easy breathing under various working conditions. Allows the user to wear the facepiece in standby position before attaching demand valve, eliminating the' waste of air, features a quick connect facepiece mounting which keeps the demand valve from interfering with users activity.

### 5. Facemask (full facepiece)

Facemasks vary slightly between different manufacturers but all have full vision and are fitted with an exhalation valve and a speech diaphragm. They are made from scratch resistant, polycarbonate lens providing a 200° field of vision. Lens fogging is kept to a minimum with the oral/nasal cup. Hypoallergenic silicone facepiece provides the wearer with superior fit and comfort, while the five strap head harness helps 'maintain face-piece-to-face seal. A speaking diaphragm allows for clear and safe communication. The facemask consists of:

a. Visor
b. Head harness
c. Exhalation valve
d. Speech diaphragm
e. Demand valve

### 6. Warning Whistle

The warning whistle is usually chest mounted. The whistle provided lets the user know when only 1/4 of air remains in the cylinder. It is positioned close to the user's hearing range. The whistle also blows when the pressure comes down to 50 to 48 bar, which indicates that the volume of the air remaining in the cylinder is enough only for 7 minutes breathing. It gives notice to the wearer to go out from the contaminated atmosphere immediately.

## 7. Reducer/Regulator

This has a fail-open design assuring the user a supply of air as long as there is air in the cylinder. It should eliminate dirt and moisture from entering into the breath. It contains only three moving parts reducing the chance of any malfunction. The regulator has a pressure gauge which indicates cylinder pressure at all times and is positioned in an easy to read location.

## 8. Harness and Backpane

The harness and back pane should be made from durable and light weight material. The harness has fully padded shoulder straps providing easy adjustment. Fully padded lower back support and back-plate system evenly distribute weight over a large area for user comfort.

SCBAs are standard accessories in all fire engines.

Compressed air used to fill SCBA must meet the following quality which is called – Grade D breathing air:

| | |
|---|---|
| Oxygen content | 19–23% |
| Balance predominantly | Nitrogen |
| Hydrocarbon (condensed) | < 5 mg/m$^3$ |
| Carbon monoxide | < 20 ppm |
| Carbon dioxide | < 1000 ppm |

### Instructions for Using Breathing Apparatus (BA Set)

1. Check breathing apparatus (BA) for any damages.
2. Check the tightness of the valve nuts, etc.
3. Check the pressure of the cylinder (200 bar)
4. Wear the apparatus and adjust the shoulder straps to support it comfortably on the back.
5. Wear the facemask and adjust the harness straps.
6. Inhale and exhale deeply 2 or 3 times to make sure that air flows freely from the demand valve.
7. Make sure that exhaling and inhaling valves are functioning properly.
8. The wearer must examine his pressure gauge occasionally to make sure that air pressure is sufficient.
9. On listening to the warning whistle, leave the contaminated atmosphere or fire scene immediately.

### Care and Maintenance of BA Sets

It is essential that every one who is required to wear a BA set is suitably trained and thoroughly understands the operating procedures. The care and maintenance of all CABA is relatively simple.

Dismantling, assembling, testing and renewal of the parts must be done in accordance with the manufacturer's specifications and recommendations. For each type of apparatus, the following points should be followed during periodical maintenance:

1. After use, the apparatus should be cleaned as necessary.
2. The facemask and breathing tube where provided must be washed and disinfected. The whole apparatus has to be thoroughly dried before re-assembling
3. If the cylinder pressure is lower than 80% of the maximum charge pressure, the cylinder has to be replaced by a fully charged one.
4. Washers or other parts, which are found defective during tests, must be renewed or replaced.
5. Excessive force should not be employed while tightening nuts as this may damage the threads.

A self-contained breathing apparatus is a self-sufficient breathing apparatus which permits freedom of movement as the wearer carries the supply of breathable air. It is used in oxygen deficient atmospheres. An oxygen deficient atmosphere is any environment where the oxygen content is less than 19.5%.

An SCBA can provide respiratory protection in an oxygen deficient environment or in situation where high or unknown concentrations of toxic gases vapour or particulates are present.

SCBAs provide the wearer with the highest level of respiratory protection available. These units utilize either a thirty or sixty minute cylinder containing pressurized breathing air which is delivered "on demand" to the facepiece. Because the cylinders are worn on the back of the worker, SCBAs are completely portable, enabling the worker the maximum mobility for special applications such as emergency response and rescue. The SCBA can protect in emergency situations. When using an SCBA, the wearer is independent of the surrounding atmosphere because he is breathing within a system admitting no outside air.

There are three different types of cylinders and two air flow supply options as follows:

1. Full wrap (high pressure only)—light weight
2. Hoop wrap (high and low pressure).
3. Aluminum (low pressure only)–heavy.

### SCBA specifications/description from a manufacturer

Self-contained breathing apparatus (SCBA) consisting of a single cylinder. It comprises of a light weight and antistatic back plate made with Kevlar; cylinder band and fully adjustable un-padded body harness; two stage pneumatic system comprising tempest automatic positive demand valve with bypass, vision-3 siliconized non-dusting face mask, shoulder mounted pressure indicator and 55 bar warning whistle; first stage pressure reducer with single high pressure cylinder connector; Harness fabricated from flame retardant polyester/polyamide materials and features half lumber pad. The apparatus is supplied with 6 litres/300 bar high pressure alloy steel cylinder with CE approved valve, along with an FRP box for storage.

## Oxygen

The air we breathe is about 21% oxygen. Fire needs an atmosphere with at least 16% oxygen.

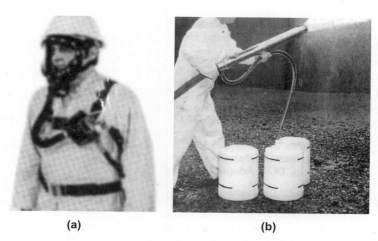

**(a)**                                    **(b)**

**Fig. 2.85:** (a) Person with SCBA, (b) Foam spray with eductor.

There are three types of SCBA:

* Demand or pressure demand–open circuit system supplied by cylinder-stored compressed air or oxygen
* Self-generating closed circuit devices.
* Liquid or compressed oxygen, closed (re-breathing) devices.

*Demand type compressed oxygen or compressed air breathing apparatus:* This is a breathing apparatus in which the user obtains respirable air or oxygen from compressed air or oxygen cylinder, which is an integral part of the apparatus and enables the wearer to breathe independently in the surrounding atmosphere. In a demand type self-contained breathing apparatus, air or oxygen is admitted in the facepiece only when the wearer inhales. Air or oxygen admitted in the facepiece is regulated to suit the rate of breathing of the person automatically. In other words, the person gets air or oxygen according to his demand. It must be remembered that an apparatus designed for air cannot be used with oxygen and vice-versa. Duration of time the wearer can remain in the irrespirable environment is dependant on the rating of the apparatus and there is an audible signal provided besides the pressure gauge to warn the wearer when to leave the area. The exhaled air passes through a non-return valve to the atmosphere. This apparatus is also called–open circuit type breathing apparatus.

*Compressed oxygen recirculating breathing apparatus:* In this type of breathing apparatus, high pressure oxygen passes through a pressure reducing and regulating valve into a breathing bag. The wearer inhales this oxygen through a one-way valve and his exhaled breath passes through a collar and then into the same breathing apparatus bag. Oxygen enters the breathing bag from the supply cylinder only when the volume of gas in the bag has decreased sufficiently to allow the supply valve to open.

*Oxygen–regenerating type breathing apparatus:* In this apparatus, moisture content from the wearer's exhaled breath reacts with granular chemical in a canister to liberate oxygen. Also, the exhaled carbon dioxide is absorbed by the chemicals in the canister. This oxygen enters a breathing bag from which the wearer inhales through a corrugated breathing tube connecting the bag to the facepiece.

**BREATHING**

KEEPS YOU
*ALIVE*

**Fig. 2.86:** Supplied air respirator.

## Fire Protective Clothing and other PPE

### Choice of the PPE

Owing to different conditions experienced in different climates, the rig to be worn by fire fighters varies. The risk to personnel, fighting fires, arises from:

a. The effect of heat on the body, face and hands.
b. The effect on lungs due to combustion products and gases used for fire extinction
c. Reduced visibility due to smoke or lighting failures.
d. Electric shock
e. Falling objects
f. The effects of wearing sweat-saturated clothing next to the skin for long periods.

**Fight fire with fire-proverb**

Specialized clothing equipment are provided to give protection against flame and radiated heat for very limited periods only. No means are available to protect human body exposed to excessively hot atmosphere for a long time. Such conditions lead to heat exhaustion and sudden collapse, which may not be preceded by a brief period of acute distress. Hence the safeguarding of men in these conditions is mainly a technical matter which depends upon correct approach, careful supervision, etc.

During exposure, heat passes through the protective clothing and the operator feels gradually his skin becoming warm. Operators must be prepared for the moment when the protective clothing and air within it become rapidly warmer than the skin. It may cause concern to an untrained operator, but it does not indicate immediate danger unless pain is felt. Training must be given to each operator so that he can recognize the approach of the point of danger. There is often a tendency to underestimate the danger to health by the prolonged wearing of wet clothing. Therefore, wet clothing must be changed as soon as possible.

Fire protective clothing are available mainly in three different specifications. They are:

   i. Approach suit
  ii. Proximity suit
 iii. Entry suit

The terms approach, proximity and entry define the garment's ability to protect the wearer. It depends on the degree of exposure to the fire.

*The approach suit:* The approach suit helps the fireman to come within a few meters of the fire. It protects the fireman from radiant heat and helps him to approach the fire safely.

*The proximity suit:* This suit enables the fire fighter to work very close to the flames. By wearing this suit, the firemen cannot enter the flames but he can reach very near to the scene of the fire for rescue operations and fire fighting operations.

*The entry suit:* This type of suits enables the fireman to work in the flame for a short period of 30 seconds. This suit helps fire fighter to enter into areas of radiant heat for a period of about 5 minutes.

In the proximity and entry suits, a pouch is made at the back of the overall, for keeping the breathing apparatus set and it should be used for such operations only.

Fire suits manufactured by firms like 'Kraps International' (Guardian brand) are fire suits approved by DRDO (Defence Research and Development Organization) fire service department and the BIS.

### Details of proximity suits

These suits are manufactured out of aluminized fabrics and suitably lined with leather or Kevlar cloth with proper insulations as inner lining. There are five-plies in the fabric. These are a layer of aluminum by vapour deposit process, protective film, another layer of aluminum, heat stable adhesive and base fabric. The outer lining is usually fiber-glass cloth.

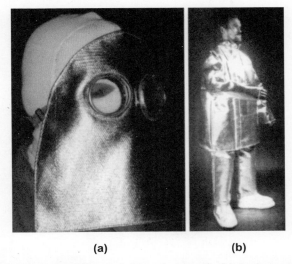

(a)            (b)

Fig. 2.87: (a) Heat-resistant clothing–facemask, (b) Jacket and trouser.

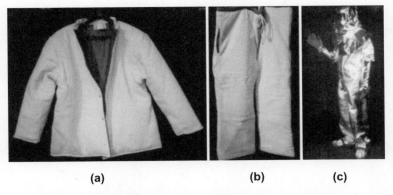

(a)        (b)        (c)

Fig. 2.88: (a) and (b) Kevlar jacket and trouser; (c) Complete fire-suit.

## Radiation

1. Transfer of heat through electromagnetic waves from hot to cold.

2. Electromagnetic waves of energy having frequency and wavelength. The shorter wavelengths (higher frequencies) are more energetic. The electromagnetic spectrum is comprised of: (a) cosmic rays, (b) gamma rays, (c) X-rays, (d) infrared rays, (e) visible light rays, (f) ultraviolet rays, (g) microwaves and (h) radio waves.

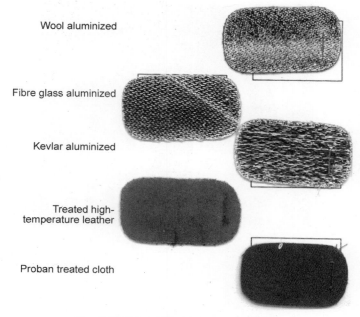

Wool aluminized

Fibre glass aluminized

Kevlar aluminized

Treated high-temperature leather

Proban treated cloth

**Fig. 2.89:** Materials used for fire suits.

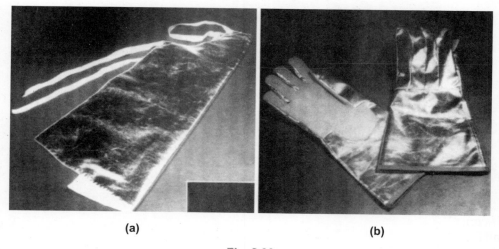

(a)                                      (b)

**Fig. 2.90**

<br>

### Hot work control

**All hot work**—welding, cutting, grinding, etc. which can provide a source of ignition should be conducted only after observing safely controlled conditions with written approval from appropriate authority (IS: 3016-1982).

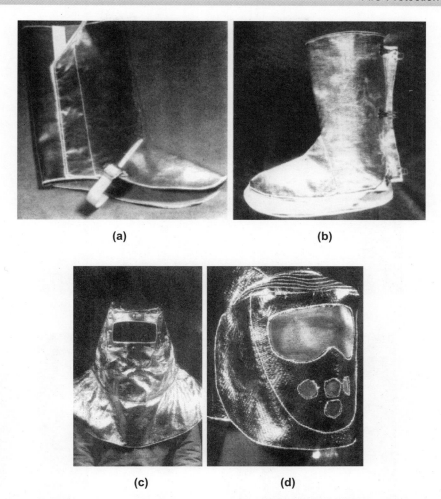

**(a)**  **(b)**

**(c)**  **(d)**

Fig. 2.91: (a) and (b) Fire suit parts; (c) Aluminized double cotton material with goggles; (d) Facemask.

Self-contained breathing apparatus [see IS: 10245 (Part 2)-1982] should be kept readily available for personal safety as polymer-based products on fire produce heavy smoke, carbon dioxide, carbon monoxide and other poisonous fumes, such as nitro fumes; chlorinated plastics evolve highly toxic and poisonous hydrocyanic gas. The fumes given off as a result of burning of fluorocarbons-polytetrafluoroethylene give traces of vapour on the depolymerization of the material into its monomer and other fluorine containing compounds which can produce polymer fume fever. The fine dust produced during disintegration may give rise to similar symptoms. Operating personnel should be trained for use of breathing apparatus.

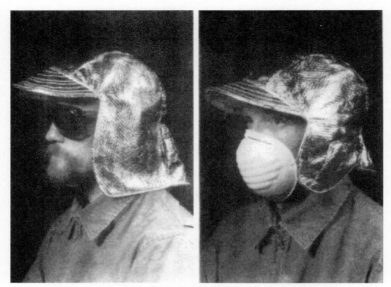

Fig. 2.92: Hood

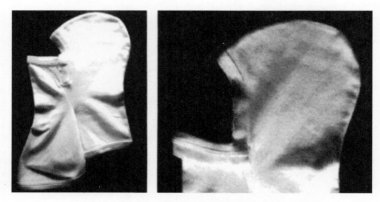

Nomex (fire-resistant) hood

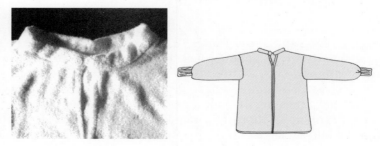

Cotton fleece

Fig. 2.93

## Classification of Personal Protective Equipment (PPE)

1. PPE worn as clothing or to cover part of the body, e.g. aprons, gloves.
2. PPE used as attachments, e.g. safety belts, goggles.

### Management of PPE

- Assessment
- Selection
- Deployment
- Control

Assessment of the risk involved in an activity is the first step. The PPE selected should provide adequate protection against the hazards involved.

### Code Designation for PPE

These codes are for easy classification and identification.

A - Safety glasses.
B - Safety glasses and gloves.
C - Safety glasses and gloves and apron.
D - Rack shield and gloves and apron.
E - Safety glasses and gloves and dust respirator.
F - Safety glasses and gloves and apron and dust respirator.
G - Safety glasses and gloves and vapour respirator.
H - Splash guard and gloves and—and vapour respirator.
I - Splash guard and gloves and dust vapour respirator.
J - Splash goggles and gloves and full suit and boots.
K - Air line head gas mask and gloves and full suit and boots.
X - Special handling instructions.

## DETECTION OF FIRES

As we have seen, all fires start small–a spark or a short circuit or a cigarette rolling on to a carpet. However, if they are not detected at an early stage and extinguished, they can spread fast and reach the blaze stage within a short period of time. At this point, they are almost impossible to extinguish with portable fire extinguishers.

Unless somebody is very near to the place, where the fire starts, the detection of the fire may not happen till the fire is close to the blaze stage. Even after it is detected, it will take some amount of time to organize the means of extinguishing it, particularly if one is using conventional fire fighting means such as water or sand or conventional extinguishers. By their very nature, these means are difficult to choose from (since these are not suited for all classes of fire), heavy and cumbersome to handle and difficult to operate.

These delays in using the means of fire fighting, combined with the delay in detection, can lead to the fire spreading and moving into blaze stage, leading to heavy losses to property and may be even life. On the other hand, if an early detection and warning is available, the fire can be detected early and if a fire extinguisher is available, the fire can be extinguished quickly and effectively with no damage to property.

The only way to attack a fire in its controllable stage is timely detection!

There are certain fuels, which when they burn, are non-luminous and do not produce smoke at all. The fuels in their pure form, when they burn, which do not give off smoke are formaldehyde, formic acid, methyl alcohol, hydrogen, etc.

Hydraulic fluids used in equipment and also vehicles, is flammable.
A dry chemical and liquid agent system is the most suitable fire extinguishing agent.

## FIRE ALARMS

### Manually-operated Fire Alarms

A fire alarm can be raised automatically by a detection system or manually by a person in the affected building. Such an alarm will generally be either wholly manual or manual-electric, not forgetting that an alarm can always be raised vocally.

### 1. Manual Systems

The purely manual means for raising an alarm involves the use of basic devices which include the following:

i. Rotary gongs which are sounded by simply leading the handle around the rim of the gong.
ii. Hand strikers, e.g. iron triangles suspended from a wall accompanied by a metal bar which is used to strike the triangle and produce a loud clanging noise.
iii. Hand bells.
iv. Whistles.

These devices are normally found in a readily available location on the walls of corridors, entrance halls and staircase landings.

They are relatively inexpensive and while they give an alarm over a limited area, are rarely adequate to give a general alarm throughout the premises; nor do they necessarily convey the alarm to a central point from which the fire brigade can be speedily summoned.

As a person is required to operate them, a continuous alarm cannot be guaranteed for as long as may be necessary. Because of this restriction in their use, it is unlikely that these devices will be the sole means of raising alarms, except perhaps, in low risk areas.

### 2. Manual Electric Alarm Systems

Manual fire alarm system consists of manual call points or pill boxes provided at various locations. These call points are connected to the zonal or central control panels. In case a fire or a hazard situation is noticed, the call points can be actuated by breaking the glass cover. To facilitate the breaking of the glass, a hammer is provided at each of the call points. The call points and the control panel are powered by emergency and also standby UPS (uninterrupted power supply) battery system. Once a call point is actuated, a red light lamp will glow at the control panel and indicate the location of the call point and also an alarm hooter will sound off.

These are systems which, although set in motion manually, operate as part of an electrical circuit. Manual call points, as they are called, are fire alarm system which allows for automatic and/or manual raising of an alarm. The call points in a manual electric system are invariably small, wall mounted boxes as shown in Fig. 2.94. They are designed to operate either.

**(a)**                                    **(b)**

**Fig. 2.94:** (a) A Manual call point, (b) Breaking the glass.

The wall mounting type manual call point, with dust and moisture proof cover, is designed to raise the alarm, once the glass is broken. An LED indicator is provided to acknowledge the operation.

Type                             :   Break glass latching type
Glass surface                    :   About 40 sq.cm
Visual indication                :   Red led.
Related accessory                :   Hammer and chain.

  i. Automatically, when the glass front is broken, or
 ii. When the glass front is broken and the button pressed in.

The majority of available models are designed to operate immediately the glass front is broken.

In Fig. 2.95 contact (1) is connected to one side of the electrical circuit and contact (2) to the other. The movement of contact (1) is governed by the spring loaded button which is maintained in the depressed position by the glass front. Normally, therefore, contact (1) is held off contact (2), but once the glass is broken, the spring forces the button outwards, allowing contact (1) to engage with contact (2), thus completing an electrical circuit and raising the alarm.

As an alterative, this type of call point can be fitted so that the electrical circuit is normally complete, a relay being incorporated to hold off the alarm. On breaking the glass, the circuit is broken, the relay de-energizes and the alarm sounds.

In either case, accidental breaking of the glass will, of course, raise a false alarm. This is most likely to happen in a situation where various goods and metals are being moved

*There is no smoke without fire–proverb*

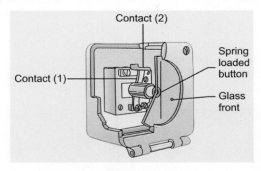

**Fig. 2.95**

about (e.g. in workshops and storage areas). This problem can be overcome by installing a call-point in which the button has to be manually pressed-in to raise the alarm after the glass has been broken. When the button is released, the alarm will continue to sound.

To help with the breaking of glass in the call-point, most manufacturers will provide, if requested, a small chromium-plated hammer for attaching by chain to the box. In the place of the hammer, some manufacturer will 'score' a little in the glass for easier breakage; this enables it to be broken by a blow with the tips of the fingers. Obviously these arrangements are desirable only when the possibility of accidental or malicious breaking is minimal.

Where neither hammer nor scored glass is available, a blow with a covered elbow, a shoe-heel or other sharp object will be effective.

In certain types of occupancy (e.g. mental wards) or in situations where vandalism is likely, a modified call point is often fitted and these points have a solid door with a keyhole. When the key is inserted and turned, the electric circuit is completed and an alarm is raised. The key will normally be in the possession of an authorized person.

Recommendations for installation of manual call points are given in BS 5839. Manual call points should be installed in prominent and conspicuous locations, especially in escape routes.

## Installation

- On each floor one or more manual call point (MCP) should be installed preferably on the exit routes.
- Call point should be installed at a height of 1.4 m above the floor at an easily accessible well-lit position free of all obstructions. The call point should form an integral part of the fire detection system.
- MCP should be of wall mounting type. The housing should be dust proof and moisture proof. Probably sealed with rubber-lining.
- The glass surface should be minimum 30 sq.cm in area and glass thickness should not exceed 2 mm. Once the glass is broken the alarm should sound on the floor as

well as on the control and indicating equipment and light should glow to indicate its operation. The alarm should be maintained by the control and indicating equipment even if someone presses the button subsequently.

## SMOKE DETECTORS

### Purpose of a Smoke Detector

A smoke detector is an excellent fire precautionary measure for residences as well as for all types of buildings.

Smoke detectors can save life. Most fatal home fires occur at night, while people sleep. Fire produces toxic gases and smoke that actually numb the senses. If somebody is asleep, or become disoriented by toxic gases, they may not even realize that there is a fire. We should not rely on the human senses to detect a fire.

Generally, all smoke detectors consist of two basic parts:

1. A sensor to sense the smoke and
2. A very loud electronic horn or speaker to wake or alert people.

Smoke detectors consume very little electric power. They can be energized with a battery or from the 240 volt mains (domestic supply).

There are two types of smoke alarms available commercially today.

1. Photoelectric smoke detectors
2. Ionization smoke detectors

Fig. 2.96: A Smoke detector installed in the roof.

### Photoelectric System or Optical Smoke Detectors

This system is especially sensitive to smouldering, gray fires. The photoelectric method relies on scattered light principle to "see" smoke and sound the alarm. A regulated pulse of infrared light is beamed into a darkened smoke chamber every ten seconds to form a precise beam. When smoke enters the chamber, it interferes with the beam and scatters its light. The amount of light scattered is monitored by a photodiode and when a predetermined amount of light strikes the photodiode, the alarm is activated. Stray light from sources such as light fixtures is blocked.

Each pulse of light actually consists of a series of micropulses of about 3 KHz. In clear air, the photodiode receives no light directly from the LED due to the angular arrangement within the detector. When visible smoke enters the detector, the light pulse emitted from the LED is scattered on to a photocell. Signals from the photocell which are synchronous with the LED pulse pattern are amplified and integrated. If the integrated photocell output reaches a certain threshold, the LED pulsing is speeded up to every 2 seconds. When two consecutive confirmations on the presence of smoke occur, the detector switches to the alarm state. Operating temperature for these types of detectors is from − 22 to + 60°C. The detector is not affected by relative humidity up to 95%.

Both ionization and photoelectric detectors are the most effective smoke sensors. Ionization detectors respond faster as compared to the photoelectric type in flaming fires with smaller combustion particles. Photoelectric detectors respond more quickly to smouldering fires or fires with smoke. The photoelectric detector uses an optical beam to search for smoke. Ionization type of detectors is less expensive than photoelectric detectors.

## Ionization Smoke Detectors

These detectors are particularly effective in the detection of combustion particles of small size, even up to 1 micron. These detectors contain a dual sided high surface area radioactive source of Americium 241(total 0.9 micro-curie). One side of the source ionizes the air in an outer chamber which is exposed to the atmosphere, while the other ionizes the air in a semi-sealed reference chamber which compensates for environmental changes. Ingress of smoke particles in the outer chamber reduces the small current which is flowing. This results in an imbalance between the two chambers which causes the detector to change to its alarm stage when its current consumption increases from 25 micro-amps to 55 milli-amps. When the detector is in the alarm stage, the indicator LED comes on and it latches until it is reset by temporarily removing the power. This type of detector is most sensitive to particles in 0.01 to 1.0 micron range and therefore senses "invisible smoke" such as is produced in the early stages of a high temperature or flaming fire as well as invisible smoke from most smouldering fires. These are very neat and unobtrusive when installed. It operates satisfactorily at temperatures between 20°C to 60°C and is immune to wind turbulences up to 20 m/sec and up to a continuous relative humidity of 90%.

## Dual Ionization System

This method of sensing smoke is especially responsive to fast developing fires. Air molecules in the control space are ionized in an air sample chamber to create a small electric flow between two charged plates. When combustion particles in the sample chamber accumulate to a preset density, they cause the current flow to drop and the alarm is activated. Dual sample chambers are also provided in some detectors to offset reduction in current flow caused by humidity and temperature changes in the space, virtually eliminating nuisance alarms.

## Answering a Fire Alarm

On receipt of an impulse from a detector or a manual call point, the indicator panel flashes a signal and also gives out a hooter sound to alert the personnel manning the control room. The hooter can be silenced by an acceptance switch which acknowledges the alarm. Smoke detectors which have been activated continue to flash an LED by which it can be identified and reset, if necessary.

## False Alarms

Sometimes, in either type of smoke detectors, steam or high humidity can lead to condensation on the circuit board and sensor, causing the alarm to sound. Some users purposely disable them because ionization detector is more likely to sound an alarm from normal cooking due to its sensitivity to minute smoke particles. However, ionization detectors have a degree of built-in security not inherent to photoelectric detectors. When the battery starts to fail in an ionization detector, the ion current falls and the alarm sounds, warning that it is time to change the battery before the smoke detector becomes ineffective. Back-up batteries may have to be used for photoelectric detectors.

## What is Required in a Smoke Detector?

- It should have a warning signal that warns when bulbs or batteries need replacing.
- The batteries and bulb should be readily available for purchase and easy to replace.
- The smoke detector's alarm must be loud enough (85 decibel or louder) to wake a sleeping person behind a closed door. Special detectors are available for hard of hearing persons.

> **Beware!**
> **Smoke has killed**
> **more people than**
> **Fire**

## Installing a Smoke Detector in Residential Premises

At the bare minimum, there should be one detector for each level in a home. A detector needs to be placed within 10 feet of sleeping areas, since most fire deaths occur at night while people are sleeping. It should never be placed in the dead-air space, such as where wall and ceiling meet or in a corner. Nor should it be placed near heating ducts or cold air returns. The air flow around these areas could prevent the smoke-filled air from collecting in the detector in sufficient amounts as to activate it. Smoke detectors should not be installed near bathrooms with showers.

**Fig. 2.97:** Ionization type smoke detectors.

Steam can sometimes cause false alarms and the moisture can rust metal components of the detector. Also, areas where nominal amounts of smoke may normally be present, such as kitchens or other cooking spaces, furnace rooms, or near fireplaces or wood-burning stoves, should be avoided.

There are other types of detectors such as the heat detector. Heat detectors are no substitute for smoke detectors. They set off an alarm in response to heat only. They do add protection and can be helpful in basements, kitchens, attics and garages. But for life safety purposes, a home should be protected by a smoke detector.

### Location of Smoke Detectors

Smoke detectors need to be placed strategically so as to cross-zone a fire before confirming the same. These alarms should be installed preferably in the ceiling. Normal flat ceiling is the best place. They should be at a minimum distance of 300 mm from any wall. If on the wall, top of the alarm should be 300–500 mm from the ceiling.

### Testing of Alarms

Fire alarm–smoke or manual type, require regular testing and cleaning. A test button is provided to ascertain the working condition of the alarm and should be tested at least once a month.

### FIRE HOOTER

The hooter is designed to provide a high sound output to warn occupants against a fire. Hooters can be powered by battery or mains. Fire hooters should have a distinctive sound to enable differentiating the hooter sound from other sounds.

| | |
|---|---|
| Sound level | : more than 85 decibel (db) |
| Type | : wall mounting type |
| Enclosure | : ms cabinet. |

In case of Fire
Do not Panic.
Call the Fire Brigade

**Fig. 2.98:** Hooter.

The hooters for fire alarm should be electronic hooters or electric bells having frequency range of 500 to 1000 Hz. If two types of alarms are used, at least one of the major frequencies should be within the frequency range of 500 to 1000 Hz.

The distribution of alarm sounders should be such that the alarm is heard at all sites which can be occupied within the protected area. A minimum sound level of either 85 db or 5 db above any other noise likely to persist for a period longer than 30 seconds, which ever is greater, should be produced by the sounder at any point which can be occupied in the building or premises.

Hooters should be suitably distributed throughout the building in regard to attenuation of sound caused by the walls, floors, and partitions. If the fire routing for the premises requires the audible alarm to arouse sleeping persons, the minimum sound level should be 85 db at the bed head.

Sound levels exceeding 120 db in areas which can be occupied may produce hearing damages.

The grouping of external fire alarm hooters can be done in either of the following way. The grouping scheme should be reflected in fire instruction issued for the use of occupancies.

*During fighting a fire, remain calm*

Sounding of alarms can be so arranged that any alarm operates all the hooters through-out the premises. This grouping is particularly suitable for smaller premises.

Sounding of alarms can be so arranged that the alarm sounds initially in the zone of the origin and then in all the adjoining areas, and in specially selected areas of high flammability.

Fixing of sounders can be at a height of 2.30 meter from the finished floor level and the distance from roof can be 150 mm (6 inches).

Fire Emergency! break glass

**Manual call point**

The striker can be used to break the glass and activate the switch.

Some panels have pre-scores on the glass for easy and safe breaking.

**Fig. 2.99**

Fire alarms should shut down the air-conditioning system as they provide ventilation. They also carry fumes, heat from a fire, to other places.

## Electrical Cabling for Detection and Alarm System

The standards prescribe the specifications for the cables and their installation and the circuit design.

The cables should conform to the following.

- PVC insulated copper conductor cables conforming to IS 694/1977 having minimum 1 mm$^2$ cross-sectional area, if stranded, at least 0.5 mm$^2$ cross-section should be used.
- Rubber insulated braided cables conforming to Indian standard 9968 (part 1) 1981.
- Armed PVC rubber insulated cables conforming to Indian standard 1554/part 1 1976.
- The cables used should be exclusively for fire detection system. The multi-core cables should not be shared for other low voltage or voltage circuit cables and wiring should be laid down in metallic or PVC conduits.
- Cables connected to detectors should be 'S' loop on the sides of the detectors which should be properly clamped to ceiling. Loop should be also be left where cables connect sounders, panels, dampers, etc. Appropriate glands should be provided where the cables enter the junction box.

- All the cables and wires should be tagged for proper identification; wires should be identified by ferrules at junction and cables by colour bands at every three meters distance.
- When connecting different buildings, etc. over-head fire alarm system should not be used. They should be laid underground according to IS: 1255/1983.

### Fire Detection/Alarm Control Panel

The detection and alarm panel should be located in areas such as the operations control room, security room at the main gate in industrial establishments, or in places which are manned round the clock from where the fire suppression and control activities can be initiated, coordinated and controlled.

**Fig. 2.100:** Fire alarm panel.

The fire detection/control panels also house the extinguisher release switch, through which the release of the extinguisher can be remotely actuated.

# 3

# Fire Risk Management

*It is with our passions as it is with fire and water; they are good servants, but bad masters.*

*—Roger L'Estrange*

## FIRE HAZARD

A fire creates three main sources of hazards–namely—heat, smoke and oxygen depletion.

Fire hazard is the risk of exposure to a fire or its products and its effects. As defined by ASTM, fire hazard is the fire risk which is greater than acceptable. The hazard is more serious, if the probability of occurrence (likelihood) is high or the severity of the possible fire is high. The hazard level is not only dependent on the characteristics of the materials in a given environment, but also on the conditions prevalent in the environment such as structural nature of the building, fire detection and protection provided, etc.

**Fig. 3.1:** Safety matches.

Fire risk management can help in risk control and risk reduction. The objective is to protect lives and preserve value. The risk management steps are:

- Identify
- Analyze
- Prioritize
- Treat

## Safety Indicators

- Positive safety culture–observable degree of effort to direct the situation and actions taken to improve level of safety.
- Measure of safety—number of accidents/injury; number of near misses; number of health issues; violation of safety rules.

## Hazards and their Control

Hazard is an attribute of a product or system that is capable of a harmful result to people or plant and equipment.

Hazard classification by effect.

- Class I–Negligible
- Class II–Marginal
- Class III–Critical
- Class IV–Catastrophic

## Hazard Mitigation

Some measures to ensure high reliability:

1. *Redundancy:* By providing more than one means to carry out an operation
2. *De-rating:* Assuring that the forces and stresses applied are lower than the stresses the part or material can withstand.
3. *Fool-proofing:* Elimination of error-prone elements in the system, designing to make human error improbable or even impossible.
4. *Warning mechanisms:* Incorporating protective systems such as fire detection and sound alarms.

## Hazard Communication

Hazard communication aims to provide users or the public with information about the hazards to which they are exposed.

The key elements of a hazard communication programme are:

- Material safety data sheet (MSDS) system
- Labelling
- Standard operating instructions or procedures.

## MSDS: Material Safety Data Sheet

MSDS is a record of the properties of a material which can indicate the known hazards arising from the material. MSDS also gives the precautionary measures to be taken while using the material and the control measures in case of incidents such as exposure, inhalation or swallowing, contact, spillage, etc.

---

### Comprehensive emergency management plan (CEMP)

A disaster plan that conforms to guidelines established by the authority having jurisdiction and is designed to address natural, technological, and man-made disasters.

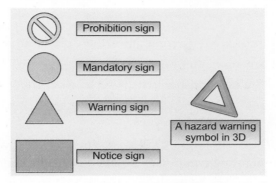

**Fig. 3.2:** Geometric scheme for safety signs.

## ACCIDENT PREVENTION

### Accident Prevention Methods

- Reduction and elimination of hazards
- Modification of energy release.
- Separation by distance
- Separation by barrier.

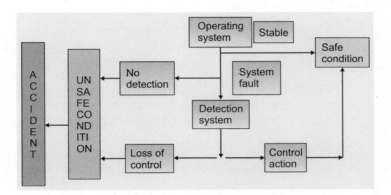

**Fig. 3.3:** Accident prevention principle (modified).

**Carbon monoxide:** A gaseous molecule having the formula CO, which is the product of incomplete combustion of organic materials. Carbon monoxide has an affinity for haemoglobin approximately 200 times stronger than oxygen and is highly poisonous. CO is a flammable gas which burns with a blue flame and has explosive limits of 12 to 75%. Carbon monoxide has approximately the same vapour density as air, 0.97 (air = 1.00).

## Points to Remember in Fire Safety Management

- Fire brigade call number is 101 (in companies, there can be an internal number also).
- Report all cases of fire–no matter how small.
- Do not keep back a fire extinguisher to its fixed location after use. Inform the fire or safety department.
- Observe strictly the fire prevention rules.
- Do not tamper with the fire fighting equipment.
- Ensure that the fire fighting equipment installed is visible and easily accessible.
- Report any defect in fire equipment immediately to the fire control.
- Do not drag fire hoses and the couplings while using the same.
- Be sure to know how to operate the fire fighting equipment.
- In fire risk zones, all hot jobs to be carried out with permits.
- In the event of fire, all hot work permits shall remain cancelled.
- Tapping of water from fire water system for other purposes should be avoided.

### FIRE PRONENESS

Statistics has revealed that communities that experience

- Low house income
- Household crowding
- Low educational achievement
- High unemployment
- Inadequate parental supervision or support.

are vulnerable and have a higher risk of fire, both internal and external fires.

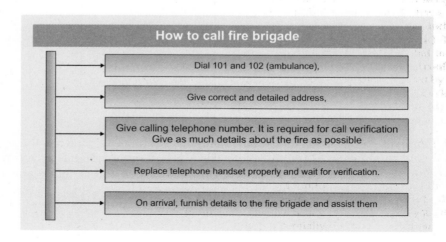

| How to call fire brigade |
| --- |
| Dial 101 and 102 (ambulance), |
| Give correct and detailed address, |
| Give calling telephone number. It is required for call verification Give as much details about the fire as possible |
| Replace telephone handset properly and wait for verification. |
| On arrival, furnish details to the fire brigade and assist them |

Minimizing the risk in either unprotected or protected facilities can be accomplished by reducing the probability of a fire. These types of "fire safety" practices are common and range from good housekeeping and other management programme controls to inherently less combustible and ignitable process and facility designs. This encompasses a broad range of elements, but all contribute to lessening the probability of a fire.

## Heat and Fire Resisting Materials

Heat and fire materials can resist high temperature, usually higher than 100° C. Examples of such materials are:

- Glass fibre materials—Usage: high temperature protection up to 650° C continuous, 900° C in flash.
- Aramid fibre materials—Usage 300° C continuous, 500° C flash.
- Aluminized tissue materials
- Kevlar
- High temperature resisting leather
- Proban® treated materials.
- Refrasil—the substitute for asbestos

## Fire Cabinets

Fire cabinets are meant for safe-keeping precious materials or documents and protect from fire and accomplish loss prevention.

**Fig. 3.4:** Fire cabinets with one to two hours of fire protection (Chubb and Godrej).

## BUILDING FIRES

Fire safety involves prevention, containment, detection and evacuation. Fire prevention basically means preventing the ignition of combustible materials by controlling either the source of heat or the combustible materials. This involves proper design, installation

*Fire prevention is better than fire protection*

and construction and maintenance of the building and its contents. Proper fire safety measures depend upon the occupancy or processes taking place in the building. Design deficiencies are often responsible for spread of heat and smoke in a fire. Spread of a fire can be prevented with design methods that limit the growth of fire and spread within a compartment and with methods that contain the fire to the compartment of origin. Early detection is essential for ensuring adequate time for escape.

In building fires, there are two levels of hazard: structurally endangering and not structurally endangering.

*Structurally endangering*: A hazardous occupancy with sufficient fire or explosion potential to defeat the basic integrity of the building frame.

*Not structurally endangering*: A hazardous occupancy with sufficient fire potential to build to full involvement and present a danger of propagating through openings or wall partitions but not possessing sufficient total potential to endanger the structural framing or flooring.

Statutory requirements for fire safety in buildings can be for material requirements or building requirements. Codes for materials are for limiting the fire growth within compartment and the codes cover requirements such as combustibility, flame spread and fire endurance. Building codes stipulate requirements such as area and height limitation, fire stops and draft stops, doors and other exits, sprinklers and fire detectors.

## Building Regulations to Prevent Fire

In building design, importance needs to be given to fire prevention and protection and also to reducing the smoke level. Ventilation provision plays an important role in smoke elimination.

Factors on which building codes evolve are:

- Type of construction (such as single storey/multi-storey)
- Use or processes carried out inside the building
- Occupancy factor

Building construction is generally classified into 5 types based on the type of materials used and the required fire resistance.

- Type I    – Fire resistant construction
- Type II   – Non-combustible construction
- Type III  – Ordinary construction using smaller wooden members.
- Type IV  – Heavy timber construction (but walls are non-combustible materials)
- Type V   – Light frame construction (wooden frames)

The above can be further classified based on the fire-resistance requirements. (Fire resistance rating is specified by the hours of protection).

A balance has to be achieved between the cost of the fire protection measures and the additional safety achieved.

*No facility is absolutely fire-safe*

### Fire Spread from one Building to Another

The spread of fire from one building to another across a vacant space can be caused by convective or radiative heat transfer or flying burning material. While flame impingement and convective heat transfer can cause a fire, ignition as a result of radiative heat transfer is the main danger. Standards prescribe minimum separation between buildings to reduce the fire spread.

### Minimum Requirements for High–rise Buildings

- Riser system and partial automatic sprinkler protection
- Smoke detection alarms
- Smoke control
- Compartmentation.

### Stages of Building Fires

*Pre-flashover (incipient stage):*  **Only** localized rise in the temperature occurs. Further growth is dependent on the air supply (ventilation). Sometimes, acts such as opening of a door increases the air supply and helps the fire to develop.

*Flashover:*  **The** fire spreads to all the available fuel within the building compartment.

*Post-flashover:*  **The** maximum temperature of over 1000° C is possible. The rate of temperature rise continues till the fuel load begins to decrease.

*Decay phase:*  **The** temperature now starts to decrease.

### Factors Affecting the Growth of Fires in Buildings

- Ventilation (air supply)—depends on the available building openings
- Fuel load—depends on the materials of construction.

The total heat output from a fire is entirely dependent on the fuel available and its rate of output is ventilation controlled.

> **Argon:** A monoatomic gas. Air contains 0.94% argon. Argon is obtained from liquid air by fractional distillation. Argon is chemically inert and is used as fire extinguishant.

### VENTILATION

Ventilation is the means of air supply into an enclosed space.

Openings at a lower level provide fresh air ventilation and thereby provide the oxygen required for a fire to continue. Ventilation is also dependent on the wind velocity and the draft available (or the pressure difference between the interior and the outside).

When smoke is predominant in a fire scene, ventilation also helps to maintain the oxygen level in a compartment. An atmosphere of less than 5% oxygen may be lethal in 5 to 10 mins.

Openings at the upper level or the roof, help to evacuate the smoke which develops from a fire.

## Roof Ventilation

Where the building is predominantly meant for occupation, such as residential buildings, ventilators should be provided at the roof level to enable egress of $CO_2$ and $CO$ produced by any fire inside the buildings.

Roof ventilation can be:

- Natural ventilation openings
- Fan-assisted or mechanical ventilation

Natural ventilation openings are definitely more reliable and should be provided in all domestic buildings. However, in crowded residential areas, exhaust fans are installed (especially in kitchens) to provide the ventilation.

In industrial establishments, special roof designs such as the "Robertson roof" ventilators are provided.

Models can be used to study and evaluate the air-flow into an enclosed space and also the fire and smoke transport and diffusion.

## BUILDING RULES FOR FIRE SAFETY

- The building should be designed to be structurally stable for a reasonable period, due to fire.
- Large buildings are subdivided into fire-resisting compartments.
- Adequate means of escape, in case of fire, to be provided
- Measures to restrict the spread of fire over internal surfaces such as walls and ceilings to be provided. The materials used must be adequately resistant to spread of flame.
- Measures to prevent fire spread form one building to another.

## Design of Buildings for Fire Safety

i. *Fire resistant materials:* Use of fire resistant materials counters the spread of fire due to destruction. Materials are designed not to fail by structural weakness, as failure can cause excessive heat transmission and the structure ceases to be an effective barrier against the spread of fire.

ii. *Compartmentation:* Countering the spread of fire due to convection.

iii. *Draft control:* Controlling the intensity and direction of drafts.

iv. *Doors and openings:* Keeping doors closed may be the most effective way of reducing fire loss. Sliding doors are better than hinged doors as pressure on hinges may keep the door open. The control of both the fire and also the smoke is necessary.

*Wind to the fire–it extinguishes the small, it inflames the great.*

*–Bussy Rabutin.*

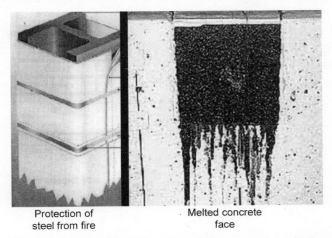

Protection of
steel from fire

Melted concrete
face

**Fig. 3.5:** Close-up of one of the faces of the concrete blocks. The outer parts of the concrete melted and a surface of glassy structures was formed in an experimental setup.

v. *Separation between buildings:* Layout design. Maintaining a minimum distance between adjacent buildings to contain the fire to the building of origin and prevent spread. Areas where the risk of fire is higher such as boiler rooms, etc. need to be isolated.

The fire resistance of building materials or components and constructions are classified as:

- R   – Load bearing
- RE  – Load bearing with integrity
- E   – Integrity
- EI  – Integrity with insulation
- REI – Load bearing with integrity and insulation.

The above criteria are followed by a time limit in minutes –15, 30, 45, 60, 90, 120, 240 or 360 minutes which signifies the time for which the fire resistance is available.

## Explanation

*Load bearing (R):* It represents the capability to bear a specified load under fire test conditions for the specified time.

*Integrity (E):* Means that no visible flame may occur on the unexposed side of the construction for more than 10 seconds. In insulated constructions, the integrity can be tested to prove if hot smoke or gases can ignite a cotton wool pad. In un-insulated, the integrity is tested with gap gauges. To pass the test, the dimension of any gaps should not be more than 25 mm in diameter or 6 mm wide for a distance of 150 mm through the construction.

*Insulation (I):* For this criterion, the temperature on the unexposed side of the construction may not, in average, exceed 140° C and the maximum temperature should not exceed 180° C.

In addition to the above, a "smoke leakage" criterion(S) can also be added.

## Common Building Materials and their Characteristics

### Concrete

Concrete is extensively used in the construction of buildings. The fire resistance of concrete depends on the type of aggregates used. The presence of moisture affects the fire resistance of concrete structures. The moisture is beneficial only if the concentration is not high as to cause spalling. Concrete made with materials such as perlite and vermiculite, lack strength and are used principally as insulating materials.

### Steel

Steel has got strength at ordinary temperatures. Ordinary mild steel used in the construction of buildings, loses half of its strength when exposed to temperatures greater than 727° C (600° C to 727° C). This is due to the phenomenon of speroidization (micro-structural changes) which cause lower strength and higher ductility.

Steel-work, whenever there is a likelyhood of exposure to fire, is encased in concrete. Though concrete loses its strength at a relatively low temperature, it acts as a good thermal insulator and considerably extends the time before steel suffers a significant loss of strength.

### Brick

Brick is a very common construction material. Brick is very stable at high temperatures also, as it does not undergo substantial changes on heating.

### Wood

Wood, today, is more used as an interior material (as furniture, etc.) than as a structural material. Wood holds more moisture than any other material.

Timber and wood products play an important role in fire development as timber is easily combustible and is also extensively used in buildings, especially residential buildings. Timber is largely used as a non-load-bearing structural member in the form of windows, doors, etc. and also for interior decoration for furniture, wall panels, etc.

More recently, plywood with fire resistant properties have been developed.

Coated wood products have also been developed for fire protection. A simple method adopted is coating wood with sodium silicate. Cement coatings are also available. Coated wood only chars when exposed to a fire and the char acts as an insulator for the inner wood.

Many fire resistant (fire-proof) paints are also in the market.

| Construction material characteristics (effect of heat) | | | | | |
|---|---|---|---|---|---|
| *Material* | *Thermal expansion* | *Density* | *Specific heat* | *Thermal conductivity* | *Strength* |
| Steel | Increases significantly | Constant | Increases up to 600° C then decreases | Decreases | Increases marginally up to 300° C then decreases significantly (by 66% at 650° C) |
| Concrete | Increases significantly | Constant (although both free and combined water is lost) | Increases | Decreases | Decreases significantly (60% at 600° C) bond strength between steel and concrete also decreases significantly |
| Wood | NA | Loses water | NA | NA | Outer layer chars and loses strength |
| Masonry | Decreases | Constant | Increases | Increases | Seldom used as structural load member (only partition) |
| Aluminium | Constant (only up to 300° C) | Constant | Constant | Constant | Decreases significantly |

## Burning of Plywood

Burning plywood releases formaldehyde. Formaldehyde, used in the adhesives is required for manufacturing plywood. Formaldehyde fumes, a colourless, pungent smelling gas, can (when more than 0.1 ppm) cause watery eyes, burning sensation in the throat and eye and nausea. Higher concentrations can trigger asthma attacks, can also cause cancer and skin diseases. The same is the case due to the burning of some textiles and polyurethane foams used in furniture upholstery.

Burning of wood, as reported by Douglas Drysdale (An Introduction to Fire Dynamics), wood requires heating, by radiation to 600° C for it to ignite spontaneously. In case the heating is by conduction, a temperature of 490° C is sufficient for the spontaneous ignition to take place.

Also, given the same ignition source, natural teak wood takes longer to ignite than plywood.

### Plastics

Plastics lack strength and is rarely considered for structural use as a load bearing material. But plastic is used as pipes, sheets, tiles, panels, doors, etc. Plastics have a very narrow temperature range of utility.

### Glass

Glass is essentially used for non-load bearing applications such as windows and partitions. Although glass is a fire resistant material up to a certain temperature, glass exhibits poor integrity due to its weak thermal shock resistance.

Windows are the openings in a building which provide the maximum ventilation. When, during a fire, the window glasses shatter and break, they fuel the fire by providing increased ventilation. Window glasses break due to stress caused by the thermal difference between the center of the glass and the edges (normally caused by the hot gases descending from the roof and the glass at the center getting heated to above 600° C). Crack initiation and a later shattering takes place. Or due to the thermal shock of the temperature difference between the inside face and the outside face (in case of thick glass panels). The breakage can be accentuated by defects in the glass or due to flaws on the surface of the glass.

*Fire resistant glass:* Fire resistant glass is made as wired-glass. Wired glass does not break-up and is shatter proof.

### Fire Doors or Fire-resisting Doors

A door which, together with its frame, is capable of satisfying the criteria of fire resistance with respect to collapse, flame penetration and excessive temperature rise. Such a door may be automatic or self-closing.

Rolling shutters are the most commonly used doors which provide fire protection, but without being self-closing.

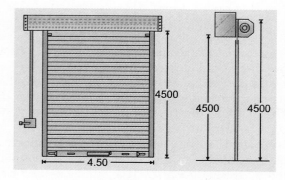

**Fig. 3.6:** Steel rolling shutters.

Fire doors are automatic and self-closing, having class (A) (B) (C) (D) or oversized certificate. 4-hour fire doors are also available.

### Class Designations

Class (A)—3-hour approved fire doors in dividing fire walls with opening not exceeding 120 sq.m in area. Class (B)—1.5 hour approved fire doors, in vertical shaft openings not exceeding 120 sq.m in area. Class (C)—0.75 hour approved fire doors in corridor or room-partition opening not exceeding 120 sq.m in area. Class (D)—0.5 hour approved fire doors in exterior wall opening not exceeding 120 sq.m in area.

### Standard Fire Tests

Standard fire tests are conducted to determine the fire resistance grading for a given method or design of building construction. It can also assist in the development of new products with superior fire resistance properties.

Such tests are described by standards like ISO: 834 (1975) and BS 476.

Even structural members such as beams, columns are also tested. The structural member is loaded so as to produce the same stress as when it is a member of a structure. The test member is heated at a standard rate under load until failure occurs (e.g. beams are heated from bottom, columns on all four sides, walls on one side). Heat is applied at a standard rate, usually 840° C in 30 mins, 945° C in 60 mins and 1050° C in 120 mins.

Failure is determined by:
- If the average temperature on the unexposed face reaches a temperature of 140° C.
- Cracks or openings occur in a separating element such that ignition can occur on the unexposed face (loss of integrity).
- No longer able to carry the load. (limit of deflection exceeds) (loss of load bearing capacity).

Structural fire resistance is graded in terms of time of 30, 60, 90,120 and 240 minutes.

### Fire Hazards and Control during Construction

The chances of fire during construction and such other temporary phase activities are very high due to:

- Lot of hot work is carried out at various locations
- Presence of combustible material like wood, packing materials, etc.
- Poor house keeping
- Lack of adequate supervision and inspection
- Inadequate fire fighting facility.

A fire watch is a necessary requirement in construction sites.

### Fire Retardants

Fire retardants reduce the likelihood of ignition and thereby prevent a fire from starting. They also reduce the rate of fire spread and thereby reduce the quantity of material burnt.

Smoke and toxic gases are as dangerous as fire. Smoke obscures vision and prevents escape. Materials with smoke suppression additives are available for increased fire protection.

## Fireproofing or Fire Stopping

Consists of using materials such as:
- Intumescent sealants
- Fire resistant mortars
- Composite sheets (fire resistant)
- Wrapping materials for pipelines.

> **"Be prepared"-Scout Motto***, meaning that we must prepare ourselves by previous thinking out and practicing how to act on any incident or emergency, so that we are never taken by surprise, knowing exactly what to do when anything unexpected happens.*

## FIRE PLANS

Fire plans help to reduce the time needed to escape (TNE) from a fire situation. A reduction of the TNE can increase the safety margin for the occupants and people exposed to a fire.
Some of the major factors affecting TNE:

- Response of the occupants to the fire
- Occupant location with respect to the fire, and the conditions of the fire scenario.
- Fire resistance property of the building structure.
- Fire protection system design
- Escape/evacuation model.

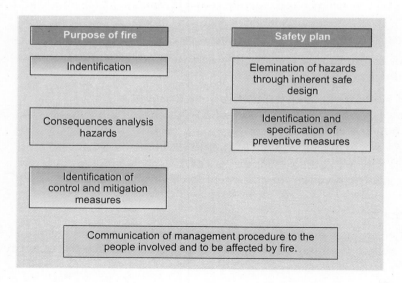

## Means of Escape

A proper design of the escape system-door, passages, lighting, instructions, etc. can reduce the TNE (time needed to escape) from a fire situation. A measure of safety is the difference between the time available for escape (TAE) and the time needed for escape.

Escape from a fire means escape from the fire affected premises to some place of safety, beyond the building, which constitutes or comprises the building, or any area enclosed by it or within it.

> **Escape plan**
> **(part of the emergency plan)**
>
> Crawl and close doors behind as you exit
> (helps to limit the spread of the fire)

In fires which grow rapidly such as flammable liquids and gases, the TAE (time available to escape) is much less than the TNE (time needed to escape).

## Emergency Plans

Emergencies, such as a fire, can occur in many shapes and at random. An emergency plan can minimize loss by helping to ensure the proper response in an emergency situation.

The features of an effective emergency plan include:

- Control point—allocation of resources and liaison will be more effective, if the emergency is controlled from a clearly defined point.
- Emergency teams
- Emergency equipment
- Provision of PPE, tools, fire fighting equipment and communication equipment.
- Staff call-in
- Action plan
- Preventive check-list.

Fire emergency response plans need to be written and made available to all employees or occupants, including visitors on premises, so that all know what is expected of them during a fire emergency.

Written fire emergency response plans should include information for occupants of a facility (especially industrial premises), regarding methods and devices available for alerting people of a fire emergency. All involved should know how the fire brigade is to be called. Even where automatic systems are expected to raise the alert, the written plan should provide for backup alerting procedures.

The addition of a non-volatile solid to a liquid in which it is soluble, lowers the vapour pressure of the solvent in proportion to the amount of substance dissolved.

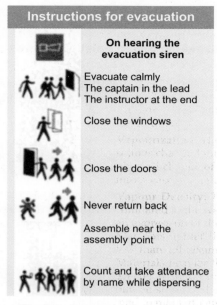

**Fig. 3.7:** Evacuation instructions.

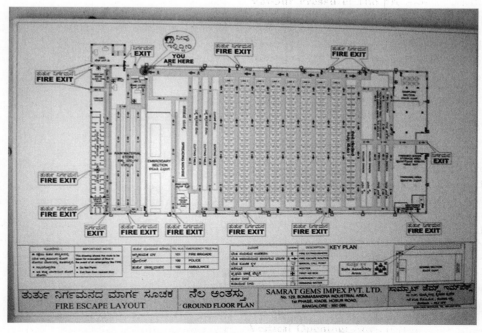

**Fig. 3.8:** Plant layout showing fire-exit points (emergency planning).

### Emergency Lights

An independent, self-energizing illumination system should be provided that will be automatically activated in the event of a major primary power failure or main lighting circuit malfunction resulting in circuit breaker interruption. If the back-up illumination system is a standby engine generator, it should provide power within 15 seconds of a failure and should be capable of sustained operation for a minimum of 72 hours. If the back-up illumination system is a standby battery system, it should provide power immediately upon failure and should be capable of sustained operation for a minimum of 4 hours.

Fig. 3.9: Emergency light.

### Emergency Door, Exit Design and Construction

Emergency doors and exits should be designed and constructed so that they are:

a. Simple to operate
b. Readily accessible
c. Clearly designated
d. Unobstructed
e. Simple to locate and operate in the dark
f. Capable of being opened in 3 sec or less
g. Capable of being opened with 44 to 133 N of operating force.

Fire barriers are probably one of the most critical and often overlooked areas of fire protection. There are many forms of fire barriers. Some are the elements of the building itself such as the walls, floors, ceilings, etc. others may be non-structural enclosures or partitions.

The principal function of a fire barrier is to prevent or reduce the spread of fire and/or smoke spread. Fire barriers are often found on primary escape routes and are vital for the safe evacuation of a building or structure.

Occupational Safety and Health Administration (OSHA) and National Fire Protection Association (NFPA) codes specify the design requirements for "ways of exit" from buildings and facilities. A designated means of emergency egress requires a minimum

*Know how to report a fire or other emergencies.*

of 45 cm (18 inches) of unobstructed "way of exit" travel from any point in a structure to an exterior safe public way. Design criteria for any unobstructed "way of exit" are:
- Functions of the nature of the building construction and contents
- The maximum occupancy capacities of its components
- Arrangement of designated "ways of exit."

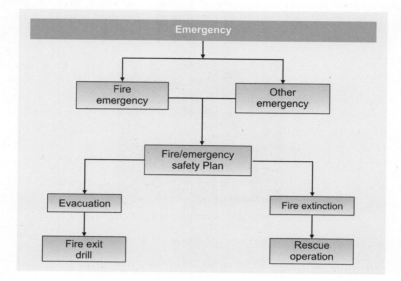

**Fire door**       **Fire alarm**

**Fig. 3.10**

**Emergency doors**
**Particular attention should be given to**
**keeping all doors unlocked.**

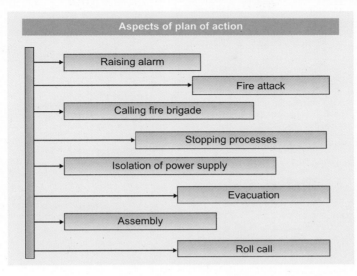

## Minimum Requirements or Duties

- *Means of escape*—with supporting provisions for the means of escape such as emergency lighting, signs, etc.
- *Means of raising the alarm*
- *Fire extinguishing appliances.*
- *Instructions*—what to do in the event of fire.

## DANGERS OF LIGHTNING

A strike of lightning can cause fires. So, tall structures and isolated buildings need to be protected from a strike of lightning. A lightning protection system routes the lightning along a known controlled path between the air and the moist earth with over 90% effectiveness in preventing fire and other damage. The power of the lightning is dissipated into the ground. The NFPA of the USA has issued a lightning protection code for guidance.

### Lightning and lightning arrestors

Lightning is a form of visible discharge of electricity between rain clouds or between a rain cloud and the earth. The electric discharge is seen in the form of a brilliant arc, stretching between the discharge points. When the electrical potential between a cloud and the earth reaches a sufficiently high value (about 10000 V), the air becomes ionized along a narrow path and a lightning flash occurs.

Buildings are protected from lightning by metallic lightning rods extending to the ground from a point above the highest part of the roof. The rod is connected to the ground with thick copper or aluminium strip. These rods form a low resistance path for the lightning discharge and prevent it from travelling through the structure itself.

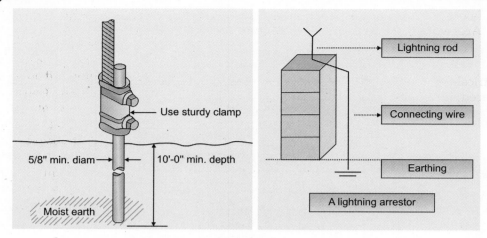

**Fig. 3.11**

Lightning protection should necessarily be provided for:

- Explosive and flammable substances manufacturing buildings
- Tall buildings and tall structures such as chimneys, where flammable gases, dust, etc. is likely to be present.
- Storage tanks containing oils, fuels and other flammable liquids.
- Electrical sub-stations.

**Fig. 3.12:** A fire-protection installation in a cable-rack.

## CAUSES OF ELECTRICAL FIRES

a. Overloading of cables by currents–cables overheat and the life of insulation is shortened.
b. Short-circuit of conductors (e.g. mechanical damage to insulation).
c. Leakage of current to earth, (e.g. due to failure of cable insulation).

d. Loose connections–resulting in localized overheating.
e. Arcs and sparks from electrical faults.

## Preventing Electrical Fires

Electrical wiring should be regularly checked and any worn or weak wire should be replaced, including cords on appliances.

**Fig. 3.13:** A carbon dioxide extinguisher being applied to an electric motor on fire.

*Water must never be used when the fire involves electrical equipment which are energized.*

**Fig. 3.14**

### Compulsory no-smoking areas

Smoking should be prohibited in any room, compartment or area where flammable or combustible liquids, combustible gases, or oxygen is used or stored and in any other hazardous location.

Fig. 3.15: Flame-proof switches.

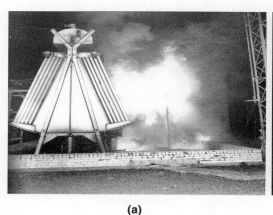 

(a)  (b)

Fig. 3.16: (a) A transformer on fire, (b) Switchgear arc flash.

The presence of electrical wires or equipment at or near the origin of a fire does not necessarily mean that the fire was caused by electrical energy. Often the fire may destroy insulation or changes in the appearance of conductors or equipment which can lead to false assumption about the cause for the fire.

For ignition from an electrical source:

- The system should have been energized.
- The heat or temperature produced by the shorting of a circuit (or can be a ground fault, parting arc, excessive current through wiring for equipment, resistance heating or ordinary sources such as light bulbs, heaters, cooking equipment and other electrical appliances) must be sufficient to ignite a combustible material, which is present in the vicinity.

*Protecting cables:* Power cables, when exposed to heat due to the operating conditions or due to heat evolved during a fire, can create further damage due to the loss of insulation of the conductors. Electrical conductors then need to be protected thermally.

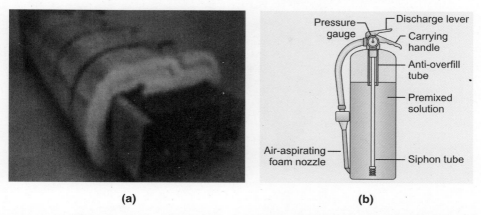

**(a)**           **(b)**

**Fig. 3.17:** (a) A cable tray protected with fire resistant material; (b) Stored pressure AFFF extinguisher (NFPA).

When electrical cables are run in open condition (without conduits), the wiring should not be concealed. For example, cords or wiring should be not run under carpets, where they might become damaged and set the carpet on fire.

Frequently electrical power outlets are over loaded and such abuse can lead to cable damage and consequent fires.

## DOMESTIC FIRES

### Causes of Domestic Fires

Human factors which are the major contributors to fire, have been identified as:
- Carelessness in work
- Lack of concentration in the activity.

Factors which are key to putting off fires are:

- The time lag between the initiation of the fire and the discovery of the fire
- The time taken to fight the fire.

Fire protection systems have largely been confined to industrial establishments and large public utility systems, while protection systems in residences is lacking. In developed countries, smoke sensors and associated alarm system is the commonly provided fire safety protection for residences which has proven to be extremely effective.

Knowledge dissemination is vital if we are to prevent fire accidents and promote fire safety. The nature of fires is now getting more complex with an increasing diversity of activities—manufacturing, storage, transport, etc. Use of fire safety equipment such as fire extinguishers is increasing (though not adequate), but the knowledge of how to use them is lacking.

**Graph 3.1:** Causes of domestic fires.

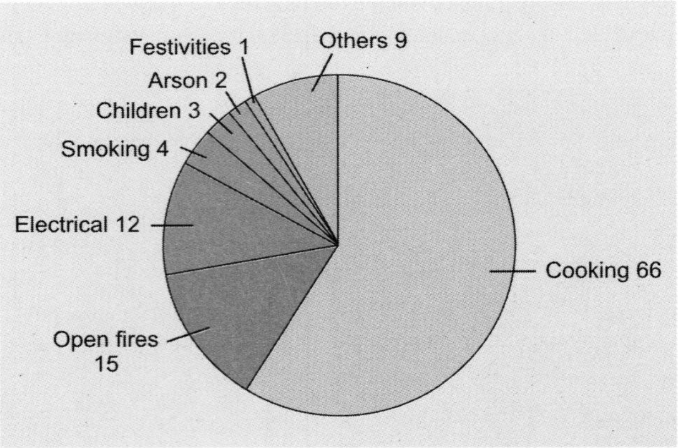

## Gaseous Fires

*Classification of gases:* Gases can be classified as

- Toxic gases
- Flammable gases
- Non-flammable gases
- Combustion supporting gases

### Toxic Gases

Toxic gases are gases which pose life hazard as they are poisonous or irritating on inhalation or when contacted. Examples of toxic gases are:

- Ammonia ($NH_3$)
- Carbon monoxide (CO)
- Chlorine ($Cl_2$)
- Hydrogen cyanide (HCN)
- Hydrogen sulphide ($H_2S$)
- Sulphur dioxide ($SO_2$)

### Flammable Gases

These are gases which when mixed with atmospheric air, form a combustible mixture which burns on the application of sufficient ignition source. Examples of flammable gases are:

- Methane ($CH_4$)
- Ethane ($C_2H_6$)
- Propane ($C_3H_8$)
- Butane ($C_4H_{10}$)

- Hydrogen ($H_2$)
- Carbon monoxide (CO)
- Acetylene ($C_2H_2$)
- Hydrogen sulphide ($H_2S$)

## Non-flammable Gases

These gases do not burn in any concentration of air or oxygen. Examples of non-flammable gases are:

- Helium (He)
- Neon (Ne)
- Argon (Ar)
- Carbon dioxide ($CO_2$)
- Sulphur dioxide ($SO_2$)
- Nitrogen ($N_2$)

## Supporter of Combustion Gases

These gases support combustion by reaction with the combustible material. Examples are:

- Oxygen ($O_2$)
- Chlorine ($Cl_2$)
- Nitrogen oxide (NO)
- Nitrogen dioxide ($NO_2$)

## Liquid Petroleum Gas (LPG)

LPG is a very common domestic fuel source. LPG is a mixture of butane and propane and is derived from processing natural gas.

LPG for domestic use is odourized (different levels possible) to enable easy detection of gas leaks.

- LPG is liquefied petroleum gas. LPG is obtained by refining of crude oil and from natural gas by oil absorption or refrigerated absorption process.
- LP gases are mixtures of hydrocarbons, which are gaseous at normal ambient temperature and atmospheric pressure, but can be liquefied at normal ambient temperature by application of moderate pressure.
- It is a clean burning, non-poisonous, dependable, high calorific value fuel.
- LPG is stored and transported in containers as a liquid, but is generally drawn out and used as a gas.
- LPG in gaseous state is nearly twice as heavy as air. Any leakage of LPG, therefore, tends to settle down at floor level.
- Liquid LPG is almost half as heavy as water. Thus, when liquid LPG gets converted to a gaseous state, it expands by about 250 times. The leakage of liquid LPG is therefore very dangerous.

LPG is chosen for use as a domestic fuel because:

- LPG contains no toxic components such as carbon monoxide and is, therefore, non-poisonous.
- Fuel gases will burn only when mixed with air in certain proportions. The minimum and maximum concentrations of a fuel gas in a gas/air mixture between which the mixture can be ignited are termed as the lower and upper limit of inflammability. The lower flammability limit for LPG is 2% and the higher limit is 8.5%.

**A domestic LPG cylinder**

Net weight—14.2 kg
Checks—Check seal on delivery;
"Switch-off regulator when not in use"

**Fig. 3.18:** LPG cylinder.

Pure LPG is colourless and odourless. LPG is distinctively odourized to give warning in case of leakage. Its smell is detectable in air at concentrations down to 1/5th of the lower explosive limit.

### Chemical Fires

Chemical fires, i.e. fires involving combustible chemicals such as fertilizers and pesticides, can release toxic combustion products. These products spread with the fire plume and can cause damage over a larger area. The residues of the combustion can be highly toxic. When these residues are carried by the fire fighting water, the effluents so created, can lead to soil and ground water contamination.

An example is the burning of PVC. Chlorinated compounds release a higher amount of CO (carbon monoxide) than non-chlorinated compounds.

### VEHICLE FIRES

Vehicles, today, have a lot of combustible materials.

The relatively small size of vehicles may result in more rapid fire growth, which may be more rapid than structural fires.

## Major Causes of Vehicle Fires

1. *Fuel leaks:* Petrol or LPG poses immediate danger. Diesel is less volatile, but as it dissipates more slowly, it remains a danger longer. Fuel leakage or spillage on hot surface or electrical components can cause a fire.
2. *Batteries:* Batteries can produce explosive gases. The battery compartment should be well ventilated. Batteries can also explode.
3. *Exhaust pipe or catalytic converter:* A hot exhaust pipe in contact with combustibles such as paper or dry grass can ignite.
4. *Tyres and brakes:* A dragging tyre can become too hot and ignite. A hot tyre can ignite the brake-fluid also. Explosion of air-inflated tyres can result from heat induced gas expansion inside the tyre which may result from an external fire or excessive use of brakes. The explosion can be very strong.
5. *Engine:* An overheated engine can cause an ignition of the vehicle interiors. It is necessary to check the engine cooling.

## Fire-Works (Fire Crackers)

Fire-work consists of a mixture of a fuel (charcoal and sulphur), an oxidizer (potassium nitrate), a metal powder (aluminum or iron or zinc or magnesium) and a binder (starch). When ignited these chemicals burn and give out a bright light. The mixture can also be made to explode with a loud sound.

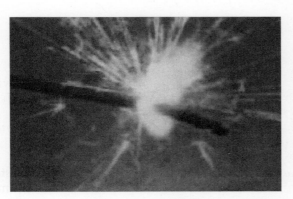

**Children should light fire-works under adult supervision**

**Fig. 3.19:** A sparkler.

## FOREST FIRES

### Types of Forest Fires

- Ground fires
- Surface fires
- Crown fires

*Ground fires:* A ground fire burns in natural litter and organic matter in the soil beneath the surface and is sustained by a glowing combustion. Once started, ground fires are very difficult to detect and control.

*Surface fires:* Surface fires burn in grasses and low shrubs, or in the low branches of trees. Surface fires move rapidly and the ease of control depends on the nature and condition of the organic fuel involved.

**Fig. 3.20:** A bush fire.

**Fig. 3.21:** An aircraft on fire.

*Crown fires:* Crown fires burn in the top foliage of trees, i.e. at the top of the trees. They are the most intense type of fires and once started, are very difficult to control. Wind plays an important role in crown fires.

Danger of wild forest fires is determined by:

- Ease of ignition
- Rate of spread
- Rate of combustion

The above factors are influenced by the fuel, the topography and the weather.

### Measures to Prevent Forest Fires

Forest fires can even be started by an empty bottle lying on dry leaves or dry grass. The bottle can, acting as a lens, concentrate the sun's rays and heat and thereby ignite the leaves to start a fire. So, polluting the forest with garbage such as plastic, empty bottles, paper, etc. can lead to a forest fire.

*Fire breaks:* Fire breaks are vacant strips of land which are made to prevent the fires from spreading. When the forest fire is uncontrollable, the fire breaks may be the only way to stop the spread of the fire. Sometimes it may be necessary to cut down trees at the boundary of a forest fire to create the fire break.

### FIRE MODELING

Fire models greatly help in effective fire risk management.

Fire models consider two distinct factors:

- The likelihood that a fire threat or a fire hazard exists and will produce a fire.
- The likelihood that an exposure of value is present when the fire occurs.

*Chlorine reacts explosively with ethylene in sunlight or ultraviolet light.*

A fire model builds a fire scenario which helps in predicting the development of fire conditions, effect of fire on structures and damage analysis (what—if analysis). All models are based on assumptions and knowledge of the assumptions will be necessary to correlate the inferences to actual conditions.

Results obtained from modeling studies are used to design fire protection systems. Models reveal the required information such as heat release and spread rate, time to flashover, combustion gas concentrations, etc. Computers can help in performing the numerous calculations.

Fire modeling is based on mathematical modeling and simulation by considering the principles of combustion of fuels, heat transfer, fluid dynamics, strength of materials, and properties of materials and other scientific phenomena, obtaining mathematical equations. The use of computers for such studies is indispensable.

Mathematical and computer modeling is applied to outdoor fires and indoor fires or confined fires in order to gain an understanding of the fire characteristics.

Four principal ways are adopted:

- Experimental fire models
- Stochastic fire models
- Field fire models
- Zone fire models.

Further studies include worst case scenarios, estimating the fuel mass (energy release), and environmental impact of fires.

Modeling of fires can also help in estimating the TAE (time available to escape) for an occupant.

*Room fire-models* (enclosed space models):

- Zone models
- Field models

## Zone Models

Zone models split rooms or enclosures into one or more zones. The most commonly used models assume a room is made up of two zones; an upper layer consisting of heated combustion products, and a lower layer which is composed of cooler air which is relatively free of combustion products. In a two-zone model, the fire forms the connection between the upper and lower layers. The layers are assumed to be well mixed, so that the conditions within each layer are constant. The predicted temperature within the hot layer, for example, is the same throughout. Many of the models include provisions for openings to the outside or to other rooms and for heat losses to the walls and ceiling. Model inputs typically include room dimensions and building materials, the sizes and locations of room openings, room furnishings characteristics, and the fire heat release rate. Outputs typically include prediction of sprinkler or fire alarm activation time, time to flashover, upper and lower layer temperatures, the height of the interface between the upper and lower layers, and combustion gas concentrations. Zone room fire models are available from several sources including the National Institute of Standards and Technology (NIST).

## Field Models

Field models, also known as computational fluid dynamics models (CFD), split a room or enclosure up into a large number of small three-dimensional boxes called cells. The enclosure may contain hundreds of thousands of cells ranging in size from centimeters to meters. Field models are based on the basic physical principles of energy, mass, and momentum conservation. The computer calculates the movement of heat and smoke between the cells, over time. At any point in time, it is possible to find the temperature, velocity, and gas concentrations within each of the cells. As in the case of the zone models, the properties within each cell are assumed to be constant. Due to the larger number of cells, however, the conditions in the enclosure can be predicted in much greater detail. Field models are capable of predicting the conditions in very large and very small spaces, in spaces with complex shapes, and in complex multiple room configurations that are not possible with zone models. Due to the complexity of field models, they require a high level of expertise to operate and are currently run on expensive computer equipment. General purpose field models are available commercially from various sources. Many computer programmes are still under development.

## Calorimeters

Calorimeters are instruments used to determine the heat output from a combustible material.

Cone calorimeter is frequently used for the fire test method.

Furniture calorimeter is used to measure the heat release rate of furniture and other materials.

These help in performing a systematic study of the fire hazard and fire modeling.

## Fire Drills

These are simulated exercises which are performed to review the evacuation system in place for fire exigencies. Clearly identifiable assembly points should be established.

Mock drills are statutory requirements for industrial units handling hazardous products. Apart from testing the level of preparedness of the operations, the drill assesses the response mechanism of the mutual aid agencies within the plant and also in the surrounding industrial area.

Translation of drill times to evacuation capability is determined as follows:

1. 3 minutes or less — prompt
2. Over 3 minutes, but not in excess of 13 minutes— slow
3. More than 13 minutes — impractical

Evacuation capability, in all cases, is based on the time of day or night when evacuation of the facility would be most difficult, such as, when residents are sleeping or fewer staff are present.

## HOT WORK

A hot work is an activity which has a potential to introduce a source of ignition.

### Hot Work Permits

The principal hazard of hot work is that it introduces a source of ignition. So, it is necessary that such work is performed only after required precautions are taken.

One-third of all fire accidents in industries have been found to be maintenance work related. And the single largest cause being a lack of adequate permit-to-work systems. Permit-to-work system is a formal written system used to control certain types of work which are considered to be potentially dangerous. This requirement is all the more necessary for work carried out by contractors or outside agencies. These activities introduce new hazards or reduce or bypass existing safeguards.

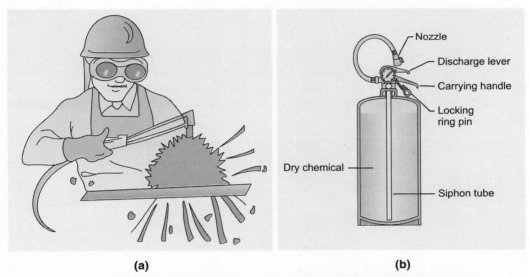

**(a)**          **(b)**

**Fig. 3.22:** (a) Welding hot work, (b) Dry chemical extinguisher (NFPA).

Arc welding is a welding process wherein coalescence is produced by heating with an electric arc, with or without application of pressure and with or without the use of filler metal.

Gas cutting is a process used for cutting steel by a flame torch using compressed gases hydrogen/acetylene and oxygen to preheat the metal and cutting it by forcing oxygen-enriched flame at higher pressure.

Welding and cutting and brazing (both gas and electric) have been responsible for many injuries, both minor and serious, and have caused many devastating fires. Acetylene used in gas welding is highly flammable and the oxygen used supports combustion. Hence, these welding and cutting flames must be handled with care at all times. Flash backs can be very serious and must be guarded against.

### Prevention of Fire from Welding and Cutting

Sparks from welding and particularly from cutting can travel a long distance. It is, therefore, recommended that such work should be carried out away from flammable materials.

Wooden floors or floors of combustible materials are a source of danger, and should be avoided. If unavoidable, these floors should be covered by sheet metal. Fire extinguishers and sand should be kept ready. In case of emergency, water may be sprayed to flood the place. Should it be necessary to carry out welding or cutting operation in the vicinity of existing wooden structures, special precautions should be taken to protect such a structure.

Cutting and welding should not be done in a place which is in proximity of a room or store containing flammable vapours, liquid, dust, etc. Welding should not be carried out in or around a painting installation.

Gas cylinders should not be allowed to come in contact with electric cables and electrical conductors.

**Fig. 3.23:** Flash back arrestor.

The use of welding, gas torch cutting, grinding and abrasive cutting equipment is a cause of many fires and explosions. They generate fires themselves or are sources of ignition. Where such work is not part of the normal routine operations, fire safety measures stipulate the use of a "hot work permit"

#### Features of the Hot Work Permit

1. The work should be authorized by a responsible person.
2. Fire and explosion hazards associated with the particular work should be identified prior to the start of the operation and necessary preventive action such as isolation, removal, protection and covering, disconnection, etc. should be carried out.
3. All combustible materials should be removed or made safe. No flammable liquids, vapours, gases or dust should be present. If necessary, shielding to be done. A test for flammables can be done.
4. The operations should be performed by trained persons.
5. Appropriate PPE should be used.

6. Relevant warning and fire fighting equipment should be readily available at the work site.
7. Fire watch should be kept, if necessary.
8. Hot work permits are always documented in writing and not just orally.
9. After completion of the work, the responsible person (who issued the permit) again inspects the work site for any signs of potential combustion or hazards.
10. Contractors and all workmen should be aware of the hot work permit system.

Hot work permits are issued only for a specific time period and if any extension is required, the permit should again be reviewed and a fresh permit should be issued.

## Backfire Arrestor

A flash back is the propagation of a flame from the burning point to the source of the fuel.

These devices provide protection against the hazards of flashback and can thus save human lives and property. A flame arrestor, installed in a fully premixed air–fuel gas distribution piping to terminate flame propagation, shuts off the fuel supply, and relieves the pressure resulting from a backfire. Flash back and flame arrestors are devices to prevent a flash back, which is a rapid flame propagation, from passing through the device and progressing into upstream equipment. Some devices have a combination of reverse flow check valve, pressure relief valve, flame barriers, etc. incorporated with the flash back and flame arrestor.

Flash back and flame arrestors can be of the wet or dry type.

## Operation of Boilers and Furnaces

A boiler is a closed vessel in which water is heated, steam is generated, steam is superheated, or in which any combination thereof takes place by the application of heat from combustible fuels, in a self-contained or attached furnace.

Boilers pose the danger of fire and explosion. For a safe operation, training and use of safety equipment is most important. Training in the correct procedures for starting, shutting down and handling possible emergency situations is necessary.

In case of pulverized coal fired boilers, they pose a greater fire and explosion hazard than any other method of coal burning. Explosions can occur during lighting–off, firing or relighting by the ignition of dangerous accumulation of unburned fuel or flammable products of incomplete combustion. Even 2 kg of pulverized coal in an air mixture can form an explosive mixture.

The prevention of the formation of inflammable mixtures is the most important protective measure against fire and explosion. Equally important is the controlling of the threshold limit values of toxic gases.

Unburnt fuel may accumulate in the boiler in a number of ways. For example:

• Through leaky main or ignition fuel inlet valves or idle combustion air chambers.
• If the fires are extinguished and the fuel is not shut-off promptly.

*Batteries produce explosive gases. Battery compartment should be properly vented.*

- If the fuel is not burning correctly, i.e. if incomplete combustion exists.
- During starting up, if there is a delay in ignition.

### Thermic Fluid Heating Systems

Thermic fluid heaters are heat transfer fluid (HTF) systems. They have the potential for releasing large quantities of heated flammable or combustible liquid. Low point drains piped to a safe location provide the ability to remove HTF from a breached piping system in order to minimize the total quantity of fluid released.

## HOUSE KEEPING

### Combustible Waste

By combustible waste, we mean the combustible or loose waste materials that are generated by an establishment or processes, and being salvageable, are retained for scrap or reprocessing on the premises where it was generated or transported to a plant for processing. These include, but are not limited to, all combustible fibres, hay, straw, hair, feathers, wood shavings, turnings, all types of paper products, soiled cloth trimmings and cuttings, rubber trimmings and buffing, metal fines, and any mixture of the above items, or any other salvageable combustible waste materials.

Waste bins should be made of metal, preferably with lids. It is preferable to segregate the combustible waste from other waste. The combustible also should be segregated based on the material type.

**Fig. 3.24:** House keeping.

## STORAGE OF FLAMMABLE LIQUIDS

Flammable liquids which are commonly used or stored at home such as LPG, kerosene, petrol, diesel, need to be stored in a proper and safe manner to avoid fires. It is not necessarily the liquid itself that one should worry about, it is the vapours that they emit which is the hazard. The vapours that are given off form a combustible mixture with the air, creating an easily ignitable medium, where the slightest spark from an electrical equipment, a light switch or static electricity can cause an explosion. Friction and high temperatures can also cause an explosion.

Flammable liquids should always be stored in suitable containers with close fitting covers. It is advisable to store only the minimum quantity in a work area (includes kitchens in domestic buildings) and store any additional quantity in special storage areas which have adequate ventilation or fire protection. A limit of 20 litres of flammable liquid with a flash point of less than 23° C is stipulated for industrial work areas.

---

### Classification of petroleum products

Petroleum products other than LPG are classified according to their closed cup flash point as follows:

Class A–Liquids which have a flash point below 23° C
Class B–Liquids which have a flash point of 23° C and above but below 65° C
Class C–Liquids which have a flash point of 65° C and above but below 93° C
Unclassified–Liquids which have a flash point of 93° C and above.

---

## Precautions in Storing Flammable Liquids

- Avoid ignition sources near the stored area.
- Keep the containers in which the liquid is stored, tightly closed.
- Flammable liquids should be stored in an area with ample ventilation which brings in a large amount of air.
- Flammable liquids should not be poured into drains or flammable liquid soaked rags should not be disposed-off into garbage bins.

Smoking or lighting a match should not be done when flammable vapours are in the air.

**No smoking**

**Smoking**

Starts fires

Fig. 3.25

---

*The best way to fight fires from flammable liquids and gases is to stop the flow of the fuel, whenever that is possible*

### Grounding and Bonding

Buildup of static electricity charges on containers and people can cause sparks that ignite flammable liquid vapours, particularly in areas where dispensing takes place. These static charges must be electrically drained off by grounding and bonding to prevent the discharge of vapour-igniting sparks. Grounding refers to the use of cables connecting an earth ground to each drum involved in dispensing. Bonding refers to connecting the containers involved in any dispensing operation with a wire to prevent spark.

### Fire Instructions

The purpose of fire instruction is to give instruction or training–what to do in the event of a fire?

**Fig. 3.26:** Training for using a fire extinguisher.

### Hazard Labelling

Hazard labelling aims at providing users with information about the hazard to which they are exposed by means of a communications programme. The hazards include any hazard which is known to be present in the environment or workplace in such a manner that people may be exposed under normal conditions of use or in a foreseeable emergency.

### Explanation of the Terms Used in CSDS (Chemical Safety Data Sheet)

#### Flash Point

The flash point of a liquid is the minimum temperature of a liquid at which it gives off vapours sufficient to form an ignitable mixture with the air near the surface of the liquid or within the vessel used. At this temperature, the vapour may cease to burn when the source of ignition is removed. The "fire point" is defined as the temperature at which the vapour continues to burn after being ignited. Neither of these parameters is related to the temperatures of the ignition source or of the burning liquid, which are much higher.

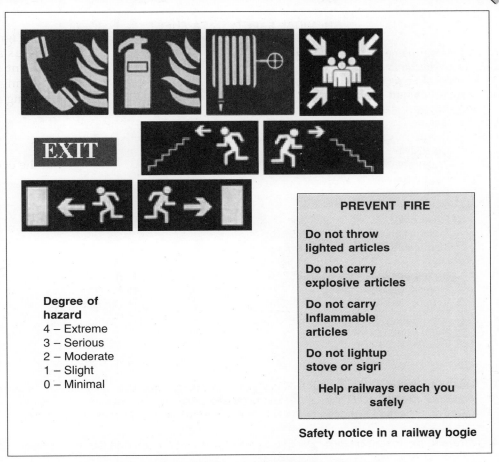

**Degree of hazard**
4 – Extreme
3 – Serious
2 – Moderate
1 – Slight
0 – Minimal

**PREVENT FIRE**

**Do not throw lighted articles**

**Do not carry explosive articles**

**Do not carry Inflammable articles**

**Do not lightup stove or sigri**

**Help railways reach you safely**

**Safety notice in a railway bogie**

**Fig. 3.27:** Examples of hazard communication.

### Measuring Flash Points

There are two basic types of flash point measurement—open cup and closed cup.

In open cup devices, the sample is contained in an open cup (hence the name) which is heated, and at intervals a flame is brought over the surface. The measured flash point will actually vary with the height of the flame above the liquid surface, and at sufficient height the measured flash point temperature will coincide with the fire point. Examples include cleveland open cup (COC) and Pensky-Martens open cup. The main difference being that the former is heated from below, while the latter is heated from the sides as well as below.

*All train-compartments should be provided with portable fire-extinguishers*

## CHEMICAL SAFETY DATA SHEET

1. **Chemical name** : **Hydrogen**
   Synonym/common name :
   Chemical Formula : $H_2$
2. **Properties and characteristics**
   A. Physical state : Liquefied gas, gas
      Colour : Colourless
      Odour : Odourless
      Corrosivity :
      Water solubility : Very slightly soluble.
      Hygroscopicity :
      Light sensitivity :
   B. Combustion property
      Flash point : Gas
      Autoignition temp. : 589° C
      Flammable limits : Lower : 4 %
                          Upper : 74 %
      Boiling point : −252° C
      Specific gravity :
      Vapour density : 0.078 at 0° C
   C. Maximum allowable concentration :
      Mode of entry into body : Inhalation
   D. Reactivity :

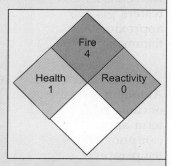

F–Fire– 4
H–Health –1
R– Reactivity–0

3. **Health hazards:** Non-toxic but suffocating in confined spaces. The liquid can cause severe frostbite or "burns" to the skin or other body tissues. Wear special protective clothing designed to prevent liquefied hydrogen or the cold vapours from coming in contact with the body.

4. **Fire and explosion hazard:** Dangerous when exposed to heat and flame, because of the danger of re-ignition.

5. **Safe handling and storage:** Keep away from heat and open flame.

6. **Fire extinguishment:** Hydrogen fires normally should not be extinguished until the supply of hydroogen has been shut-off. If the hydrogen gas has ignited, use water to keep the fire-exposed containers cool and to protect men who are stopping the source of a leak. For small hydrogen fire, use dry-chemical, $CO_2$ extinguishers.

7. **First aid:** In case of burn, report the nearest first aid centre.

8. **Specific information and reference:** Hydrogen gas is odourless and may escape unnoticed until ignited, and/or before entering into affected areas.

*Chemical safety data sheet for hydrogen. (See appendix for CSDS of more materials)*

Closed cup testers, of which the Pensky-Martens closed cup is one example, are sealed with a lid through which the ignition source can be introduced periodically. The vapour above the liquid is assumed to be in reasonable equilibrium with the liquid. Closed cup testers give lower values for the flash point (typically 5–10° C) and are a better approximation to the temperature at which the vapour pressure reaches the lower flammable limit (LFL).

## Autoignition Temperature

Autoignition temperature of a substance is the minimum temperature required to initiate or cause self-sustained combustion independent of the heating or heated elements. This term applies mostly to flammable liquids, although there are certain solids such as camphor and naphthalene that evaporate slowly or volatilize at ordinary room temperatures and therefore have flash point still in the solid state.

Autoignition temperature observed under one set of conditions may be changed substantially by a change of conditions. For this reason, autoignition temperatures should be looked upon as only approximations. As there are as many differences in the test methods such as size and shape of containers, method of heating and ignition source, different ignition temperatures are obtained.

Data for some domestic fuels:

| Fuel | Flash point temperature | Autoignition |
|------|-------------------------|--------------|
| Petrol | < –40° C | 246° C |
| Diesel | > 62° C | 210° C |
| Kerosene (paraffin oil) | 38–72° C | 220° C |
| Vegetable oil | 327° C | – |
| Bio-diesel | > 130° C | – |

Autoignition temperatures of some common chemicals (° C).

| Chemical | Autoignition temperature ° C |
|----------|------------------------------|
| Methane | 601 |
| Benzene | 560 |
| Ethylene | 490 |
| Acetone | 465 |
| Propane | 450 |
| n-Butane | 405 |
| Hydrogen | 400 |
| Methanol | 385 |
| n-Heptane | 280 |

## Flammable (Explosive) Limits and Flammability Range

In the case of flammable vapours, there is a minimum concentration of vapour in air or "oxygen" below which propagation of flame does not occur on contact with a means of ignition. There is also a maximum concentration of vapour in air above which propagation of flame does not occur. These limit mixtures of vapour with air, are known as the "lower and higher limits of flammability" and are usually expressed in terms of volume of the gas or vapour in air.

With insufficient vapour, the mixture is said to be too lean, with an excess of vapour, the mixture is said to be too rich. Thus, the flammable range of a gas lies between the two limits–too lean and too rich. For example, if a mixture of methanol vapour in air contains less than 6% (too lean) or more than 36.5% of methanol vapour (too rich) cannot propagate a flame. But mixtures containing from 6 to 36.5% methanol vapour will propagate flame. These figures represent the range of flammability of methanol vapour.

## Boiling Point

The boiling point of a flammable liquid is the temperature at which a continuous flow of vapour bubbles occurs in a liquid being heated in an open container. The boiling point may be taken as an indication of the volatility of the material. All materials, which are composed solely of carbon and hydrogen, are combustible and to some degree flammable. If they are liquids with low boiling point, they can be assumed to be fire hazards.

## Vapour Pressure

When a liquid evaporates, the molecules leave the liquid and fill the space above it. If the liquid evaporates in a closed container, the number of molecules in the space above the liquid will eventually reach a maximum at a given temperature. The pressure exerted by the vapour is called vapour pressure of the liquid at that temperature.

Every flammable liquid has a vapor pressure, which is a function of that liquid's temperature. As the temperature increases, the vapour pressure increases. As the vapour pressure increases, the concentration of evaporated flammable liquid in the air increases. Hence, temperature determines the concentration of evaporated flammable liquid in the air under equilibrium conditions.

## Vapour Density

Vapour density of a gas in relation to air is the simplest ratio of the molecular weight of the gas to the composite molecular weight of air. Vapour density is the weight of a volume of pure gas compared to the weight of an equal volume of dry-air at the same temperature and pressure. It is calculated as the ratio of the molecular weight of the gas to the average molecular weight of air, (29). The vapour density of a gas indicates how much its vapour is heavier/lighter than the equal volume of air under the same temperature and pressure. A vapour density value of less than 1 indicates that the gas is lighter than air and will tend to rise in a relatively calm atmosphere. A value of greater than 1, indicates that the gas is heavier than air and may travel at low level for a considerable distance to a source of ignition and flash back (if the gas is flammable).

## Maximum Allowable Concentration

The maximum value is an exposure level under which people can work consistently for 8 hours a day with no harmful effects.

## Hazard Warning Labels and Transport Labels

Flammable materials in transport pose a serious hazard today. The volume of such products being transported by various means–road, rail, sea has increased considerably. Since the materials are in movement, it is not possible to provide adequate fire protection measures to prevent fire incidences. The vehicles can meet with accidents which endanger the vicinity.

The purpose of hazard warning labels is to communicate information on:
- Chemical identification
- Chronic and acute health hazard
- Degree of flammability and physical hazard
- Proper PPE

## HAZARD DIAMOND

Hazard diamonds are displayed in locations storing hazardous materials or on containers transporting hazardous materials. Hazard diamonds serve as visible notices indicating the level of hazard. The diamond is made up of four contrasting colours–blue, red, yellow and white. The numbers superimposed over the colours rank the severity or danger, ranging from one to four, with four being the highest rating. The hazard diamond is usually complemented by the MSDS (material safety data sheet), which is a complete detailed description of the chemical, including the hazards posed by the material. Even though the display of the hazard diamond is still not statutory, (statutory on transport vehicles carrying hazardous material), storage facilities for hazardous materials should, as a good safety practice, post the hazard diamond. Apart from warning anybody in the vicinity of the location, hazard diamonds help in providing information about the material, controlling and fighting any spill or fire due to the hazardous material.

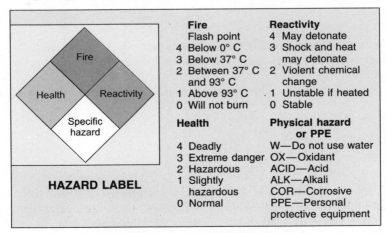

**Fire**
Flash point
4 Below 0° C
3 Below 37° C
2 Between 37° C and 93° C
1 Above 93° C
0 Will not burn

**Health**
4 Deadly
3 Extreme danger
2 Hazardous
1 Slightly hazardous
0 Normal

**Reactivity**
4 May detonate
3 Shock and heat may detonate
2 Violent chemical change
1 Unstable if heated
0 Stable

**Physical hazard or PPE**
W—Do not use water
OX—Oxidant
ACID—Acid
ALK—Alkali
COR—Corrosive
PPE—Personal protective equipment

**HAZARD LABEL**

Fig. 3.28: Hazards diamond with explanation.

## Explanation of the Hazard Diamond

The hazard diamond for a material pictorially conveys the nature and level of hazard posed by the material. It can be easily understood with very little explanation. The colour depiction helps to clearly comprehend the dangers which can arise from a material. The selection of the appropriate rating for each material is carefully done and conveys the results of the hazard evaluation of the material. It is useful to become familiar with the hazard diamond designations to be able to recognize the hazards involved. The rating for each of the categories is expressed on a scale of 0 to 4.

*Fire:* (Flammability).

(Colour code Flammability-RED, Health-BLUE, Reactivity-YELLOW).

### Flammability–4

Very flammable gases, very volatile flammable liquids and materials that in the form of dusts or mists readily form explosive mixtures. When dispersed in the air, as a part of the protection from this category, shut off the gas or liquid and keep cooling water streams (fog) on exposed tanks or containers. Use water supply in the vicinity of the dusts, so as not to create dust clouds.

### Flammability–3

Liquids that can be ignited under all temperatures and conditions. Water may be ineffective on these liquids because of their low flash point. Solids which form coarse dust, solids or fibrous form that create flash fires. Solids that burn rapidly, usually because they contain their own oxygen and any material that ignites spontaneously at normal temperatures in air.

### Flammability–2

Liquids which must be moderately heated before ignition can occur and solids that readily give-off flammable vapours. Water spray may be used to extinguish the fire because the material can be cooled to below its flash point.

### Flammability–1

Materials that must be preheated before ignition can occur. Water may cause frothing of the liquid which has flammability rating number if it gets below the surface of the liquid and turns to steam. However, water spray gently applied to the surface will cause frothing which will extinguish the fire. Most combustible solids have a flammability rating-1.

### Flammability–0

Materials that will not burn.

### Reactivity–4

Materials which in themselves are readily capable of detonation or explosive decomposition or explosive reaction at normal temperatures and pressures. This includes materials which are sensitive to mechanical or localized thermal shock. If a chemical with this hazard rating is in an advanced or massive fire, the area should be evacuated.

### Reactivity–3

Materials which in themselves are capable of detonation or of explosive decomposition or of explosive reaction but which require a strong initiating energy or must be heated under confinement before initiation. This includes materials which are sensitive to thermal and mechanical shock at elevated temperatures and pressures. Also includes materials which may react explosively with water without requiring heat or confinement. Fire fighting should be done from an explosion resistant location.

### Reactivity–2

Materials which in themselves are normally unstable and readily undergo violent chemical change but do not detonate. This includes materials which can undergo chemical change with rapid release of energy at normal temperatures and pressure or which can undergo violent chemical change under elevated temperature and pressures. This also includes those materials which may react violently with water or may form potentially explosive mixtures with water in advanced or massive fires. Fire fighting should be done from a protected location.

### Reactivity–1

Material which in themselves are normally stable, but which may become unstable at elevated temperatures and pressure or which may react with water with some release of energy but not violently. Caution must be exercised in approaching the fire and applying water.

### Reactivity–0

Materials which are normally stable even under fire exposure conditions and are not reactive with water. Normal fire fighting procedures may be used.

### Health–4

A minor exposure of vapour or gas could cause death or the gas/vapour/liquid could be fatal on penetrating the fire fighters' normal full protective clothing which is designated for resistance to heat. For most chemicals having a health-4 rating, the full protective clothing available to the average department will not provide adequate protection against skin contact with these materials. Only special protective clothing designed to protect against the specific hazard should be worn.

### Health–3

Materials extremely hazardous to health, but areas may be entered with extreme care. Full protective clothing, including self-contained breathing apparatus, rubber gloves, and boots should be used.

### Health–2

Material hazardous to health, but areas may be entered freely with self-contained breathing apparatus.

### Health-1

Material only slightly hazardous to health. It may be desirous to wear self-contained breathing apparatus.

### Health-0

Materials which on exposure under fire conditions would offer no health hazard beyond that of an ordinary combustible.

## Hazard Warning Label for Transport

| | |
|---|---|
| A. Hazardous chemical technical name: X1, X2, X3 | D. Hazard diamond |
| B. UN number: | |
| C. Specialist advice: | |
| Telephone number | E. Company/consignor's name |

## Explanation of the Hazard Warning Label

A. Indicates the correct technical name of the hazardous chemical. (The commercial name can also be indicated)

X1 – 1 – Water jet application
2 – Water fog application (or fine spray)
3 – Foam application
4 – Dry agent

X2 – W – Full body protective clothing needed with breathing apparatus.
Y – Full chemical protective suit with breathing apparatus
R – Protective gloves with breathing apparatus

X3 – E – Consider evacuation

B. Indicates the UN number identifying the product carried.
Example: 1011–Butane (2 WE)
1075–Petroleum gases, liquefied (2 YE)
1965–Hydrocarbon gases, liquefied (2 WE)
1978–Propane (2 YE)
1977–Liquid nitrogen (2 RE)

C. Telephone number of the consignor from which specialist advice can be obtained. Can also indicate instructions as:

*Contain*–prevent by any available means spillage from entering drains.
*Dilute*–may be washed to drain with large quantities of water.

D. Pictorial hazard diamond indicating contents are dangerous and nature of the hazard (a single diamond with picture).

E. Address of the consignor.

## Pictorial Hazard Diamonds

Pictures convey messages and help in visualizing the message. A set of standard pictorial symbols have been developed to indicate the hazardous material and its nature or characteristics.

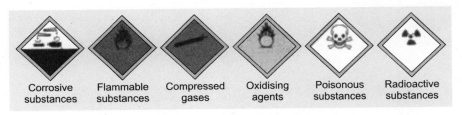

**Fig. 3.29:** Pictorial hazard diamonds.

---

### Fire watch

The fire watch is an individual appointed to play an important role in hot work: to prevent fire and to respond to fire if and when it occurs. The fire watch must stay near the person doing the hot work making sure all the procedures are followed. After the work has ended, the fire watch should continue to check the area for a period of no less than one hour.

---

## Hazard Definitions

1. *Combustible liquid:* Any liquid having a flash point at or above 37° C, but below 93.3° C.
   An example is fuel oil.
2. *Compressed gas:*
   (a) A gas or mixture of gases having, in a container, an absolute pressure exceeding 2.8 bar at 21° C, or
   (b) A gas or a mixture of gases having, in a container, an absolute pressure exceeding 7 bar at 54.4° C.

---

*Chlorine (with heat) plus ammonia explodes due to formation of extremely sensitive nitrogen trichloride*

**Fig. 3.30:** Other pictorial hazard diamonds and signs.

(c) A liquid having a vapour pressure exceeding 2.8 bar at 37.8° C as determined by ASTM-D- 323-72.

An example of a compressed gas is argon.

3. *Explosive:* A chemical that causes a sudden, almost instantaneous release of pressure, gas, and heat when subjected to sudden shock, pressure or high temperature.

An example is dynamite.

4. *Flammable:* A chemical that meets one of the following definitions:
   (a) A gas that forms a flammable mixture in air.
   (b) A liquid with a flash point below 37.8° C.
   (c) A solid that can cause fire through friction, absorption of moisture, spontaneous chemical change, or which can be ignited readily, and when ignited burns so vigorously and persistently as to create a serious hazard.

Examples of flammables are hydrogen gas, petrol and aluminium metal powder.

5. *Organic peroxide:* A chemical with an organic compound that contains the bivalent -O-O structure and which may be considered to be a structural deviation of hydrogen peroxide where one or two of the hydrogen atoms have been replaced by an organic radical.

An example of an organic peroxide is benzyl peroxide.

6. *Oxidizer:* A chemical other than a blasting agent or explosive that initiates or promotes combustion in other materials, thereby causing fire, either of itself or through the release of oxygen or other gases.

An example is concentrated nitric acid.

7. *Pyrophoric:* A chemical that will ignite spontaneously in air at a temperature of 54.4° C or below.

An example is butyl lithium.

8. *Unstable:* A chemical which in the pure state, or as produced or transported, will vigorously polymerize, decompose, condense, or will become self-destructive under conditions of shock, pressure or temperature.

**Fig. 3.31:** DOT regulation for tankers showing hazard label areas.

It should also be kept in mind that any gas, when contained in any portable metal container in a compressed or liquefied state comes under the purview of the Gas Cylinder Rule, 1981 framed under Indian Explosive Act, 1884.

9. *Water reactive:* A chemical that reacts with water to release a gas that is either flammable or presents a health hazard.

An example is potassium metal.

## Role of Heat Release Rate (HRR)

HRR is the primary fire hazard indicator. The rate of heat release, especially the peak amount, is the primary characteristic determining the size, growth, and suppression requirements of a fire environment.

HRR is used to calculate the available escape time and the HRR data is utilized in predictive modelling for hazard predictions.

The rate of heat release from an unwanted fire signifies the risk of the fire to life and property. The rate at which heat is released is more important than the total amount of heat released, in determining the effect of a fire. The HRR is measured in terms of Joules per second, which is also equal to watts. Since a fire puts out much more heat than a watt, HRR is also expressed in KW or MW.

HRR is also a good indicator of the other effects of fire (other than heat release)– smoke, toxic gases, room temperature or other fire hazard, as they increase with an increasing HRR.

So, a high HRR is an indication of the extent of fire hazard. Control of the HRR leads to a control on the fire.

**Graph 3.2:** HRR and escape time available in an interior fire of burning wood.

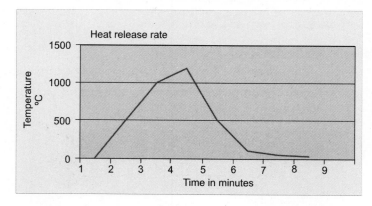

## FIRE DISASTERS

A fire incident which involves a large number of people, who are affected by the effects of the fire, can be called a fire disaster. Numerous and frequent disasters are indicative of systemic deficiencies.

## The Bangalore Circus Fire in Feb. 1981

The fire at the Venus Circus left 109 dead and 400 injured. The sufferers were mostly school children who had come to watch the circus show.

## The Sirsa School Disaster of 1995

The fire took more than 350 lives. Children of a local school had gathered to celebrate the annual day. The tent in which the celebrations were taking place caught fire and fell over. The tent with more than 1500 people inside had only one exit. Fire is suspected to have been started by a short circuit in the lights. Stampede due to terrified crowds trying to escape the flame through the small exit of the huge tent caused the majority of the fatalities.

### Reasons for the Disaster

1. Poor design and strength of the structures.
2. Inadequate exits—there was a stampede when the fire broke out due to the crowds trying to escape the flames, through a small and narrow exit to the huge tent.
3. Poor cables/substandard electrical equipment and faulty cabling.
4. Materials were not fire-resistant—the tent was made of nylon.
5. No provision of emergency lights.
6. Inadequate fire protection
7. No emergency management plans—no facility for burns treatment.
8. Failure of the regulatory system.

### 1989 Railway Catastrophe in Russia

The incident involved the release of hydrocarbon fuel into the atmosphere. A pipeline carrying LPG at 28 bar pressure ruptured, releasing a dense vapour cloud of the fuel over the nearby railway tracks. When two trains passed the area, the vapour cloud ignited and exploded, causing a huge fire which killed/injured more than a thousand people.

### The Srirangam Marriage Hall Fire–Jan 2004

More than 50 people perished in the marriage hall fire. A temporary thatched roof hall had been erected on the terrace where the ceremonies were being held. Bare wires had been inserted into power outlets for the lighting required for taking video pictures. The wires loosened and short-circuited, the sparks ignited the dry shelter and the burning roof collapsed on the assembly. There was only one two foot staircase-exit to the hall. Panic and resulting stampede killed most of the aged, women and children.

## FIRE INVESTIGATION

Fire investigation may be carried out to:

1. Establish the origin and cause of a fire.
2. Find out whether the fire was a case of arson.
3. Find out whether the fire was accidental or was caused intentionally.
4. Study the effect of the fire on the building and its contents in order to formulate building codes and standards.
5. Suggest preventive measures to avert recurrence.

## Steps in Fire Investigation

1. *Fire cause analysis or fire analysis:* Study of the origin and cause
   - Specific ignition source
   - Point or area of origin
   - Means of occurrence
2. *Fuel analysis and characterization:*
   - Burning rates of different items
   - Effect of configuration and ignition source on burning.
3. *Fire debris analysis:*
   - Validation of fire pattern analysis
   - Residues, marks and fire indicators.
4. *Fire incident reconstruction:*
   - Reburning at actual site.
   - Lab reburning

Fire investigation studies include—Fire cause analysis or fire analysis and fire incident reconstruction.

### Fire Cause Analysis or Fire Analysis

Determination of origin (point or area of origin) and the cause of the fire which will indicate the specific ignition source and the means of occurrence.

### Fire Incident Reconstruction

Means reconstructing all or certain aspects of a fire incident scene and actually burn it under lab conditions with environmental control and instrumentation to either determine how the fire ignited and developed or to proof test a hypothesis. Sometimes this involves actual reburning at the site of the fire incident. Modelling studies can provide useful information.

*Fire pattern analysis:* Fire pattern analysis involves study of the residues, marks and fire indicators and debris analysis. It is a type of forensic examination.

*Ignition studies:* In order to conduct ignition studies, a knowledge of the various data pertaining to the fuel which is involved in the fire is a necessary prerequisite such as—fuel analysis and characterization; ignition studies; effect of the configuration at the fire site and the ignition source on the burning; burning rates of different materials and effects of configuration and ignition source on burning.

*Rapid fire growth and extensive damage is usually indicative of incendiary fires*

# *Appendix*

*It is with our passions as it is with fire and water, they are good servants, but bad masters*

— *Roger L"Estrange (1616-1704)*

## FIRST AID

First aid is an emergency action and service.

Emergency can occur at any time, even when every precaution has been taken to prevent and avoid them. It is a situation which requires immediate action.

First aid is the immediate care given to the victim of injury or sudden illness until more advanced care can be provided.

### Aims of First Aid

- To save life
- To protect the victim
- To prevent the condition from worsening and to relieve pain.
- To promote recovery.

Fire produces three types of hazards—heat, smoke and oxygen depletion.

The skin feels pain, if the temperature rises above 45° C. Inhalation of hot gases from a fire can damage the pharynx and the upper airway.

*Smoke inhalation:* Smoke is a product of incomplete combustion. Contents of smoke can be soot particles, gases (like CO), aerosols, volatilized organic molecules and free radicals. Smoke can also irritate the throat which can go into a spasm and close the airway. Some combustible materials can give-off toxic fumes which can become fatal.

A fire uses up oxygen in the atmosphere. So the oxygen level in room with a fire is low and can result in asphyxia to persons in the vicinity. The brain suffers irreversible damage, if oxygen supply is interrupted for more than about 3 minutes.

### First Aid Treatment

It is important to have a knowledge of the injuries caused by fire, heat or smoke to be able to provide first aid treatment.

### Action during an Emergency

The basic principles of first aid apply to all injuries regardless of the severity. It is the first aider's responsibility to act quickly, calmly and correctly in order to preserve life, prevent detioration in the victim's condition and promote recovery. These objectives are met by:

- A rapid but calm approach
- A quick assessment of the situation and the victim
- A correct diagnosis of the condition based on the history of the incident, symptoms and signs
- Proper disposal of the victim according to the injury or condition.

General symptoms and signs of asphyxia

- Casualty may be scorched and burned
- Symptoms and signs of shock due to burns.

For entering a gas or smoke filled room, use of a proper respirator will be necessary.

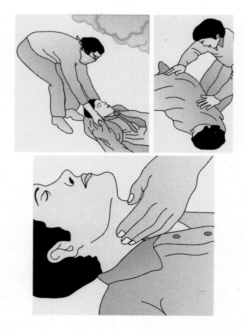

**Fig. A1.1:** Handling a smoke affected victim.

### Treatment

1. Remove the casualty to safety
2. Extinguish any clothing that is on fire or smouldering
3. If the casualty is unconscious but breathing normally, place the casualty in the recovery position.

4. If breathing is difficult or has stopped, begin artificial respiration immediately.
5. Treat any burns and scalds.

**The recovery position** – ensures optimum ease of breathing ➡

**Fig. A1.2**

## Mouth to Mouth Ventilation

The most efficient way to breathe for a casualty is to transfer air from one's own lungs in to the casualty's, by blowing into them through the mouth (mouth to mouth ventilation).

The air that we exhale contains about 16% oxygen which is sufficient to sustain life. In mouth to mouth ventilation, you blow air from your lungs into the casualty's mouth or nose (or mouth and nose together in a child) to fill the casualty's lungs. When you take your mouth away, the casualty will breathe out as the elastic chest wall resumes its shape at rest. Mouth to mouth ventilation enables you to watch the casualty's chest for movement, indicating that the lungs are being filled or that the casualty is breathing again.

Mouth to mouth ventilation can be performed in most cases and is easiest to carryout, if the casualty is lying on his back, but it should be started immediately whatever the

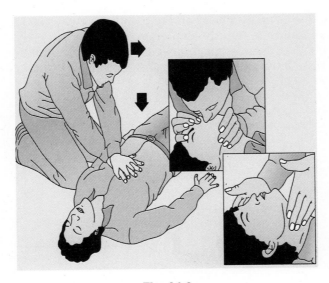

**Fig. A1.3**

position the casualty happens to be in. The first four inflations should be given swiftly. The casualty may start breathing again at any stage, but may need assistance until breathing settles down into a normal rate.

Mouth to mouth ventilation may be unsuitable, if there are serious facial injuries, if there is recurrent vomiting or if the casualty is pinned face down.

If mouth to mouth ventilation is not possible, one of the manual methods may have to be used.

## External Chest Compression

If mouth to mouth ventilation by itself is unsuccessful and the casualty's heart stops, external chest compression must be performed in conjunction with mouth to mouth ventilation. This is because without the heart to circulate the blood, oxygenated blood cannot reach the casualty's brain.

### Method

1. Lay the casualty on the back on a firm surface. Kneel alongside the casualty facing the chest and in line with the heart. Locate the lower half of the breast bone; find the sternal notch at the top and the intersection of the rib margins at the bottom. Place your thumbs midway between these two landmarks to find the center.
2. Place the heel of one hand on the center of the lower half of the breast bone, keeping your fingers off the ribs. Cover this hand with the heel of the other hand and lock your fingers together.
3. Keep your arms straight and move forwards until your arms are vertical. Press down on the lower half of the breast bone (about 4–5 cm for the average adult). Move backwards to release pressure. Complete 15 compressions at the rate of 80 compressions per minute (to find the correct speed count one and two and three and so on).
4. Move back to the casualty's head and reopen the airway. Seal the airway and give two breaths of mouth to mouth ventilation.
5. Continue with 15 compressions followed by two full ventilations, repeating heart check after the first minute. Thereafter, check the heart beat after every 12 cycles or 3 minutes.

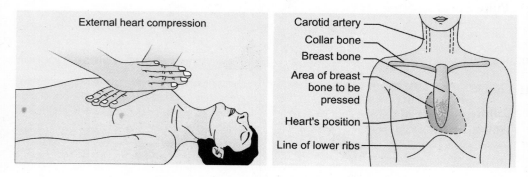

**Fig. A1.4:** External chest compression.

6. As soon as the heart beat returns, stop compressions immediately. Continue mouth to mouth ventilation until natural breathing is restored.
7. Place the casualty in the recovery position.

## Cardiopulmonary Resuscitation (CPR)

| | | |
|---|---|---|
| FIRST | 4 Quick respirations | |
| **1 Rescuer** | 15 Compressions | 80 per minute |
| | 2 Respirations | |
| (.... 1 and 2 and 3 and 4 ........... and 15) | | |
| **2 Rescuers** | 5 Compressions | 60 per minute |
| | 1 Respiration | |
| (1 one thousand, 2 one thousand ........ 5 one thousand) | | |

## HEAT STROKE

A high temperature environment such as that encountered during a fire incidence, can raise the body temperature and lead to a heat stroke. Heat stroke develops when the body can no longer control its temperature by sweating and can occur quite suddenly. It can develop in people of any age who have been exposed to heat or from prolonged confinement in a hot atmosphere such as a burning room.

### Symptoms and Signs

The casualty complains of headache, dizziness and of feeling hot. Casualty becomes restless.

Unconsciousness may develop rapidly and become very deep. Casualty will be hot with a body temperature of 40° C or more and will look flushed although skin remains dry. Pulse is full and pounding, the breathing may be noisy.

### Aim

Reduce the casualty's temperature as quickly as possible and seek medical aid.

---

**If a house is on fire**

• Call for help immediately
• Do not enter the house, but wait for the firemen

---

*Brass burns spontaneously in gaseous chlorine*

- If somebody is burnt, pour water on the burn and then cover it with a loose clean bandage.
- If burns are large or if the burnt person is a child, send the person to hospital.

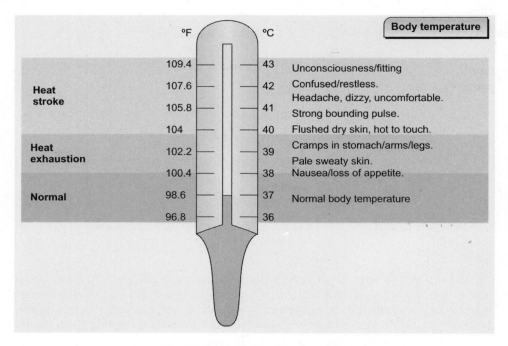

**Fig. A1.5:** Body temperature chart.

### Treatment

Move the casualty to a cooler environment and remove the casualty's clothing.

If the casualty is conscious, place in a half-sitting position with the head and shoulders supported.

If the casualty is unconscious but breathing normally, place in the recovery position

Wrap the casualty in a cold, wet sheet and keep it wet. Direct currents of air on to the casualty by fanning until the casualty's body temperature drops to 38° C.

### THE BASIC LIFE SUPPORT

First assess the casualty's responsiveness.

Check the ABC

- A–Airway
- B–Breathing
- C–Circulation

### Airway

The air way is necessary to be able to take oxygen into the lungs, which is then distributed to all the parts of the body. While it is possible for some parts of the body to survive for a time without oxygen, vital parts of the body such as the brain can be affected, if oxygen supply is stopped for more than 3 minutes. The airway may be obstructed due to:

- Lack of breathing and/or heart beat.
- A state of unconsciousness, which as it develops, is likely to interfere with the open airway and eventually the breathing.

So, to preserve the airway it is necessary to:

- Open the airway to allow unobstructed passage of air to the lungs.
- Lay the casualty in the recovery position to maintain an open airway and thereby preventing an unconscious casualty from becoming asphyxiated.

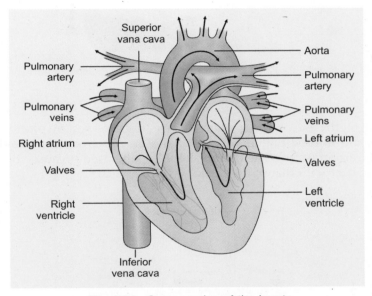

**Fig. A1.6:** Cross-section of the heart.

### How to Open the Airway?

If a casualty is unconscious, the airway may be narrowed or blocked, making breathing difficult or impossible. This occurs due to several reasons.

- The head may tilt forward narrowing the air passage.
- Muscular control in the throat will be lost, which may allow the tongue to slip back and block the air passage.
- Because the reflexes are impaired, saliva or vomit may lie at the back of the throat and block the airway.

The method used to open the airway is known as the "head-tilt and chin-lift" Method.

- Place one hand under a casualty's neck and your other hand on the casualty's forehead and tilt the head backwards. This will extend the head and neck and open the air passage.

• Transfer your hand from the neck and push the chin upwards. The tilted jaw will lift the tongue forward, clearing-off the airway.

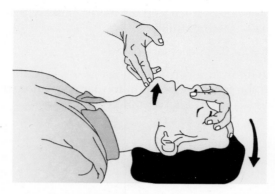

**Fig. A1.7:** Head–chin lift (American Heart Association).

## Circulation

Feel for a pulse on the side of the casualty's neck closest to you by placing the first two fingers (index and middle fingers) of your hand on the groove beside the casualty's Adam's apple (carotid pulse). The thumb should not be used because you may confuse your own pulse beat with that of the casualty's.

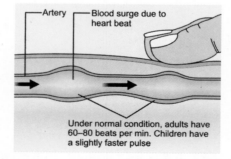

**Fig. A1.8:** Check pulse by feeling the carotid artery.

## FIRST AID FOR BURNS

A burn injury is one of the most painful injuries. Burns damage the nerve endings and cause severe pain. Burn injury is also the one of the major accidental injury contributors, causing even death in many cases.

Burn injuries are complex injuries. Burns, apart from damaging the skin, affects a number of other organs and functions in the body. The muscles, bones, blood vessels and nerves can be affected. The respiratory system, with possible airway obstruction, respiratory failure and respiratory arrest, can also be affected.

Burns injure the skin and thereby impair the body's normal fluid/electrolyte balance, body temperature, body thermal regulation, joint function, manual dexterity and physical appearance. Burns make the skin open to infection and this could lead to serious skin infection complications. Burn casualties also experience severe trauma and emotional and psychological problems. Burns cause shock, which can be very perilous.

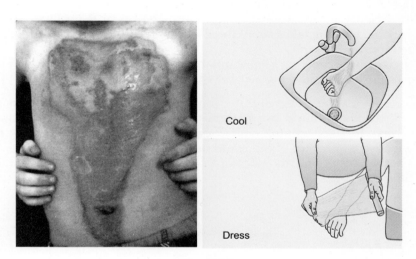

Cool

Dress

Fig. A1.9

## Classifying Burns

Source and degree of burn.

### Source

Burns due to fire can be due to the flame or radiation or due to liquids heated up by the fire, like steam, or due to hot objects.

*Types of burns:*
- Dry heat–example is as from flames or hot objects—most common.
- Wet heat–also called a scald, example is as from steam or hot liquids
- Friction-example is as caused by rubbing
- Chemical–from acids and alkalies.
- Electrical–by contact with live wires.
- Radiation–from radioactive radiation. Also sun burns.

### Degree of Burn

*First degree burns:* **First degree burns are superficial injuries that involve only the epidermis or outer layer of skin. They are the most common and the most minor of all burns. The skin is reddened and extremely painful. The burn usually heals on its own without scarring within 2–5 days. There may be peeling of the skin and some discolouration.**

These burns are also known as superficial burns.

*Second degree burns:* Second degree burns occur when the first layer of the skin is burnt through and the second layer, the dermal layer, is damaged, but the burn does not pass through to underlying tissues. The skin appears moist and there will be deep intense pain, reddening and blisters. Second degree burns are considered minor, if they involve less than 15% of the body surface in adults and less than 10% in children. When treated with reasonable care, second degree burns will heal themselves and produce very little scarring. Healing takes about three weeks. The risk of shock is high and needs medical attention.

Second degree burns covering a substantial percentage of the body can lead to fatality. These burns are also known as partial thickness burns.

*Third degree burns:* Third degree burns involve all the layers of the skin. They are known as full thickness burns and are the most serious of the burns. While a third degree burn may be very painful, some casualties feel little or no pain as the nerve endings have been destroyed. There could be damage to underlying tissues. The skin is usually charred black and includes areas of dry and white. Final treatment may involve plastic surgery.

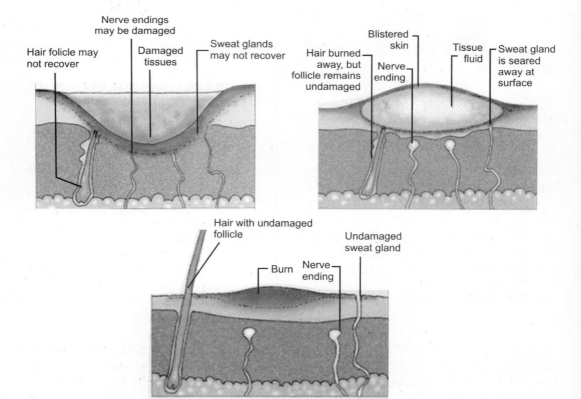

**Fig. A1.10:** Treatment of burns (3, 2, 1 degree).

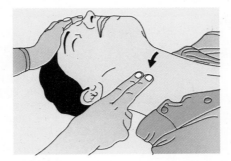

**Fig. A1.11:** Detecting the pulse (American Heart Association).

## Treating Burns

The concerns in treating burns are:

- Exclusion of air from the burned area
- Reducing pain
- Minimizing the onset of shock
- Prevention of infection

### First Aid

The first aid given depends on the cause of the burn and the degree of severity.

Clothing can be removed from the injured area. If there is a tendency to chill, the other parts can be covered.

Special dressings are to be used but should not be wrapped too tight.

| Fire loading of floor area | Classification of fire severity |
|---|---|
| kg/sq.m | |
| 0 to 34 | Light |
| 35 to 73 | Moderate |
| More than 74 | Severe |

Excluding any appreciable quantities of rapidly burning materials such as foamed plastics or flammable liquids. When such materials are present the class is considered moderate or severe.

### Treating for Shock

Shock stuns and weakens the body. When the normal blood flow in the body is upset, death can result.

- When treating a casualty, assume that a shock is present or will occur shortly.
- Elevate the casualty's feet higher than the level of the heart (unless a broken leg, head injury or abdominal injury is present).

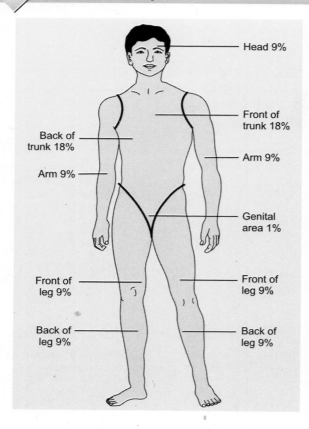

Head 9%

Front of trunk 18%

Back of trunk 18%

Arm 9%

Arm 9%

Genital area 1%

Front of leg 9%

Front of leg 9%

Back of leg 9%

Back of leg 9%

It is important to know the percentage of the skin surface involved in the burn. The adult body is divided into regions, each of which represents 9% of the total body surface. These regions are head and neck, each upper limb, the chest, the abdomen, the upper back, the lower back and the buttocks, the front of each lower limb, and the back of each lower limb. This makes up 99% of the human body. The remaining 1% is the genital area. With an infant or small child, more emphasis is placed on the head and the trunk.

Any partial thickness burn of 1% or more needs medical attention.

Any partial thickness burn of 9% or more, or a full thickness burn needs urgent medical treatment, as the risk of shock is high. An ambulance should be called.

**Fig. A1.12:** Rule of nines for burns.

- Loosen clothing at neck, waist.
- Prevent chilling or over heating. The key is to maintain body temperature.
- Carefully watch the casualty's airway and be prepared to turn the casualty to the recovery position, if necessary, or even to resuscitate, if there is a stoppage of breathing.
- Try to prevent fluid loss.
- Calm and reassure the casualty.

These are represented by **WARMTH**–**W**armth, **A**ir, **R**est, **M**ental rest, **T**reatment, **H**elp.

## FIRST AID ROOM

Minimum equipment required for a first aid room are:
- A glazed sink with hot and cold water always available.
- A table with a smooth top at least 180 cm × 105 cm.
- Means for sterilizing instruments.
- A couch.
- Two stretchers.

- Two buckets or containers with close fitting lids.
- Two rubber hot water bags.
- A kettle and spirit stove or other suitable means of boiling water.
- Twelve plain wooden splints 900 mm × 100 mm × 6 mm.
- Twelve plain wooden splints 350 mm × 75 mm × 6 mm.
- Six plain wooden splints 250 mm × 50 mm × 12 mm.
- Six woollen blankets.
- Three pairs of artery forceps.
- One bottle of spiritus ammoniac aromaticus (120 ml).
- Smelling salts (60 gm).
- Two medium size sponges.
- Six hand towels.
- Four "kidney" trays.
- Four cakes of toilet, preferably antiseptic soap.
- Two glass tumblers and two wine glasses.
- Two clinical thermometers.
- Two tea spoons.
- Two graduated (120 ml) measuring glasses.
- Two mini measuring glasses.
- One wash bottle (1000 cc) for washing eyes.
- One bottle (one litre) carbolic lotion 1 in 20.
- Three chairs.
- One screen.
- One electric hand torch.
- Four first aid boxes or cupboards stocked to standards.
- An adequate supply of anti-tetanus toxoid.
- Injection — morphine, pethidine, atropine, adrenaline, coramine, novocain (6 each).
- Coramine liquid (60 ml).
- Tablets — 25 each of antihistaminic, antispasmodic.
- Syringes with needles — 2 cc, 5 cc, 10 cc and 50 cc.
- Three surgical scissors.
- Two needle holders, big and small.
- Suturing needles and materials.
- Three dissecting forceps.
- Three dressing forceps.
- Three scalpels.
- One stethoscope.
- Rubber bandage — pressure bandage.
- Oxygen cylinder with necessary attachments.

## ACCIDENT REPORTING, INVESTIGATION AND RECORD-KEEPING

### Accident Definition

An accident is an undesirable event that results in personal injury and/or property damage or loss to process or even an incident with near miss (or near hit or narrow escape). An event is considered an accident, if it is unexpected, unavoidable and unintended.

Another definition of accident: "An accident is an unplanned event that interrupts the completion of an activity, and that may (or may not) include injury or property damage."

Definition by Suchman (1961)—Definition is based on three distinguishing characteristics:

- Degree of expectedness
- Degree of avoidability
- Degree of intention.

These are measured by the degree of warning, duration of occurrence, degree of negligence, degree of misjudgment.

### Accident Terms

#### Accident Proneness

It is the susceptibility of an individual or condition to accidents. This factor is difficult to establish.

#### Accident Liability

This a broad term which also includes other factors such as work situations, personal stress (temporary factors).

#### Major Accident

Major accident means an occurrence (including in particular, a major emission, fire or explosion) resulting from uncontrolled developments in the course of an activity or owing to natural events, leading to a serious danger to persons, whether immediate or delayed, inside or outside the installation or damage to property or adverse effects on the environment.

#### Public Accident

Any accident, other than motor vehicle accidents, that occurs in the public use of any premises. Includes deaths in recreation (swimming, hunting, etc.), transportation, except motor vehicle, public buildings, etc. and deaths from widespread natural disasters even though some may have happened on home premises. Excludes accidents to persons in the course of gainful employment.

#### Temporary Total Disability

An injury which does not result in death or permanent disability but which renders the injured person unable to perform regular duties on one or more full calendar days after the day of the accident.

#### Work Injury

Those injuries which arise out of and in the course of gainful employment regardless of where the accident occurred.

#### Fatal Accident

An accident which results in one or more deaths.

*Responsibility of Employers*

All employers are required to provide evidence to show that in the work environment:
- All major accident hazards have been identified
- Adequate steps have been taken to:
  - Prevent such major accidents and to limit their consequences to persons and the environment.
  - Provide the persons working on the site with the information, training and equipment including antidotes necessary to ensure their safety.

## Causes of Accidents

Accidents are caused by unsafe acts, unsafe conditions or a combination of both. It is seldom due to a single cause. Therefore, it is necessary to unearth all the root causes. It is necessary to determine why an accident occurred–what caused it? Accidents do not simply happen. They occur because of human error, failure of mechanical or electrical systems and failure in systems of work. Accidents and incidents are evidence of the failure of control measures and a forecast of future events.

The objective of accident reporting, accident investigation and accident analysis is to:
1. Provide information needed to determine injury rates, identify problem areas, satisfy statutory requirements, etc.
2. Identify, without placing any blame, the basic causal factors that contributed directly or indirectly to each accident.
3. Identify deficiencies in the management system.
4. Suggest corrective action alternative for a given accident.
5. Suggest corrective action alternative for the management system.

Some important reasons for accidents are:
- Faulty system design such as bad roads, bad ergonomics
- Faulty construction/installation
- Defective equipment/tools
- Faulty operational methods such as wrong procedure, unsafe practices, etc.
- Administrative/management failure
- Human behavioural failure such as physically unfit (weak), intoxicated, etc.
- Sabotage, malicious intent, crime, etc.

Accidents also happen because of ignorance, carelessness, and faulty supervision. Reasons for human error can be distraction, unsafe work procedures, improper safety devices, poor supervision and training.

*Effect of Accidents (Short-term and Long-term)*

- Death–immediate and delayed
- Physical injuries—disabling and non-disabling
- Mental injuries
- Social trauma
- Disruption of people's way of life
- Environmental damage
- Financial loss

- Property damage
- Consequential losses

## Accident Investigation

Accident investigation is primarily concerned about injury to personnel. The study of the total accident situation falls under accident analysis and can be termed as root-cause analysis.

Accident investigation system can provide important feedback into the processes of risk assessment, prevention and control.

The purpose of accident investigation can be summarized in one word–PREVENTION. An accident is an undesirable event that results in personal injury and/or property damage or loss to process or even an incident with near miss (or near hit). The difference in the effect of an accident and near miss is only a matter of chance, but the potential to harm remains the same. Accidents are always preceded by numerous incidences of 'near misses'. Being undesired makes it something that must be prevented whenever possible. The purpose of accident investigation, then, is accident prevention or in simple terms, to establish all relevant facts and opinions as to how and why an accident occurred, so conclusions can be drawn about what must be done to prevent recurrence which can possibly be with more serious consequences and not for just fulfilling statutory requirements.

### Investigation Procedure

Investigation is intended to determine what happened, why it happened and how it can be prevented from happening again.

Before accident investigation is carried out, if injury is involved, it is necessary to provide first aid and medical care to the injured persons. However, accident investigation should be carried out as immediately as possible after the occurrence of the accident or near miss as this will help in being able to observe the conditions as they were at the time of the accident, prevent disturbance of evidence, and identify witnesses.

The thorough investigation of accidents should be properly assigned to a responsible, qualified team. For practical purposes, a prompt and immediate investigation can be assigned to the operating level managing the process. Accident prevention can result in considerable cost-savings also.

'Accident investigation is like peeling an onion. Beneath one layer, of causes and recommendations, there are other, less superficial ones' —Trevor Kletz'.

Perhaps the only positive aspect of an accident is the opportunity to learn how to prevent or lessen the impact of an accident happening again.

Operating personnel should be encouraged to suggest ways and means of preventing accidents. Without active cooperation from all levels, accident prevention plans are bound to fail.

Organised effort to prevent accidents and concern for safety must be demonstrated.

### Immediate and Underlying Causes

Basic causes are indicative of management and administrative failure.

> *Accidents are caused, they do not just happen*

In railway accidents taking place in various parts of the country–in every case, a detailed investigation is carried out, the entire workmen and officials are detected and some are punished, again without looking into the common underlying causes. It has been proved that this exercise of fault detection and punishment does not lead to improvement. The basic fault here is the assumption that accidents occur due to special causes, i.e. the person on the job is responsible, wherein reality the actions required are on the system. The causes of the accident are common to a majority of the cases and that the accident could have happened anywhere, to anybody.

Accident investigation is primarily concerned about injury to personnel. The study of the total accident situation is known as accident analysis.

### When Investigation should be made?

Each accident should be investigated as soon as practical after its occurrence. The injured person's version of the accident should be obtained as early as practical and witnesses should be interviewed with as little delay as possible. The scene of the accident should be appraised promptly before any changes are made that might interfere with determining what happened. A proper damage assessment should be carried out. However, medical treatment, even for minor injuries, gets preference and must never be delayed in order for the investigation nor is questioning advisable when the victim is in pain or emotionally upset.

### Re-enacting Accidents

Re-enacting can be carried out only in certain specific cases, when additional information is needed or when some of the facts have to be checked. Re-enacting is not necessary for the majority of the accidents and should not be attempted unless additional information or verification of the facts is required.

### Getting the Facts

It is recommended to use the "W" questions–why, what, where, when, who, and how.

1. Why did the condition exist?
2. Why had no one noticed and corrected the condition?
3. What caused the condition to exist?
4. What caused the condition to be involved?
5. Where was the condition?
6. Where was source of the condition?
7. Where else does the condition exist?
8. How can we find out?

If the accident was caused by an unsafe act or work practice, these questions can be explored.

1. When did the practice act occur?
2. When do similar conditions occur?
3. Who was responsible for it? Who should take corrective actions?
4. How should it be corrected?
5. How can it be avoided in future?

In the event of an action and if the accident was caused by the action, ask;
1. Why was it being done?
2. Why was it being done this way?
3. Why was it necessary?
4. What was its purpose?
5. What other way could it be done?
6. What instructions were not followed?
7. Where should it be done?
8. Where else is it being done?
9. When should it be done?
10. Who is best qualified to do it?

### Accident Facts

All available sources of information, including the injured employee, witnesses, the accident scene and even re-enactment of the accident, should be checked. The investigation must be based on facts avoiding speculations and judgement. The accuracy of all information should be carefully checked. Asking direct questions implying a yes or no answer are very helpful.

### Interviewing the Injured

Before interviewing the injured persons, explain briefly that the purpose of the interview is to learn what happened and how it happened so that recurrence of similar accidents can be prevented. An assurance should be given that the purpose of the interview is not to blame or make the injured person look bad. Questions should be asked to establish:
1. What the injured person was doing?
2. How he was doing it?
3. What happened?

Such questions as to why he did, what he did should be left for later after what happened is definitely established. Do not phrase questions so that they are likely to be antagonistic, and do not try to corner the individual, even though his version contradicts itself. Explain the point of conflict tactfully.

To bring the interview to a close, discuss how to prevent recurrence of the accident. Get the injured person's ideas and discuss them with him. Emphasis should be placed on the precautions that will prevent recurrence and the interview should be closed on a friendly note. No attempt should be made to write up the report while the interview is being made. The report should be written as soon as possible after the interview and after all the sources of information have been checked.

### Interviewing Witnesses

Witnesses are important sources of information. Included are eyewitnesses to the accident or any other persons who know about the circumstances related to the accident even though they are not direct witnesses. Establish if the witness actually saw the accident. Be sure to gather the facts as distinguished from speculation and conjecture on what might have happened. Circumstantial evidence is also important. If the witness saw the accident site immediately before the accident, this evidence may be of value to show that the machine was guarded or unguarded, etc.

Interview witnesses as promptly as possible after an accident. Get their version of the accident first. Then ask specific questions to clarify your understanding or to obtain additional information. No questions should be asked that imply answers wanted or not wanted. Refrain from asking questions or saying anything that blames or threatens the person who met with the accident. Do not badger witnesses or give them a bad time and do not resort to sarcasm, open skepticism, or accusations. Handle all discrepancies with tact and let the witnesses feel they are also partners in the investigation.

Who can give the answers?
Who can show what is being done?
How is the best way to do it?
How can the job or detail be improved?

*Taking a Witness Statement*

Written signed statements from the injured person(s) or witnesses should always be taken when the accident type is permanent partial disability, permanent total disability, or a fatality. The statement should be in the words of the witness and should be signed by the witness and the person taking the statement. All the principal of the five questions (what, where, why, when, and how) should be kept in mind when taking a statement. A copy of the statement could be given to the person, if he requests it (page 254).

Mere slogans or signs do not prevent or reduce accidents. Sometimes they may in fact have the contrary effect, as they may instil a false impression of a safe environment. The need is an adequate understanding of the situation and the problem by proper evaluation and implementation of safety measures in totality.

> When Clothing is on
>
> Fire
>
> Stop
> Drop
> and
> Roll

Dangerous Occurrences

Examples of dangerous occurrences:
- Bursting of a pressure vessel/pressure pipeline.
- Collapse of a building/structure, temporary or permanent, including roof, gallery, tunnel, bridge, etc.
- Road accidents involving dangerous substances
- Escape of hazardous or flammable substances
- Communicable diseases
- Mass poisoning
- Failure of dams, nuclear reactors.

*"The measure of a good accident report is quality, not quantity"*

## WITNESS STATEMENT

Place of interview _____

Date _____ Time _____

I, (Mr./Ms) _____ of _____
(complete address)

_____

Was/am employed by_____

_____

_____

Whose address is _____

_____

_____

from _____ to _____ . My occupation
(is/was) _____

in _____ department.

In reference to the accident involving injury/death to _____

On _____ at approximately _____ the exact location of the accident

Was _____ , I was located in the area of _____

Approximately_____ meters away. I was/were not a supervisor involved in this
accident. I make the following statement of my own free will:

_____

_____

_____

_____

_____

I have read the above or have had the above read to me and it is true and correct to the
best of my knowledge.

Signature _____

Witness signature_____

Date _____ 20 __ Time _____

Page _____ of _____ pages.

## A Brief Accident Report Format

---

### INCIDENT REPORT

Department/location of accident          Employee/persons involved

Type of incident                         Description
                                         Give details of nature, extent, location, etc.

Injury                    [   ]          _____

Minor (first aid)         [   ]          _____

Major                     [   ]          _____

Fatal                     [   ]          _____

Damage                    [   ]          _____

Near miss                 [   ]          _____

---

Incident Date                    Time                    Place

---

Key points to check:

Equipment/tools          _____

Static                   _____

Powered                  _____

PPE                      _____

Access                   _____

Lighting                 _____

Training                 _____

Instruction              _____
Enclose sketches/photo if any
(continue on a separate sheet, if necessary)

---

Contributory causes

---

Actions already taken

---

Further actions recommended

### Hazardous Condition Reporting

Report no: _____

Region/area _____

Specific location _____

Date of report _____

Date received _____

Reported by _____

Phone _____

Determined by _____

Phone _____

Reason for reporting _____

_____

Description of condition _____

_____

Discovery of circumstances _____

_____

Significant effects on safety _____

_____

Corrective actions taken _____

_____

_____

_____

Effect on the condition _____

_____

Planned follow-up/corrective action initiated date _____

Started by _____

To complete by _____

Follow-up action taken _____

Date _____

Any incidents after report _____

Enforcement action taken _____

Case-closed/further follow-up/investigation required.

Remarks/notes _____

_____

_____

_____

## Properties of Flammable Liquids

Generally, a **flammable liquid** means a liquid which may catch fire easily. There is a precise definition of **flammable liquid** as one with a flash point below 37° C. Less flammable liquids (with a flash point between 37° C and 93° C) are defined as **combustible liquids**.

The properties—flash point, ignition-temperature, flammability-limits, specific-gravity (relative density), vapour-density, boiling-point, and water-solubility, determine the relative fire risk of a given chemical.

## Properties of Common Chemicals

| Chemical | Health | Fire | Reactivity | S/N |
|---|---|---|---|---|
| Acetic acid (glacial) | 2 | 2 | 2 | |
| Acetone | 1 | 3 | 0 | |
| Acetylene | 1 | 4 | 3 | |
| Ammonia (anhydrous) | 3 | 1 | 0 | |
| Butane | 1 | 4 | 0 | |
| Calcium carbide | 1 | 4 | 2 | W |
| Calcium oxide | 1 | 0 | 1 | |
| Carbon monoxide | 2 | 4 | 0 | |
| Carbon tetrachloride | 3 | 0 | 0 | |
| Castor oil | 0 | 1 | 0 | |
| Chlorine | 3 | 0 | 0 | OX |
| Chloroform | 2 | 0 | 0 | |
| Coconut oil | 0 | 1 | 0 | |
| Cod liver oil | 0 | 1 | 0 | |
| Creosote oil | 2 | 2 | 0 | |
| Denatures alcohol | 0 | 3 | 0 | |
| Dimethyl-phalate | 0 | 1 | 0 | |
| Ethane | 1 | 4 | 0 | |
| Ethyl alcohol | 0 | 3 | 0 | |
| Ethylene | 1 | 4 | 2 | |
| Ethylene glycol | 1 | 1 | 0 | |
| Formaldehyde (water solution) | 2 | 2 | 0 | |
| Gas (natural) | 1 | 4 | 0 | |
| Gasoline | 1 | 3 | 1 | |
| Glycerine | 1 | 1 | 0 | |
| Hydrazine(anhydrous) | 3 | 3 | 2 | |
| Hydrochloric acid | 3 | 0 | 0 | W |
| Isopropyl alcohol | 1 | 3 | 0 | |
| Lubricating oil (mineral) | 0 | 1 | 0 | |

*(Contd.)*

| Chemical | Health | Fire | Reactivity | S/N |
|---|---|---|---|---|
| Methane | 1 | 4 | 0 | |
| Methyl alcohol | 1 | 3 | 0 | |
| Methyl ethyl ketone | 1 | 3 | 0 | |
| Mineral spirits | 0 | 2 | 0 | |
| Naphtha | 1 | 3 | 0 | |
| Nitrogen (liquefied) | 3 | 0 | 0 | |
| Oxygen (liquid) | 3 | 0 | 0 | OX |
| Paraffin oil | 0 | 1 | 0 | |
| Phenol | 3 | 2 | 0 | |
| Pine oil | 0 | 2 | 0 | |
| Potassium | 3 | 1 | 2 | W |
| Propane | 1 | 4 | 0 | |
| Silane | 1 | 4 | 2 | |
| Silver nitrate | 1 | 0 | 0 | OX |
| Sodium | 3 | 1 | 2 | W |
| Sodium hydroxide (lye) | 3 | 0 | 1 | ALK |
| Sulphur | 2 | 1 | 0 | |
| Sulphur dioxide | 2 | 0 | 0 | |
| Sulphuric acid | 3 | 0 | 2 | W |
| Toluene | 2 | 3 | 0 | |
| Trichloroethylene | 2 | 1 | 0 | |
| Trinitrotoluene | 2 | 4 | 4 | |

CSDS for some important hazardous chemicals:

- LPG
- Chlorine
- Methyl alcohol
- Naphtha
- Sulphuric acid
- Ammonia

### Panic hazard

Experience indicates that panic seldom develops even in the presence of potential danger, so long as occupants of buildings are moving towards exits which they can see within a reasonable distance and with no obstructions or undue congestion in the path of travel. However, any uncertainty as to the location or adequacy of means of egress, the presence of smoke or fumes and the stoppage of travel towards the exit, such as may occur when one person stumbles and falls on stairs, may be conducive to panic. Danger from panic is greater when a large number of people are trapped in a confined area.

## CHEMICAL SAFETY DATA SHEET

1. **Chemical name**          :    **LPG**
   Synonym/common name    :    Bottle gas
   Chemical formula         :    $C_4H_{10} + C_3H_8$

2. **Properties and characteristics:**
   A. Physical state        :    Gas, liquified gas
      Colour               :
      Odour                :    Odourless(mercaptan added)
      Corrosivity          :
      Water solubility     :    Very slightly soluble.
      Hygroscopicity       :
      Light sensitivity    :
   B. Combustion property:
      Flash point          :    Gas
      Autoignition temp.   :    589° C
      Flammable limits     :    Lower : 1.9%
                                Upper : 10.5%
      Boiling point        :    0° C                    F–Fire–3
      Specific gravity     :                            H–Health–1
      Vapour density       :    0.078 at 0° C           R–Reactivity–1
   C. Maximum allowable concentration : 1000 ppm or 1000 mg/cu.m air

      Mode of entry into body        :  Inhalation
   D. Reactivity:

   Fire 3 / Health 1 / Reactivity 1

3. **Health hazards:** Unknown, may act as a simple asphyxiant.

4. **Fire and explosion hazard:** Dangerous when exposed to heat and flame. It can react with oxidizing materials.

5. **Safe handling and storage:** Keep away from heat and open flame; keep away from oxidizing material; protect container from physical damage; avoid inhalation.

6. **Fire extinguishment:** Use dry powder, $CO_2$ extinguisher or water spray. Cool the container or bullet.

7. **First aid:** In case of inhalation, remove person to fresh air and report to first aid center for treatment.

8. **Specific information and reference:** NFPA code.

## CHEMICAL SAFETY DATA SHEET

1. **Chemical name** : **Chlorine**
   Synonym/common name :
   Chemical formula : $Cl_2$

2. **Properties and characteristics:**

   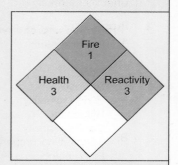

   A. Physical state : Gas or crystals
      Colour : Greenish yellow
      Odour : Acrid odour, bitter taste
      Corrosivity : yes
      Water solubility : soluble.
      Hygroscopicity :
      Light sensitivity :
   B. Combustion property
      Flash point : Gas
      Autoignition temp. :
      Flammable limits : Lower :
                          Upper :
      Boiling point : 35° C
      Specific gravity : 1.47 at 0° C
      Vapour density : 2.419
      Vapour pressure : 4800 mm at 20° C

      F–Fire–1
      H–Health–3
      R–Reactivity–3

   C. Maximum allowable concentration : 1 ppm or 3 mg/cu.m air
      Mode of entry into body : Contact or Inhalation
   D. Reactivity: Reacts with ammonia, acetylene, ether, fuel gases, hydrogen, turpentine, most hydrocarbons, organic matter and finely divided metals.

3. **Health hazards:** Chlorine causes lung irritation and damage. The gas irritates the eyes and can cause conjunctivitis, in high concentrations the gas irritates the skin. Concentrations of 50 ppm are dangerous even for short exposures. 1000 ppm may be fatal, even where exposure is brief.

4. **Fire and explosion hazard:** Non-combustible, but most combustible materials will burn in chlorine as they do with oxygen. When heated it emits highly toxic fumes, will react with water or steam to produce toxic and corrosive fumes of hydrogen chloride.

5. **Safe handling and storage:** Keep away from chemicals especially acetylene, turpentine, ether, ammonia, fuel gas, hydrocarbons, hydrogen and finely divided metals. Use goggles, rubber apron, gloves and respiratory protection.

6. **Fire extinguishment:** Wear full protective clothing with respiratory protection. Keep surrounding areas under water fog for absorbing released chlorine.
   Do not use water directly on the leak.
   Use dry chemical powder and $CO_2$ extinguisher.

7. **First aid:** Remove person to fresh air, immediately wash the affected part with water, remove contaminated clothing, report to the first aid center.

8. **Specific information and reference:** For neutralization of chlorine, caustic soda or soda ash and hydrated lime solution can be used to absorb chlorine.

## CHEMICAL SAFETY DATA SHEET

1. **Chemical name**            :   **Methyl alcohol**
   Synonym/common name      :   Methanol
   Chemical formula         :   $CH_3OH$

2. **Properties and characteristics:**

   A.  Physical state       :   Liquid-volatile
       Colour               :   Clear, colourless
       Odour                :
       Corrosivity          :
       Water solubility     :   Soluble
       Hygroscopicity       :   Yes-slight
       Light sensitivity    :
   B.  Combustion property
       Flash point          :   11.1° C
       Autoignition temp.   :   475° C
       Flammable limits     :   Lower  :  7.3%
                                Upper  :  36.5%
       Boiling point        :   65° C
       Specific gravity     :   0.7913
       Vapour density       :   1.11
       Vapour pressure      :   100 mm at 21.2° C

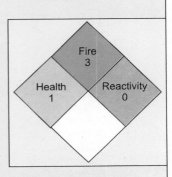

F–Fire–3
H–Health–1
R–Reactivity –0

   C.  Maximum allowable concentration: 200 ppm or 262 mg/cu.m air
       Mode of entry into body  :  Contact, ingestion and inhalation
   D.  Reactivity           :   Unknown

3   **Health hazards:** Methanol poisoning causes injury to the central nervous system particularly to the optic nerve. Sudden weakness, headache, nausea, vomiting, abdominal pain, diminution of vision and a state of unconsciousness may result. Failure of respiration may occur followed by coma and death. Continued skin contact can cause dermatitis.

4.  **Fire and explosion hazard:** Highly flammable.

5.  **Safe handling and storage:** Keep container away from heat and source of ignition. Use chemical goggles.
    Faceshield and rubber gloves.
    Adequate ventilation should be provided around methanol storage and handling area.

6.  **Fire extinguishment:** Water spray, foam, $CO_2$, or dry chemical type extinguisher.

7.  **First aid:** Remove from exposure, give rest and keep warm, wash the affected part with water, remove contaminated clothing, report to the first aid center for treatment.

8.  **Specific information and reference:** NFPA-code.

## CHEMICAL SAFETY DATA SHEET

1.  **Chemical name**          :    **Naphtha**
    Synonym/common name        :    Petroleum benzene
                                     Petroleum ether
    Chemical formula           :

2.  **Properties and characteristics:**
    A.  Physical state         :    Liquid-volatile
        Colour                 :    Colourless
        Odour                  :    Like petroleum product
        Corrosivity            :
        Water solubility       :    Nil
        Hygroscopicity         :
        Light sensitivity      :
    B.  Combustion property
        Flash point            :    32° C
        Autoignition temp.     :    287.7° C
        Flammable limits       :    Lower: 1.1%
        Upper                  :    5.9%
        Boiling point          :    40–80° C
        Specific gravity       :
        Vapour density         :    2.5 at 0° C
    C.  Maximum allowable concentration    :    500 ppm
        Mode of entry into body           :    Skin and inhalation
    D.  Reactivity                        :    Reacts with oxidizing materials

Fire 3
Health 1 | Reactivity 1

F–Fire–3
H–Health–1
R–Reactivity–1

3.  **Health hazards**:  Direct contact of naphtha with skin will lead to skin diseases like dermatitis. Vapours inhaled or absorbed through skin in extreme cases may act as irritant, asphyxiants or anaesthetic.

4.  **Fire and explosion hazard**: Naphtha is a highly flammable liquid giving rise to fire and explosion risk.

5.  **Safe handling and storage**: Avoid skin contact and exposure.
    Use rubber boots, aprons, rubber gloves, goggles and gas mask with organic vapour canister while handling.
    Adequate ventilation should be provided around naphtha storage and handling area.

6.  **Fire extinguishment**: Use foam, $CO_2$, or dry chemical type extinguisher. Water spray may be used to reduce the rate of burning and for cooling containers.

7.  **First aid:** In case of contact, wash with water.

8.  **Specific information and reference**: NFPA-code.

## CHEMICAL SAFETY DATA SHEET

1.  **Chemical name** : **Sulphuric acid**
    Synonym/common name :
    Chemical formula : $H_2SO_4$

2.  **Properties and characteristics**
    A.  Physical state : Liquid only
        Colour : Colourless oily liquid
        Odour : Acidic
        Corrosivity : Highly corrosive
        Water solubility : Soluble
        Hygroscopicity : Yes
        Light sensitivity :
    B.  Combustion property
        Flash point : Non-flammable
        Autoignition temp. :
        Flammable limits : Lower:
        Upper:
        Boiling point : 333° C
        Specific gravity : 1. 34
        Vapour density : 0.078 at 32° C
        Vapour pressure : 1 mm at 146° C

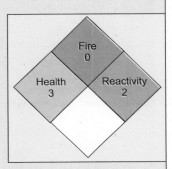

F–Fire–0
H–Health –3
R–Reactivity –2

C.  Maximum allowable concentration : 1 mg/cu.m of air
    Mode of entry into body : Inhalation, contact, ingestion
    D.  Reactivity: Reacts violently with water, exothermic reaction, powerful oxidizer.

3.  **Health hazards:** It is highly corrosive and toxic to tissue. Contact with skin causes rapid destruction of tissue causing severe burns. Repeated contact with dilute solution can cause dermatitis and repeated or prolonged inhalation of mist of sulphuric acid can cause inflammation of the upper respiratory tract.

4.  **Fire and explosion hazard:** It is very powerful acidic oxidizer which can ignite or even explode on contact with many materials such as $NH_4OH$, HCl, $H_2$, Fe, water, etc.

5.  **Safe handling and storage:** While handling sulphuric acid, protective safety gear–gum boots, rubber gloves, faceshield or goggles and rubber apron should be worn.

6.  **Fire extinguishment:**

7.  **First aid:** If splashed with acid, wash the affected part with copious quantity of water. Remove contaminated clothing.
    Do not attempt to neutralize the acid in contact with the skin.

8.  **Specific information and reference:** Manual of hazardous chemical reactions (NFPA-1966).

## CHEMICAL SAFETY DATA SHEET

1. **Chemical name**          :   **Ammonia**
   Synonym/common name    :   Ammonia gas
   Chemical formula       :   $NH_3$

2. **Properties and characteristics**
   A.  Physical state     :   Gas liquefied by compression
       Colour             :   Colourless
       Odour              :   Extremely pungent odour
       Corrosivity        :
       Water solubility   :   Soluble
       Hygroscopicity     :
       Light sensitivity  :
   B.  Combustion property
       Flash point        :   Gas
       Autoignition temp. :   651° C
       Flammable limits   :   Lower  :  16%
                              Upper  :  25%

       Boiling point      :   −72° C
       Specific gravity   :   0.897
       Vapour density     :   0.6
       Vapour pressure    :   10 bar at 27.7° C
   C.  Maximum allowable concentration  :  50 ppm
       Mode of entry into body          :  Contact and inhalation
   D.  Reactivity: Reacts with oxidizing gases, chlorine, mercury, acids, bromine, iodine, copper, zinc, aluminium, silver and their alloys. Galvanized surfaces can be corroded by ammonia.

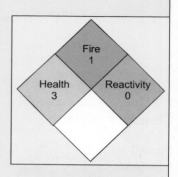

F-Fire–1
H–Health–3
R-Reactivity–0

3. **Health hazards:** Highly irritant to skin, eyes and mucous membranes of the respiratory system.

4. **Fire and explosion hazard:** Moderate when exposed to heat and flame. Forms explosive compounds in contact with silver or mercury.

5. **Safe handling and storage:** Use chemical goggles/face-shield, rubber gloves, apron, gum boots and gas-mask. In case of high exposure, use SCBA.

6. **Fire extinguishment:** Shut-off the source of supply
   Use dry chemical or $CO_2$ fire extinguishers or water fog.
   Keep surrounding area cool under water fog.

7. **First aid:** In case of contact, immediately wash the affected part with water; remove contaminated clothing; report to the first aid center.

8. **Specific information and reference:** National fire code by NFPA.

# 2

# *Appendix*

## CHEMISTRY

Chemistry is defined as the science of substances, their properties and reaction with each substance. Substances may be solid, liquid or gas in living or non-living systems. But all have a common factor. All of them consist of chemical substances. Chemical substances can be a single chemical or can be mixtures.

## Molecules

Irrespective of being a single chemical or mixture, every substance is made up of millions of very tiny particles. A mixture of different substances contains different types of molecules, while a single chemical has only one type of molecule. In their properties and behaviours, all molecules of the same substance are exactly alike. One may define a molecule as the smallest particle of a substance capable of existing without depending on any other substance.

## Atoms

Molecules are made up of still smaller particles, called atoms. Molecules of all substances consist of different combinations of atoms. Atoms form the foundation of all substances. Molecules are broken down or changed during chemical reactions. Atoms can never be chemically split into anything smaller. (But they may be split through nuclear fission which is a non-chemical action.) An atom includes a dense nucleus comprised of a specific number of positively charged protons and uncharged neutrons surrounded by orbits of negatively charged electrons.

## Matter and the Atomic Structure

Matter can be defined as something which possesses mass and occupies space.

## Dalton's Atomic Theory

Postulated by Dalton (1807) states that each element is composed of extremely small particles called atoms. All atoms of a given element are identical. Atoms of different elements have different properties. Atoms of an element are not changed into different types of atoms by chemical reactions. Atoms are neither created nor destroyed in a chemical reaction.

Compounds are formed when atoms of more than one element combine. In a given compound, the relative number and kinds of atoms are constant.

### Atomic Number

The number of protons always equals the number of electrons in an atom, and that number is equal to the atomic number. For instance, carbon has an atomic number of six and therefore has six protons and six electrons.

### Atomic Mass

The weight of an atom is determined by the number of neutrons and protons that are present in the nucleus. The proton and neutron, which are similar in mass, each weighs approximately 1,836 times greater than a single electron. Thus the mass contributed by electrons is insignificant when determining atomic weight or atomic mass. The atomic mass is the sum of the protons and neutrons in the nucleus. Carbon has an atomic mass of twelve. Since there are six protons in carbon (remember, it has an atomic number of six and, therefore, must have six protons), it must have six neutrons.

Atomic mass = No. of protons + no. of neutrons

The atomic mass of carbon = 12

The atomic no. of carbon = 6 = the no. of protons

Number of neutrons = Atomic mass–no. of protons

Number of neutrons = 12–6 = 6

### Atomic Weight Scale

An atom or one molecule has a very small weight. In order to compare the weight of one atom with that of another, an atomic weight scale is used. The atomic weight of oxygen is 16, while that of hydrogen is 1.

Comparison of the weight of other atoms can be made with that of hydrogen to find out how many times heavier they are.

$$\text{Atomic weight is} = \frac{\text{The weight of one atom of the element}}{\text{The weight of one atom of hydrogen}}$$

*Example:* The atomic weight of sodium is 23. This shows that one atom of sodium is 23 times heavier than one atom of hydrogen.

### Molecular Weight

The atomic weight scale is used to compare the weight of one molecule with the weight of one atom of hydrogen.

$$\text{Molecular weight} = \frac{\text{The weight of one molecule of the given substance}}{\text{The weight of one atom of hydrogen}}$$

For example, atomic weight of sulphur (S) is 32. Molecular weight of sulphur dioxide ($SO_2$) is 64.

Molecular weight can also be calculated as below.

Molecular weight = atomic weight of sulphur + 2 × atomic weight of oxygen

$$= 32 + (2 \times 16) = 64$$

### Charge of an Atom

Since the number of protons (positive charges) always equals the number of electrons (negative charges) in an atom, positive charges equal negative charges and atoms in the elemental state have no charge. Only when an atom takes an electron from another atom does the particle become charged. This charged form of the atom is known as an *ion*. Positively charged ions are called cations and negatively charged ions are called anions. For instance, when chlorine accepts an electron from sodium, the sodium ion that is formed will have one more proton than electrons. It will therefore have a positive charge and be called a *cation*. The chlorine (or chloride) ion will have one more electron than protons. It will take on a negative charge and be called an *anion*. The compound formed by this transfer of electrons is sodium chloride or table salt, which is nothing like the highly reactive sodium or extremely poisonous chlorine from which it was formed.

### Energy Levels

The electrons surround the nucleus of the atom. Electrons are located in energy levels. There are seven energy levels. Each has a specific maximum number of electrons that can exist in it. The number of electrons, which an energy level can hold is equal to $2n^2$ where, n = energy level. The letter *n* represents the principal quantum number that specifies the energy level of the atom in which an electron is located. The chart below identifies the various energy levels and maximum number of electrons possible. The energy level closest to the nucleus is represented by energy level 1.

| Electron energy levels |
|---|
| **Possible number of electrons** |
| 2 |
| 8 |
| 18 |
| 32 |
| 50 (theoretical, not filled) |
| 72 (theoretical, not filled) |
| 98 (theoretical, not filled) |

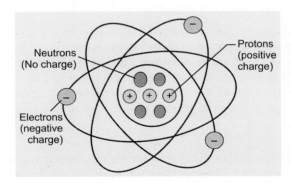

Fig. A2.1

## Energy Sublevels

Within each principal energy level, is one or more energy sublevels (orbitals) or sub-shells. The number of sublevels possible for any one principal energy level is equal to the value of the quantum number (n) for that energy level. While there are theoretically 7 possible sublevels, only four are actually used for the known elements. The others are not currently needed. Sublevels are numbered with consecutive whole numbers. The first sublevel is 0 followed by numbers 1 through 6. These numbers are the azimuthal quantum numbers, L. The value of L can never be greater than n-1. Based on the 112 known and verified elements, the following table represents possible sublevels in the atom.

| Energy sublevels (orbitals) | | |
|---|---|---|
| Principal quantum number | Sublevel number, L | Sublevel letter |
| 1 | 0 | s |
| 2 | 0,1 | s, p |
| 3 | 0,1,2 | s, p, d |
| 4 | 0,1,2,3 | s, p, d, f |
| 5 | 0,1,2,3 | s, p, d, f |
| 6 | 0,1,2,3 | s, p, d |
| 7 | 0 | s |

As more elements are identified, the sublevels of 5, 6 and 7 will fill.

## Elements

Substances consisting of only one type of atoms are called elements. There is an element corresponding to each different type of atom. For example, oxygen is an element which consists of oxygen molecules. Each molecule is consisting of two oxygen atoms combined together like $O_2$.

## Compounds and Mixtures

A compound is formed when two or more atoms of different elements are chemically bound together to form molecules.

Two or more different kind of molecules come together to form a mixture. With the help of physical or mechanical means, a mixture can be separated into the substances which make up the mixture. Only a re-arrangement of the atoms can breakdown or change the compound, which is made up of identical molecules.

## Radicals

There are certain groups of atoms, which are common to families of related compounds. They are known as radicals. Radicals can be defined as a group of atoms present in series of compounds that maintain their identity through chemical changes affecting the molecule. Radicals are bracketed while writing formulae. For example, formula of 1 molecule form of calcium hydroxide is Ca $(OH)_2$. Here OH stands for the radical and the number 2 signifies that the radical appears twice. Two atoms of hydrogen and two atoms of oxygen are present in every calcium hydroxide molecule.

## Density

The density of a substance is its mass per unit volume. It can be calculated by dividing the mass of a body by the volume.

$$\text{Density} = \frac{\text{Mass}}{\text{Volume}}$$

$$\text{Mass density} = \frac{\text{Mass}}{\text{Volume}}$$

$$p = \frac{m}{V} \left(\frac{Kg}{M^3}\right)$$

Unit of density in SI unit is kilograms per cubic meter ($kg/m^3$) or gram per cubic centimeter ($g/cm^3$). Water has a density of 1000 $kg/m^3$ or 1 $gm/cm^3$.

### Relative Density or Specific Gravity

Relative density is the ratio of the mass of any volume of the substance to the mass of an equal volume of water. It has no unit.

In SI system, relative density is a comparison of mass density to a standard. For solids and liquids, the standard is fresh water.

---

### Combustible dust

Any time a combustible dust is processed or handled, a potential for deflagration exists. The degree of deflagration hazard will vary depending on the type of combustible dust and processing methods used. For a combustible dust explosion to occur, has the following four requirements:

1. A combustible dust
2. A dust dispersion in air or oxygen at or exceeding the minimum combustible concentration
3. An ignition source such as an electrostatic discharge, an electric current arc, a glowing ember, a hot surface, welding slag, frictional heat, or a flame
4. Confinement.

---

### Vapour Density

The ratio of the weight of a gas or vapour to the weight of an equal volume of hydrogen under the same conditions of temperature and pressure decides the weight of the gas or vapour. The vapour density of air when compared with hydrogen is 14.5, carbon dioxide is 11 times heavier than oxygen and its vapour density is 22.

| Relative density(specific gravity) of various substances | | | |
|---|---|---|---|
| Water (fresh) | 1.00 | Magnesium | 1.74 |
| Water (sea) | 1.03 | Manganese | 8.0 |
| Aluminium | 2.56 | Mercury | 13.6 |
| Antimony | 6.70 | Mica | 2.9 |
| Bismuth | 9.80 | Nickel | 8.6 |
| Brass | 9.40 | Oil (linseed) | 0.94 |
| Brick | 2.10 | Oil (olive) | 0.92 |
| Calcium | 1.58 | Oil (petroleum) | 0.76–0.86 |
| Carbon (diamond) | 3.40 | Oil (turpentine) | 0.87 |
| Carbon (graphite) | 2.30 | Paraffin | 0.86 |
| Carbon (charcoal) | 1.80 | Platinum | 21.5 |
| •Chromium | 6.5 | Sand (dry) | 1.42 |
| Clay | 1.90 | Silicon | 2.60 |
| Coal | 1.36–1.40 | Silver | 10.57 |
| Cobalt | 8.60 | Slate | 2.1–2.80 |
| Copper | 8.77 | Sodium | 0.97 |
| Cork | 0.24 | Steel (mild) | 7.87 |
| Glass (crown) | 2.50 | Sulphur | 2.07 |
| Glass (flint) | 2.80 | Tin | 7.3 |
| Gold | 19.3 | Tungsten | 19.1 |
| Iron (cast) | 7.21 | Wood (pine) | 0.56 |
| Iron (wrought) | 7.78 | Wood (oak) | 0.7–1.0 |
| Lead | 11.40 | Wood (teak) | 0.8 |

Vapour density (VD) formula:

1. Vapour density $= \dfrac{\text{Molecular weight}}{2}$

2. $VD = \dfrac{\text{Molecular weight of gas}}{\text{Molecular weight of air (29)}}$

To determine the vapour density of air, if molecular weight of air is 29.

Example: Given,

Molecular weight of air = 29

$$VD \text{ of air} = \dfrac{\text{Molecular weight}}{2} = \dfrac{29}{2}$$

Therefore,

VD of air = 14.5

To determine the vapour density of butane ($C_4H_{10}$).
if (C=12, H=1).

Chemical formula of butane = $C_4H_{10}$

Molecular weight of butane:

$$= 12 \times 4 + 1 \times 10 = 48 + 10 = 58$$

$$\text{VD of butane} = \frac{\text{Molecular weight}}{2}$$

$$\text{VD of butane} = \frac{58}{2} = 29$$

Therefore, VD of butane = 29 (air =14.5)
Therefore, butane is two times heavier than air.

## Classification of Elements

The known elements are classified based on the principle–the properties of elements are the periodic function of their atomic numbers. The classification table is known as the periodic table.

*Periodic table:* Periodic table is the classification or arrangement of elements.

The modern periodic table is designed based on "the properties of elements are the periodic functions of their atomic numbers".

The properties of elements depends on the electronic configuration of their atoms.

The long form of the periodic table has 7 horizontal rows called periods and 18 vertical columns called groups.

Based on the valence electron, the elements are classified as given below.

| Block | Group | Electronic configuration |
|---|---|---|
| s– block elements | 1 to 2 group | $ns^1$ and $ns^2$ |
| d–block elements | 3 to 12 group | $(n-1)\, d^{1-10}\, ns^2$ |
| p–block elements | 13 to 17 group | $ns^2\, np^{1-5}$ |
| Zero group elements | 18th group | $ns^2\, np^6$ |

Evaluation of the hazard of a combustible dust should be determined by the means of actual test data. The following list represents factors that can be considered when determining the deflagration hazard of a dust:

1. Minimum dust concentration to ignite
2. Minimum energy required for ignition (joules)
3. Particle size distribution
4. Moisture content as received and dried
5. Deflagration index
6. Layer ignition temperature
7. Maximum explosion pressure, at optimum concentration
8. Electrical volume resistivity measurement
9. Dust cloud ignition temperature
10. Maximum permissible oxygen concentration (MOC) to prevent deflagration

| Group period | 1 | 2 | 3 | 4 | 5 | 6 | 7 | 8 | 9 | 10 | 11 | 12 | 13 | 14 | 15 | 16 | 17 | 18 |
|---|---|---|---|---|---|---|---|---|---|---|---|---|---|---|---|---|---|---|
| 1 | 1 H | | | | | | | | | | | | | | | | | 2 He |
| 2 | 3 Li | 4 Be | | | | | | | | | | | 5 B | 6 C | 7 N | 8 O | 9 F | 10 Ne |
| 3 | 11 Na | 12 Mg | | | | | | | | | | | 13 Al | 14 Si | 15 P | 16 S | 17 Cl | 18 Ar |
| 4 | 19 K | 20 Ca | 21 Sc | 22 Ti | 23 V | 24 Cr | 25 Mn | 26 Fe | 27 Co | 28 Ni | 29 Cu | 30 Zn | 31 Ga | 32 Ge | 33 As | 34 Se | 35 Br | 36 Kr |
| 5 | 37 Rb | 38 Sr | 39 Y | 40 Zr | 41 Nb | 42 Mo | 43 Tc | 44 Ru | 45 Rh | 46 Pd | 47 Ag | 48 Cd | 49 In | 50 Sn | 51 Sb | 52 Te | 53 I | 54 Xe |
| 6 | 55 Cs | 56 Ba | 71 Lu * | 72 Hf | 73 Ta | 74 W | 75 Re | 76 Os | 77 Ir | 78 Pt | 79 Au | 80 Hg | 81 Tl | 82 Pb | 83 Bi | 84 Po | 85 At | 86 Rn |
| 7 | 87 Fr | 88 Ra | 103 Lr ** | 104 Rf | 105 Db | 106 Sg | 107 Bh | 108 Hs | 109 Mt | 110 Ds | 111 Rg | 112 Cn | 113 Uut | 114 Uuq | 115 Uup | 116 Uuh | 117 Uus | 118 Uuo |

| *Lanthanoids | 57 La | 58 Ce | 59 Pr | 60 Nd | 61 Pm | 62 Sm | 63 Eu | 64 Gd | 65 Tb | 66 Dy | 67 Ho | 68 Er | 69 Tm | 7 Yb |
|---|---|---|---|---|---|---|---|---|---|---|---|---|---|---|
| **Actinoids | 89 Ac | 90 Th | 91 Pa | 92 U | 93 Np | 94 Pu | 95 Am | 96 Cm | 97 Bk | 98 Cf | 99 Es | 100 Fm | 101 Md | 102 No |

## Organic and Inorganic Chemistry

Chemistry is mainly divided into two branches—organic chemistry and inorganic chemistry. Carbon is present in a very large number of compounds, many of which have a small number of other elements such as hydrogen, nitrogen and halogens. Organic chemistry deals with the chemistry of carbon compounds while inorganic chemistry deals with all the other elements.

## Chemical Equation

Chemical equation is the representation of a chemical reaction by writing the symbol and molecular formula of the reactants and the products of the reaction. Sometimes the physical states of the reactants and the products are also indicated in a chemical equation.

## Balancing a Chemical Equation

Based on the 'law of conservation of energy' which states that matter can neither be created nor destroyed, the number of atoms found in the products must match the numbers in the reactants. This process of equalizing the number of atoms on both sides of a chemical equation is called 'balancing'.

For example

$$(2)\ H_2 + O_2 \longrightarrow (2)\ H_2O$$

Here, (2) is included for $H_2$ and for $H_2O$ to balance the equation. The equation now has equal number of hydrogen and oxygen atoms on both sides of the equation. We can now state—two molecules of hydrogen combine with one molecule of oxygen to yield two molecules of water.

## Types of Chemical Bonds

### Ionic Bond

A bond formed when electrons are transferred between atoms of the reacting elements.
Example of an ionic bond

$$Na + Cl \longrightarrow NaCl$$

Characteristics of compounds formed by ionic bonds are:

- They are generally solids at room temperature
- The melting points and boiling points are high
- They are hard
- They are good conductors of electricity
- Reactions of ionic compounds are fast.

### Covalent Bond

A bond formed when a pair of electrons is being shared by atoms, e.g. the bond formed by two hydrogen atoms H–H. The process of pairing of electrons takes place in covalent bonds. Example of a covalent bond.

$$C + O_2 \longrightarrow CO_2$$

Compounds formed by covalent bonds show:

- Low melting and boiling points
- They are usually insoluble in water, but soluble in organic compounds
- They are poor conductors of electricity
- Reactions of covalent compounds are slow.

## Types of Chemical Reactions

1. Chemical combination
2. Chemical decomposition
3. Chemical displacement
4. Chemical double decomposition or double displacement.

### Chemical Combination

A chemical combination is a type of chemical reaction in which two or more reactants combine to form a single product. For example:

$$2\,Mg + O_2 \longrightarrow 2\,MgO$$
$$C + O_2 \longrightarrow CO_2.$$
$$NH_3 + HCl \longrightarrow NH_4Cl.$$

### Chemical Decomposition

A chemical decomposition is a type of reaction in which a single reactant decomposes to form two or more products. For example:

$$CaCO_3 \text{ (in the presence of heat)} \longrightarrow CaO + CO_2$$

### Chemical Displacement

A chemical displacement is type of reaction in which an element present in a compound is displaced by another element. For example:

$$CuSO_4 + Zn \longrightarrow ZnSO_4 + Cu$$

### Chemical Double Decomposition

A chemical double decomposition is a reaction in which the reactants mutually exchange their radicals to form two new compounds. For example:

$$AgNO_3 + NaCl \longrightarrow AgCl + NaNO_3$$

## Acids

An inorganic acid is derived from one or more inorganic compounds. Inorganic acids do not contain carbon atoms.

The word acid is derived from the Latin word which means "sour".

An acid is a compound, which when dissolved in water yields hydronium ion as the only positive ion.

For example:

$$HCl \quad + \quad H_2O \longrightarrow \quad H_3O^+ \quad + \quad Cl^-$$

Hydrochloric + Water      Hydronium    Chlorine
    acid                      ion         ion

An inorganic acid is derived from one or more inorganic compounds. Inorganic acids do not contain carbon atoms.

Examples:
- Sulphuric acid ($H_2SO_4$)
- Hydrochloric acid (HCl)
- Nitric acid ($HNO_3$)
- Hydrofluoric acid (HF)

Most of the inorganic acids are treated as strong acids because they are extensively dissociated into ions (effectively 100%).

Organic acids–An organic acid is an organic compound with acidic properties.

Examples:
- Acetic acid
- Citric acid
- Tartaric acid
- Oxalic acid
- Formic acid, etc.

Organic acids are treated as weak acids because they are less than 100% dissociated into ions in solvents.

- Acids are widely used in industry and are amongst the most frequently encountered chemical hazards.
- Acids are often corrosive, attacking the eyes and the skin.
- All acids have sour taste in dilute solution.
- Acids are substances which contain hydrogen which may be replaced by a metal.

Example:
Zinc reacts with dilute sulphuric acid to produce zinc sulphate and liberates flammable hydrogen gas.

$$Zn \quad + \quad H_2SO_4 \longrightarrow \quad ZnSO_4 \quad + \quad H_2$$

Zinc     Dil. sulphuric     Zinc sulphate     Hydrogen
           acid                                 gas

Acids will liberate carbon dioxide from a carbonate or bicarbonate.

Example:

$$NaHCO_3 \quad + \quad H_2SO_4 \longrightarrow \quad NaHSO_4 \quad + \quad H_2O \quad + \quad CO_2$$

sodium      Sulphuric     Sodium bi-    Water    Carbon
bicarbonate     acid           sulphate                     dioxide

Acids are neutralized by bases to give salt and water. This reaction is referred to as neutralization reaction.

$$\text{Acid} \quad + \quad \text{Base} \quad \longrightarrow \quad \text{Salt} \quad + \quad \text{Water}$$

For example:

$$\underset{\substack{\text{Hydrochloric}\\\text{acid}}}{\text{HCl}} \quad + \quad \underset{\substack{\text{Sodium}\\\text{hydroxide}}}{\text{NaOH}} \quad \longrightarrow \quad \underset{\substack{\text{Sodium}\\\text{chloride}}}{\text{NaCl}} \quad + \quad \underset{\text{Water}}{\text{H}_2\text{O}}$$

### Properties of Sulphuric Acid

- Chemical formula: $H_2SO_4$
- Specific gravity: 1.84
- Boiling point: 330° C

The properties of sulphuric acid may be considered under three headings:
1. Reactivity towards water
2. Behaviour as an acid
3. Behaviour as an oxidizing agent

1. *Reactivity towards water:*
   - Sulphuric acid in concentrated solution is a very powerful dehydrating agent.
   - That is, it has the ability to absorb large quantities of water.
   - This it does with the evolution of a large quantity of heat, sufficient to boil the added water. Hence dilution of such acid with water is a dangerous process.
   - Organic compounds containing carbon, hydrogen and oxygen may be dehydrated by concentrated sulphuric acid.

Sugar is dehydrated to sugar charcoal.

$$\underset{\text{Sugar}}{\text{C}_{12}\text{H}_{22}\text{O}_{11}} \quad \longrightarrow \quad \underset{\text{Carbon}}{12\text{C}} \quad + \quad \underset{\text{Water}}{11\text{H}_2\text{O}} + \text{HEAT}$$

2. *Behaviour as an acid:*
   - Concentrated sulphuric acid (98%) contains little water and consequently shows little reaction towards metals at room temperature.
   - For this reason, concentrated sulphuric acid can be transported in unlined tankers.
   - When, however, the concentration is reduced to a lower level, say 60%, the reactivity of sulphuric acid increases considerably.
   - Tankers lined with rubber, PVC, etc. are needed to transport acid of this concentration.
   - These facts suggest that if a spillage of concentrated sulphuric acid occurs, dilution will produce a more corrosive acid unless sufficient water can be added to make the acid very dilute, when the reactivity would again decrease.

3. *Behaviour as an oxidizing agent:* Hot concentrated sulphuric acid acts as a powerful oxidizing agent. When heated, sulphuric acid decomposes and produces nascent oxygen, which helps in the oxidation process. It oxidizes metals as well as non-metals, and in turn gets reduced.

## Neutralization

Neutralization is the reaction between an acid and a base, producing a salt and a neutralized base; e.g. hydrochloric acid and sodium hydroxide form sodium chloride and water.

$$HCl + NaOH \longrightarrow H_2O + NaCl$$

| HCl | NaOH | H₂O | NaCl |
|---|---|---|---|
| Hydrochloric acid | Sodium hydroxide | Water | Sodium chloride |

## Acids and Alkalies (pH)

The pH of any fluid is a measure of the hydrogen ion concentration. The pH values are represented on a scale of 1 to 14 with 1 to 7 considered as acidic and 7 to 14 considered as basic or alkaline. A value of exactly 7 is considered is considered to be neutral.

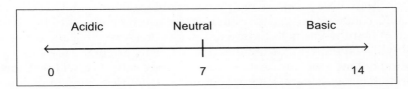

**pH range of values**

## Hydrocarbons

Compounds formed by two elements–namely carbon and hydrogen are called hydrocarbons. Petrol, diesel, kerosene, LPG, natural gas are all hydrocarbons which are combustible.

Hydrocarbons can be mainly divided into two classes:
  i. Aliphatic compounds which contain chains of carbon atoms.
  ii. Aromatic compounds, which contain a special kind of ring of six carbon atoms known as a benzene ring.

Most of the organic chemicals are capable of burning. Many important fuels we use, such as natural gas, petrol, paraffin and diesel oil are organic compounds.

Aliphatic compounds also known as paraffin hydrocarbons.

The paraffin hydrocarbons are compounds made up of carbon and hydrogen only. The simplest compound available in these series is methane ($CH_4$). Natural gas has methane as the main constituent. It reacts readily with chlorine, hydrochloride, bromine, etc. There are also substances known as "acetylenes". The gas acetylene ($C_2H_2$) is liable to explode when exposed to heat or mechanical shock, even when air or oxygen is absent. Acetylene is flammable and forms mixture in air with a very wide explosive range. Acetylene is used for manufacturing plastics, other chemicals and in oxygen-acetylene welding and acetylene is also widely used in other areas.

## Aromatic Hydrocarbons

There is a series of hydrocarbons known as "aromatics". The base of compounds is a ring structure known as benzene ring. Benzene ($C_6H_6$) is a flammable compound. There is a high proportion of carbon in its molecule. So there is not enough oxygen to oxidize all the carbon. Naturally, a good portion of carbon is released as thick black smoke. Aromatic compounds usually burn with a very smoky flame. Some aromatic compounds like toluene and xylene are important solvents. Some aromatic compounds, like benzene are highly toxic.

## Liquefied Petroleum Gases (LPG)

LPG is produced from naturalgas. LPG contains mainly propane and butane. Propane ($C_3H_8$) and butane ($C_4H_{10}$) are gases at room temperature and pressure. As the critical temperatures of these gases are well above room temperature, they can be liquefied by the application of pressure alone. A very small amount of liquid will produce a great volume of gas. Both these gases are highly flammable and are widely used as fuel gases. Inside each LPG container we get, for domestic fuel use, there is a liquid with pressurized gas above it. When the gas is withdrawn, the gas pressure will fall; at the same time, more liquid will evaporate to restore the pressure to its original value.

Propane and butane gases are heavier than air. They naturally, will seek lower ground on release from their containers. They are odourless, colourless but flammable. A stinging agent known as mercaptan is added so critical temperature. Above the critical temperature, substances can only exist as gases. So if the cylinders containing gases are heated above the critical temperature, they are likely to explode. The critical temperature of propane is 96.7° C and that of butane 152° C.

> Gas fires normally should not be extinguished unless the source of burning gas can be shut off, because an explosive mixture can be formed with air that, if ignited, can cause greater damage than the original fire.

| Methane | $CH_4$ | |
|---------|--------|---|

$$\begin{array}{c} H \\ | \\ H-C-H \\ | \\ H \end{array}$$

| Ethane | $C_2H_6$ | |
|--------|----------|---|

$$\begin{array}{c} H\ \ H \\ |\ \ \ | \\ H-C-C-H \\ |\ \ \ | \\ H\ \ H \end{array}$$

| Propane | $C_3H_{2\times3+2}$ | |
|---------|---------------------|---|

$$\begin{array}{c} H\ \ H\ \ H \\ |\ \ \ |\ \ \ | \\ H-C-C-C-H \\ |\ \ \ |\ \ \ | \\ H\ \ H\ \ H \end{array}$$

Butane $\qquad$ $C_4H_{2 \times 4 + 2}$

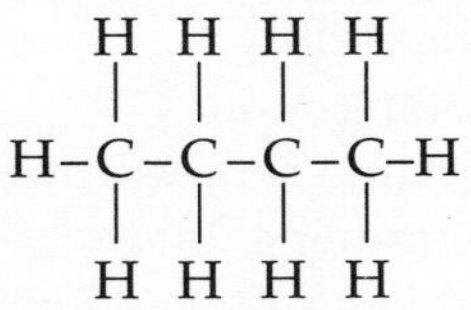

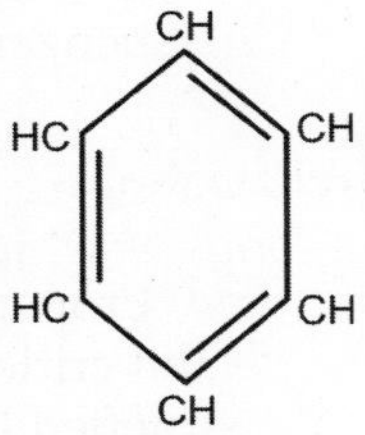

$$
\begin{array}{c}
\text{H} \quad \text{H} \quad \text{H} \quad \text{H} \\
| \quad | \quad | \quad | \\
\text{H}-\text{C}-\text{C}-\text{C}-\text{C}-\text{H} \\
| \quad | \quad | \quad | \\
\text{H} \quad \text{H} \quad \text{H} \quad \text{H}
\end{array}
$$

Benzene $C_6H_6$

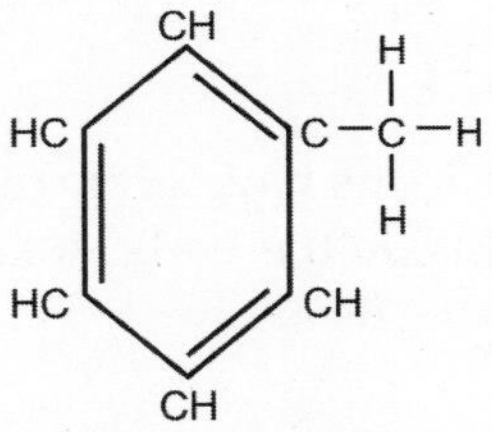

Toluene $C_6H_5CH_3$

## Organic Solvents

Organic solvents are organic compounds ranging through liquid hydrocarbon, alcohol, aldehydes, ketones, esters, ethers, carbon disulphide, turpentine oil, etc. which are commonly used as solvents.

### Characteristics of Organic Solvents

- They have natural dissolving power.
- They are volatile.
- They are easily removed from the articles with which they have been in contact.
- They evaporate at normal temperature and pressure.
- They are highly flammable in nature and have their flash point below the ordinary temperature.
- They are potential fire hazards. Therefore, these compounds should be stored in close containers, in well-ventilated room and away from any source of ignition.
- Organic compound like methyl alcohol, ethyl alcohol, and acetone are miscible in water in all proportions.
- Organic compound like gasoline, benzene, hexane, toluene, xylene, etc. are immiscible in water and being lighter than water, when mixed with water, float upon it.
- Organic compounds like carbon disulphide (specific gravity = 1.3) is heavier than water and being immiscible in water, when mixed with water, sink to the bottom.

### Properties of Some Organic Solvents

*Methyl alcohol ($CH_3OH$):*

- Methyl alcohol is a clear colourless, volatile, flammable liquid.
- Flash point $\qquad$ : 11°C
- Specific gravity $\qquad$ : 0.8 (water = 1)
- Vapour density $\qquad$ : 1.1

- Miscibility    : Miscible in water
- Boiling point   : 64° C
- Ignition temperature  : 464° C
- Flammable limits  :
  Lower limit = 7.3%
  Upper limit = 36%

*Health hazards of methyl alcohol:*

- Deaths have occurred from drinking even a small amount of methyl alcohol.
- Inside the body, it is oxidized into Formaldehyde and Formic acid, both of which are poisonous.

*To fight fires of methyl alcohol:* Use following extinguishing agents:

- Alcohol resistant foam.
- Dry chemical powder.
- Carbon dioxide.
- Water spray in case of a spill fire.
- Dilution techniques are less effective on tanks, where they can cause an over flow.

*Ethyl alcohol ($C_2H_5OH$):*

- Pure ethyl alcohol is a clear, colourless, volatile flammable liquid.
- It burns with pale blue flame.
- Flash point    : 13° C
- Specific gravity   : 0.8
- Vapour density   : 1.6
- Boiling point    : 78° C
- Ignition temperature  : 423° C
- Miscibility with water : Miscible
- Flammable limits  :
  Lower limits = 4.3%
  Upper limits = 19%

*Turpentine oil:*

- Turpentine oil is a major liquid hydrocarbon.
- It is used as a solvent for oil, resins, varnishes, paints, rubbers, polishes and waxes.
- It reacts vigorously with chlorine and iodine.

*Carbon disulphide:*

- Colour: colourless, volatile flammable liquid.
- Specific gravity: 1.3 (water = 1)
- Vapour density: 2.6 (Air = 1)
- Ignition temperature: 120° C
- Flash point: –25° C

- Toxicity: Highly toxic
- Limits of flammability:
    Lower limit flammability = 1.25%
    Upper limit flammability = 50%
- Carbon disulphide burns with bluish flame. It produces carbon dioxide and sulphur dioxide gases as a product of combustion.

Reaction

$$CS_2 \quad + \quad 3O_2 \quad \longrightarrow \quad CO_2 \quad + \quad 2SO_2$$

| Carbon disulphide | Oxygen dioxide | Carbon dioxide | Sulphur dioxide |

*To fight fire of carbon disulphide ($CS_2$):*

- Use BA set for self-protection
- Use water
- Use carbon dioxide
- Use dry chemical powder
- Use foam

## Halogens

Fluorine, chlorine, bromine and iodine form a group of elements known as halogens. They, more or less, resemble each other and the compounds formed by them also have somewhat similar properties.

Fluorine is the most reactive non-metal substance. It reacts with most other elements. The gas is highly toxic and dangerous because of its very high reactivity with compounds which can be oxidized. It can react with water to produce great heat giving off toxic and corrosive flames.

*Chlorine* is the most common and most important of all halogens. The critical temperature is 144° C. It can be easily liquefied under pressure at room temperature. Chlorine is an extremely toxic gas. It was used as a poisonous gas in World War-I. It reacts with water to form hydrochloric acid and hydrochlorous acid. Chlorine is a highly reactive element and an oxidizing agent. It can react and even cause explosions with turpentine, ether, ammonia gas, hydrocarbons, hydrogen and powdered metals. Its reaction with acetylene is very violent. Large quantity of chlorine is used in the bleaching of wood pulp, cotton and linen fabrics. It is also used in order to manufacture chlorine compounds like as carbon tetrachloride (CTC), hydrochloric acid, chlorates and different insecticides.

*Bromine* at room temperature is a dark red liquid, which readily gives out toxic vapours. This liquid can cause serious bums, if it comes into contact with the skin. Bromine is an oxidizing agent and can ignite combustible substances with which it comes into contact.

*Iodine* is a dark-violet almost black solid, which sublimes readily to give off a violet toxic vapour. To a certain extent, it can react with oxidizable materials. It is almost insoluble in water but can dissolve in alcohol to give tincture of iodine, which is used as an antiseptic.

## Properties of Chlorine Gas

*Physical properties:*

- Chlorine is highly toxic gas.
- Chemical formula: $Cl_2$
- Colour: Greenish yellow gas
- Odour: Pungent and irritating odour
- Molecular weight: 70
- Solubility in water:
- Maximum 1% soluble in water at $10°$ C. Solubility decreases with rise in temperature.
- Specific gravity of liquid: 1.468
- Vapour density: 2.48 (Air = 1)
- TLV: 1.O PPM

*Chemical proerties:*

- Reaction of chlorine with water: When chlorine reacts with water, a weak solution of hydrochloric acid and hypochlorous acids are formed.

$$Cl_2 + H_2O \longrightarrow HOCl + HCl$$
Chlorine   Water          Hypo-          Hydro-
                          chlorous acid   chloric acid

- Hypochlorous acid further dissociates to give hydrochloric acid and nascent oxygen.

$$HOC \longrightarrow HCl + O$$
Hypo-              Hydro-      Nascent
chlorous acid      chloric acid   oxygen

Note: Nascent oxygen will increase the rate of corrosion.

- Reaction of chlorine with sodium hydroxide solution: Chlorine reacts readily with sodium hydroxide solution to form sodium hypochlorite.

$$2NaOH + Cl_2 \longrightarrow NaOCl + NaCl + H_2O$$
Sodium     Chlorine      Sodium      Sodium   Water
hydroxide                hypochlorite   chloride

- Reaction of chlorine with calcium hydroxide: Chlorine reacts with calcium hydroxide to form bleaching powder.

$$2Ca(OH)_2 + 2Cl_2 + 2H_2O \longrightarrow Ca(OCl)_2.4H_2O + CaCl_2$$

- Reaction of chlorine with hydrogen: In presence of sun light, chlorine forms explosive mixture with hydrogen.
  Reaction: Sun light

$$H_2 + Cl_2 \longrightarrow 2HCl$$
Hydrogen   Chlorine          Hydrogen chloride

* Reaction of chlorine with turpentine oil:

$$C_{10}H_{16} \quad + \quad 8Cl_2 \quad \longrightarrow \quad 10C \quad + \quad 16HCl$$

| Turpentine<br>oil | Chlorine | Carbon | Hydrogen<br>chloride |

## Some Combustible Solids

*Wood:* It is a very useful non-metallic material. There is high water content in wood. Wood used for construction purpose is known as timber. Considerable quantity of heat is required to dry timber. The properties of wood are such that a long flaming fuel is produced, which burns steadily and spreads fire quickly.

*Coal:* Coal is formed as a result of the action of pressure and temperature on the decayed vegetable products of ancient forests. It is a very complex mixture of carbon and a variety of organic compounds. Coal is mainly divided into two types, i.e. hard coal and soft coal. The harder coals contain more carbon and are used in special furnaces for producing steam. The softer coals are more useful for producing gas. The surface of coal has reactive centers where combustion with oxygen can easily occur. Coal can produce a spontaneous ignition due to direct oxidation. Higher the oxygen content in the coal, greater the chances of danger. Pulverized coal with more than 10% oxygen content is very risky in storage. A small proportion of the coal dust is enough to form an explosive mixture in air.

**(a)**                                                                 **(b)**

Fig. A2.2: Combustible solids (a) Wood, (b) Coal.

## Heat Output of Fuels

Chemical heating value, MJ per kg of fuel

$$= 33.7 \, C + 144 \, (H_2 - (O_2)/8) + 9.3 \, S$$

C is the mass of carbon per kg of fuel
$H_2$ is the mass of hydrogen per kg of fuel
$O_2$ is the mass of oxygen per kg of fuel
S is the mass of sulphur per kg of fuel

## Theoretical Air Required to Burn Fuel

Air (kg per kg of fuel) $= ((8/3)C + 8 \, H_2 - (O_2)/8 + S) \, 100/23$

### The Hazards of Hydrogen Sulphides (H₂S) Gas

Hydrogen sulphide gas ($H_2S$) is found associated with:

- Biological and industrial processes
- Often associated with gas, oil and water.

So, prevalent in oil-wells, marshes, sewage lines, water wells, etc.
Hydrogen sulphide is both toxic and flammable.
Fire and explosion hazard of $H_2S$

- Explosive between 4.3 to 46%
- Autoignition temperature 260° C

$H_2S$ fire can be extinguished using $CO_2$ or dry powder. But because of the toxicity of $H_2S$, respiratory protection (SCBA) should be used.

## Metals

Three quarters of all the elements are metals. Metals have a set of properties which are distinctive. If an element possesses most of these properties, it is considered as a metal.

### Important Properties of Metals

1. All metals except mercury are solids.
2. Metals can be hammered into shapes (malleable)
3. Metals can be drawn into wire (ductile)
4. Metals have high melting and boiling points.
5. Metals can form alloys (alloys are eutectic mixtures of two or more metals)
6. Many metals combine with oxygen easily
7. Metals usually dissolve in mineral acids, often releasing hydrogen.

### Thermal Properties of Metals

All metals expand in dimensions when heated, i.e. their temperature gets raised. The increase in length is linear with temperature–as temperature increases the length increases proportionately. For metals, if the temperature is raised by 100° C, we can assume an expansion of 1 mm per metre.

Increase in length = $L \alpha (T_2 - T_1)$

Where,

$L$ = original length, metres
$\alpha$ = coefficient of linear expansion
$(T_2 - T_1)$ = rise in temperature

Increase in volume = $V \beta (T_2 - T_1)$

Where

$V$ = original volume
$\beta$ = coefficient of volumetric expansion
$(T_2 - T_1)$ = rise in temperature

Coefficient of volumetric expansion = coefficient of linear expansion × 3

$\beta = 3\alpha$

*Typical Expansion Data*

| | |
|---|---|
| Aluminium, lead, zinc | 3 to 4 mm/m/100° C |
| Copper | 2 mm/m/100° C |
| Tunsten | 0.4 mm/m/100° C |
| Invar steel | 0.3 mm/m/100° C |
| Platinum | 1 mm/m/100° C |

*Reaction of Metals with Water or Steam*

1. *Potassium, sodium and calcium:* These metals react immediately with water and cause the release of flammable hydrogen gas and leave a metal hydroxide residue. In a few cases, the hydroxide thus formed is itself a corrosive alkali. In the case of potassium, the reaction is so vigorous that the metal seems to ignite on contact with water. Sodium moves rapidly on the surface of water and, if prevented from doing so, it will ignite. Large pieces of these metals pose a danger of explosion when they come into contact with water. Calcium reacts steadily with cold water but vigorously with hot water.
2. *Magnesium:* Magnesium shows little reaction with cold water even if magnesium is powdered. At higher temperature, the rate of reaction increases and a steady flow of hydrogen is produced by its reaction with steam. If the metals are already burning, the reaction with cold water becomes very high. If cold water is added, magnesium gives rapid production of hydrogen with subsequent explosions.

*Pyrophoric Materials*

When a metal is very finely divided, the surface area gets increased. Then the chances of combustion also get increased. Some of the metallic powders and dusts can burn or explode spontaneously in air. When this takes place at ordinary temperature, the material is said to be pyrophoric. A good number of flammable metal powders such as magnesium, calcium, sodium, potassium, etc. are pyrophoric.

Some metal powders burn in carbon dioxide, nitrogen or under water. There are some metal powders which when damp, can cause fires and explosions, even in the absence of air, without a warning, and even in the absence of heat.

## PHYSICS

### Energy

Energy is nothing but the ability to do work. It is available in different forms such as heat, electricity, kinetic energy, potential energy and so on. Energy cannot be created. Energy cannot be destroyed. Energy can only be converted into a different form, e.g. mechanical energy can be converted to heat energy by friction. Joule is an energy unit. It is used for all forms of energy like mechanical energy, heat energy, electrical energy, etc.

### Heat and Temperature

Heat can be produced mechanically, electrically or chemically. Heat can be converted into other forms of energy as steam pressure energy in steam boiler. Heat energy cannot be measured directly. The temperature changes when heat is applied to a substance.

Temperature may be defined as a measure of the heat content of a body. It is not the amount of heat in a body. Heat always moves down from a hot body to a cold one. Temperature can be measured using an instrument called thermometer. A thermometer has a small stem with a tiny bore, containing a suitable liquid, usually mercury. It is sealed at the end.

Temperature can be measured using either a fahrenheit scale (° F), (now obsolete) or a centigrade scale (° C). The latter is known as celsius (° C) scale. An instrument known as "pyrometer" is used in order to measure temperature above the range of a mercury thermometer.

### Conversion of Temperature Scales

While a fahrenheit scale thermometer has 180 graduations, the celsius thermometer has only 100 graduations for the same length.

To convert degrees centigrade to fahrenheit, multiply degree centigrade by 9/5 and add 32.

$$° F = (° C \times 9/5) + 32$$

To convert degree fahrenheit to centigrade, subtract 32 and then multiply by 5/9

$$° C = (° F\text{-}32) \times 5/9$$

### Absolute Temperature Scale

Calculation involving gases is done with the help of a different temperature scale. It is named after Lord Kelvin.

The lowest attainable temperature of a gas is considered to be 0 Kelvin (0°K) on the absolute scale.

$$0° K = –273° C$$
$$0° C = 273° K, 100° C = 373° K$$

### Heat Units

*The joule:* This is a unit of energy and this unit is used for all forms of energy like heat energy, mechanical energy, electrical energy and so on. "joule" is the SI unit of heat. The 'joule' is the work done when the application of a force of one Newton moves through 1 metre in the direction of the force.

$$1000 J = 1 kJ \text{ (kilojoules).}$$

*The calorie:* Calorie can be defined as the quantity of heat required for raising the temperature of 1 gram of water through 1° C.

$$1 \text{ caloric} = 4.18 \text{ joules.}$$

*The British thermal unit:* This may be defined as the quantity of heat required to raise the temperature of one pound of water through 1° F.

$$1 BTU = 1005 J = 1.005 kJ$$

*Heat capacity:* The heat necessary for raising the temperature of a body by 1° C is known as heat capacity.

The SI unit of heat capacity is J/° C.

*Specific heat capacity:* We make use of specific heat capacity to compare one substance with another. The specific heat capacity is the heat needed for raising the temperature of unit mass of the substance through 1° C. In SI system, the unit is joules per kilogram per degree centigrade (J/kg per ° C).

| Specific heats of gases | | | |
|---|---|---|---|
| Gas | Specific heat at constant pressure kJ/kg ° K or kJ/kg ° C $C_p$ | Specific heat at constant volume kJ/kg ° K or kJ/kg ° C $C_v$ | Ratio of specific heat $C_p/C_v$. |
| Air | 1.005 | 0.718 | 1.40 |
| Ammonia | 2.060 | 1.561 | 1.32 |
| Carbon dioxide | 0.825 | 0.630 | 1.31 |
| Carbon monoxide | 1.051 | 0.751 | 1.40 |
| Helium | 5.234 | 3.153 | 1.66 |
| Hydrogen | 14.235 | 10.096 | 1.41 |
| Hydrogen sulphide | 1.105 | 0.85 | 1.30 |
| Methane | 2.177 | 1.675 | 1.30 |
| Nitrogen | 1.043 | 0.745 | 1.40 |
| Oxygen | 0.913 | 0.652 | 1.40 |
| Sulphur dioxide | 0.632 | 0.451 | 1.40 |

## Change of State and Latent Heat

### Latent Heat of Vapourization

Latent heat of vapourization is the quantity of heat needed to change unit mass of a substance from liquid state to vapour without changing the temperature. Latent means hidden.

Water boils at 100° C. At this temperature, the boiling water absorbs more heat from the source in order to convert into steam. Though the water absorbs more heat, the temperature does not exceed the boiling point. The additional heat goes into the vapour, but does not reveal its presence by producing a rise in temperature. 22,66,000 joules are required to convert 1 kg of water at its boiling point into steam at the same temperature. As the steam condenses to form water, the same quantity of latent heat is given out. 86,000 joules are needed in order to convert, at the same temperature, 1 kg of alcohol into vapour. The SI unit of latent heat is joule/kilogram (J/kg) or kilojoules/kilogram (kJ/kg).

It is the amount of heat needed to convert unit mass of a substance from solid state to liquid state without any change in temperature.

Water freezes into ice at 0° C. When ice melts into water some latent heat is absorbed. In order to convert 1 kg of ice at 0° C at the same temperature 3,36,000 joules are required. Water and other substances take in latent heat when they melt and give out latent heat when they solidify. This is known as latent heat of fusion.

Latent heat of fusion of ice = 336 kJ/kg

Latent heat of steam from and at 100° C = 2266 kJ/kg

1 tonne of refrigeration = 336 000 kJ/day = 233 kJ/min

## Volatile Liquids

Some liquids have a low boiling point so they change from liquid to vapour very easily at ordinary temperatures.

These liquids are known as "volatile liquids", e.g. ether, spirit.

## Thermal Expansion of Solids, Liquids and Gases

A substance irrespective of it being solid, liquid or gas expands when heated. Thus expansion resulting from heat is known as thermal expansion.

### Solids

A solid, when heated expands in all three dimensions. As a result it increases in length, breadth and thickness. Expansion of materials in one direction is called "linear expansion".

A solid substance expands when its temperature is raised. This amount of expansion is called "linear expansion". The coefficient of linear expansion is defined as the increase in the length of a unit length when its temperature is raised by one degree. The coefficient of linear expansion for steel is 0.000012 per ° C. For copper it is 0.000017 per ° C. It is 0.000023 per ° C for aluminum. When metal structures are subjected to large variations in temperature, they should be given enough allowance for linear expansion of the parts. "Invar" is an alloy of nickel (36%) and iron (64%). This alloy has a very low coefficient of linear expansion 0.0000001 per ° C. This alloy is used in order to manufacture measuring rods and tapes, which have to be very accurate over a range of temperatures. This alloy is also used for manufacturing watch and clock parts or parts of other mechanisms, which have to remain unchanged in size and shape over a range of temperature.

### Liquids

When temperature rises, the volume of a given mass is increased and the density decreases. As liquids have no shape, they therefore have no fixed dimensions other than volume. The coefficient of cubical expansion of liquids is much greater than that of solids (approximately 30 times higher). In a fire situation, if a sealed container containing liquid is heated, the liquid will expand and cause a rise in pressure, if there is no air space or relief valve to bring down the pressure.

### Water

*Properties of water:*

*Vapour pressure:* Water boils at 100° C at atmospheric pressure.

## Vapour pressure of water at different temperatures

| Temperature ° C | Pressure bar | Temperature ° C | Pressure bar |
|---|---|---|---|
| 5 | 0.008 | 190 | 12.552 |
| 10 | 0.012 | 200 | 15.551 |
| 20 | 0.023 | 210 | 19.081 |
| 30 | 0.042 | 220 | 23.021 |
| 40 | 0.073 | 230 | 27.979 |
| 50 | 0.123 | 240 | 33.480 |
| 60 | 0.199 | 250 | 39.780 |
| 70 | 0.311 | 260 | 46.94 |
| 80 | 0.473 | 270 | 55.05 |
| 90 | 0.701 | 280 | 64.19 |
| 100 | 1.013 | 290 | 74.45 |
| 110 | 1.432 | 300 | 85.92 |
| 120 | 1.985 | 310 | 98.70 |
| 130 | 2.701 | 320 | 112.90 |
| 140 | 3.614 | 330 | 128.65 |
| 150 | 4.760 | 340 | 146.08 |
| 160 | 6.180 | 350 | 165.37 |
| 170 | 7.920 | 360 | 186.74 |
| 180 | 10.027 | 370 | 210.52 |

## Density and viscosity of water (At atmospheric pressure)

| Temperature ° C | Density kg/cu. m | Viscosity Kinematic ($m^2$/sec) |
|---|---|---|
| 0 | 999.87 | $1.79 \times 10^{-6}$ |
| 10 | 999.73 | $1.31 \times 10^{-6}$ |
| 20 | 998.23 | $1.01 \times 10^{-6}$ |
| 30 | 995.67 | $0.804 \times 10^{-6}$ |
| 40 | 992.24 | $0.661 \times 10^{-6}$ |
| 50 | 988.07 | $0.556 \times 10^{-6}$ |
| 60 | 983.24 | $0.473 \times 10^{-6}$ |
| 70 | 977.21 | $0.415 \times 10^{-6}$ |
| 80 | 971.83 | $0.367 \times 10^{-6}$ |
| 90 | 965.34 | $0.328 \times 10^{-6}$ |
| 100 | 958.38 | $0.296 \times 10^{-6}$ |

| Specific heat and volume expansion for liquids | | |
|---|---|---|
| *Liquid* | *Specific heat (at 20° C)kJ/kg ° K or kJ/kg° C* | *Coefficient of volume expansion (× 10⁻⁴)* |
| Alcohol (ethyl) | 2.470 | 11.0 |
| Ammonia | 0.473 | |
| Benzine | 1.738 | 12.4 |
| Carbon dioxide | 3.643 | 1.82 |
| Mercury | 0.139 | 1.80 |
| Olive oil | 1.633 | |
| Petroleum | 2.135 | |
| Petrol | 2.093 | 12.0 |
| Turpentine | 1.00 | 9.4 |
| Water | 4.183 | 3.7 |

## Gases

Every gas expands by the same amount with the same temperature rise. Temperature and pressure affect the change of volume of a gas. In order to study the effect of one, the other must be maintained at its constant value. There are three gas laws, namely, Boyle's law, Charles law and the law of pressures. While Boyle's law explains the change, if temperature is kept constant, Charles' law explains the changes when pressure is kept constant. Finally, the law of pressures explains the changes, if the volume is kept constant.

> **NTP–Normal temperature and pressure**—is defined as air at 0° C (293.15 K, 32° F) and 1 atm (101.325 kN/m², 101.325 kPa, 14.7 psia, 0 psig, 30 in Hg, 760 torr).
>
> **STP–Standard temperature and pressure**—is defined as air at 21° C (273.15 K, 68° F) and 1 atm (101.325 kN/m², 101.325 kPa, 14.7 psia, 0 psig, 30 in Hg, 760 torr).

*Boyle's law:* Boyle's law states that the volume of a gas is inversely proportional to the pressure upon it, if the temperature is kept constant. The volume of gas gets halved, when the pressure applied to a given mass of gas is doubled.

When the pressure applied to a given mass of gas is doubled, the volume of gas is halved. If $P_1$, and $V_1$, are initial pressure and volume and if $P_2$ and $V_2$ are final pressure and volume

$$P_1 V_1 = P_2 V_2$$

Initial pressure × initial volume = final pressure × final volume.

*Charles' law:* According to this law, the volume of a given mass of gas is directly proportional to its absolute temperature provided that its pressure is kept constant.

If the pressure is kept constant, the volume of a given mass of gas increases by 1/273 of its volume at 0° C for every ° C raise in temperature.

If $V_1$ and $T_1$ are the initial volume and temperature and $V_2$ and $T_2$ are final volume and temperature.

$$V_I.V_2 = T_1. T_2$$

*The law of pressures*: It states that the pressure of a given mass of gas is directly proportional to its absolute temperature, if its volume is kept constant. These conditions exist when a cylinder of gas undergoes a rise of temperature with its valve kept closed.

### The General Gas Law

The three gas laws may be combined into a single expression as shown below.

$$P_1 V_1/T_1 = P_2 V_2/T_2$$

These gas laws are applicable to all gases, if they remain as gas over the temperature and pressure range involved. When the temperature and the pressure levels, at which liquefaction occurs, are approached, the gas laws are no longer applicable.

### Liquefaction of Gases

The boiling point of a liquid gets raised when there is an increase in pressure. Many substances, which are in the form of gases at atmospheric temperature and pressure, can be compressed to raise the boiling points above atmospheric pressure and then the gas turns into liquid. The gases, even after increasing the pressure, which cannot be liquefied at atmospheric temperature are called, "permanent gases". These gases can be liquefied only if the temperature is lowered sufficiently.

> Substances are classified into two groups–pure substances and mixtures. Mixtures can be separated by simple physical methods.

### Critical Temperature

It is the highest temperature beyond which a gas cannot be liquefied with increase of pressure alone.

### Gas and Vapour

The critical temperature of $CO_2$ is 31.1° C. Below this temperature, it can be liquefied by increased pressure and it is described as vapour. Above the critical temperature, it cannot be liquefied and it is known as gas.

### Critical Pressure

The pressure needed to liquefy a gas at its critical temperature. The critical pressure of $CO_2$ is 73.1 bar.

### Sublimation

The change of state directly from solid to gas is called sublimation. When a very low pressure is produced, the boiling point of water can be brought down to 0° C or below.

When this takes place, ice does not melt into water but vaporizes into steam totally, on a rise in temperature.

## Humidity

The air in the atmosphere contains water in the form of vapour, which is invisible. The air can hold up to 19 gm of water per cubic metre at 20° C, when it is said to be saturated. As the temperature increases, the air can hold more water vapour.

Humidity is the ratio of the quantity of water vapour which is present in the air at a particular temperature, to the quantity of air which it can hold when saturated.

Practically any combustible when in the form of dust and mixed with air in proper proportion, will burn so rapidly as to cause a severe explosion if ignited by heat or spark or flame. Ignorance of this fact has caused many serious disasters. Grain flour, coal dust, metal powders wood dust, rubber dust, leather dust–all constitute a hazard in this regard. Dust prevention and good house keeping is very important. All equipment must be dust proof and maintained so. Explosion vent should lead outdoors to a safe location. Vacuum cleaning is better than sweeping.

For example: At 20° C, if the air contains 9.5 gm of water vapour, we know that air, at saturation, at 20° C, can hold up to 19 gm of water vapour, then 9.5/19 = 50%, relative humidity is 50%.

| Temperature ° C | −20 | −10 | 0 | 10 | 20 | 30 | 40 |
|---|---|---|---|---|---|---|---|
| Quantity of water maximum (saturation) gram/cu.m | 1 | 2.5 | 5 | 10 | 19 | 35 | 50 |

Hygrometers or psychrometers are the instruments used to measure the relative humidity.

| Melting point of some materials (at normal atmospheric pressure) | | |
|---|---|---|
| *Material* | *Chemical formula* | *Melting temperature ° C* |
| Aluminium | Al | 658 |
| Argon | Ar | −190 |
| Cobalt | Co | 1495 |
| Gold | Au | 1064 |
| Helium | He | −271 |
| Lead | Pb | 327 |
| Magnesium | Mg | 651 |
| Mercury | Hg | −39 |

*(Contd.)*

| Material | Chemical formula | Melting temperature ° C |
|----------|------------------|--------------------------|
| Nickel | Ni | 1435 |
| Nitrogen | N2 | –210 |
| Platinum | Pt | 1755 |
| Silver | Ag | 961 |
| Sodium chloride | NaCl | 804 |
| Tin | Sn | 232 |
| Water | $H_2O$ | 0 |
| Zinc | Zn | 419 |

| Piping colour code (conventional) | |
|------------------------------------|--------|
| *Material* | *Colour* |
| Water | Red |
| Steam | Silver grey |
| Combustible liquids | Brown |
| Gas | Yellow |
| Acid and base | Violet |
| Air | Clear blue |
| Other liquids | Black |

| Specific colour code for piping | |
|----------------------------------|--------|
| *Material* | *Colour* |
| Hydrocarbons; flash point < 55° C | Clear red |
| Hydrocarbons; flash point > 55° C | Dark blue |
| Acetylene | Maroon |
| Ammonia | Clear red |
| Argon | Yellow |
| Nitrogen | Black |
| Helium | Brown |
| Hydrogen | Dull red |
| Oxygen | White |

## Properties of some gases (At 1.013 bar and 0° C)

| Gas | Chemical formula | Specific weight kg/cu.m | Liquefaction temperature ° C |
|---|---|---|---|
| Air, pure and dry | | 1.29 | −195 |
| Argon | Ar | 1.78 | −185.7 |
| Nitrogen | $N_2$ | 1.026 | −195.8 |
| Butane | $C_4H_{10}$ | 2.68 | −0.5 |
| Chlorine | $Cl_2$ | 3.21 | −34.5 |
| Ethylene | $C_2H_4$ | 1.26 | −103.8 |
| Fluorine | $F_2$ | 1.70 | −187 |
| Hydrogen | $H_2$ | 0.09 | −252.8 |
| Helium | He | 0.18 | −268.9 |
| Carbon monoxide | CO | 1.25 | −190 |
| Oxygen | $O_2$ | 1.43 | −183 |
| Ozone | $O_3$ | 2.14 | −112 |
| Propane | $C_3H_8$ | 2.02 | −44.5 |

## Ideal air composition

| | By weight | By volume |
|---|---|---|
| Nitrogen | 75.8% | 78% |
| Oxygen | 23.2% | 21% |
| Other gases Argon, carbon dioxide and other rare gases constitute | 1% | 1% |

## Density of liquids

| Material | Specific weight kg/cu.m | Material | Specific weight kg/cu.m |
|---|---|---|---|
| Actone | 791 | Glycol | 1115 |
| Alcohol at 90° C | 834 | Vegetable oil | 910–930 |
| Ethyl alcohol | 789 | Mineral oil | 910–940 |
| Methyl alcohol | 792 | | 780–810 |

## Density/viscosity and temperature

| Material | Temperature ° C | Density kg/cu.m | Viscosity in centipoises |
|---|---|---|---|
| Water @ atm pr. | 20 | 998 | 1.004 |
| | 90 | 964 | 0.315 |
| Boiling water | 100 | 958 | 0.283 |
| | 200 | 866 | 0.136 |
| Fuel oil kerosene | 20 | 840 | 6 |
| | 50 | 840 | 6 |
| Fuel oil heavy | 20 | 940 | 215 |
| | 50 | 940 | 60 |
| Mercury | 0 | 13600 | 1.685 |
| | 200 | 13100 | 1.01 |
| Sodium | 200 | 904 | 0.45 |
| | 600 | 809 | 0.21 |

## Density of common materials

| Material | Density kg/cu.m | Material | Density kg/cu.m |
|---|---|---|---|
| Bitumen | 840–1160 | Paper | 700–1160 |
| Wood heavy | 650–1000 | Glass | 2530 |
| Wood light | 350–700 | Nylon | 1090–1150 |
| Limestone | 2000–2200 | Plexiglass | 1170–1200 |
| Ice@0° C | 920 | Polyethylene | 910–970 |
| Sand | 1710 –1800 | Polystyrene | 980–1100 |

## Specific heat and linear coefficient of expansion of solids

| Solid | Mean specific heat between 0° C and 100° C kJ/kg° C | Coefficient of linear expansion (0° C to 100° C ($\times 10^{-6}$) |
|---|---|---|
| Aluminium | 0.909 | 23.8 |
| Antimony | 0.209 | 17.5 |
| Bismuth | 0.125 | 12.4 |
| Brass | 0.383 | 18.4 |
| Carbon | 0.795 | 7.90 |
| Cobalt | 0.402 | 12.3 |
| Copper | 0.388 | 16.5 |
| Glass | 0.896 | 9.0 |
| Gold | 0.130 | 14.2 |

*(Contd.)*

| Solid | Mean specific heat between 0° C and 100° C kJ/kg° C | Coefficient of linear expansion (0° C to 100° C (× 10⁻⁶) |
|---|---|---|
| Ice (–20° C–0° C) | 2.135 | 50.4 |
| Iron (cast) | 0.544 | 10.4 |
| Iron (wrought) | 0.465 | 12.0 |
| Lead | 0.131 | 29.0 |
| Nickel | 0.452 | 13.0 |
| Platinum | 0.134 | 8.6 |
| Silicon | 0.741 | 7.8 |
| Silver | 0.235 | 19.5 |
| Steel (mild) | 0.494 | 12.0 |
| Tin | 0.230 | 26.7 |
| Zinc | 0.389 | 16.5 |

### What to do on discovery of fire

Remain calm—think quickly to accomplish certain functions to be performed by you on seeing fire.

Size—up the situation. Quickly determine whether the fire is small enough for you to handle or if help is required. Spot the closest fire fighting equipment. Weigh possibilities of personal danger.

Extinguish the fire—if the fire has not developed beyond a stage where you can handle alone, first action should be to extinguish.

## ELECTRICITY

Electrical energy is the energy possessed by charged particles or electrons by virtue of their motion. The work done in bringing a unit positive charge from infinity to a point is called the potential difference at that point. The potential difference (V) is expressed in units of volts. Electric current (I) is the flow of positive or negative charges under the influence of a potential difference. The electric current is expressed in units of amperes(A). Charged particles can move from one point to another only if there is a potential difference between the points. The level of potential difference at any point depends on the deficiency of electrons at that point.

An electric circuit is a continuous path along which electric current flows. The property of a conductor to oppose the flow of electrons through it called the electrical resistance of the conductor. Resistance (R) is expressed in units called ohms. Electric resistance causes heating effect in a conductor. Due to the resistance offered by the conductor, a part of the electrical energy gets converted into heat energy.

The resistance of a conductor depends on:

- Nature of the material of the conductor
- Length and thickness of the conductor
- Temperature of the conductor.

## Ohm's Law

'At constant temperature, the current flowing through an electric conductor is directly proportional to the potential difference across its ends,' i.e. $V = I. R$.

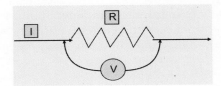

**Fig. A2.3:** Voltmeter.

### Application of Ohm's Law

A current of 4 mA (milli-amperes) flows through the resistance of 625 ohms shown above. What voltage will the voltmeter show?

From Ohm's law,

$$V = I \times R = 625 \times (4/1000) = 2.5 \text{ V}.$$

Electric power–the rate at which work is done by electrical energy is called electric power. The unit of power is watt {or horse power (HP)}.

One horse power = 746 watts.

Electricity can be easily converted to other forms of energy. Therefore, electric energy is the most convenient and widely employed form of energy.

### Protective Devices used in Electrical Circuits

### Fuses

Fuses provide short circuit and overload protection in electrical circuits. Fuses are an intentional weakened link of an electrical circuit, acting as a safety link in case of current overloads. An excessive current is interrupted by the melting of the fuse wire through which the current flows.

- Fuse wires are made of tin-lead alloy (low melting alloy).
- Fuses should have proper rating for its desired application.
- Fuses are connected in series in the live wire.
- Fuses act based on the heating effect of electricity.

Fuses are available in different physical dimensions, shapes and constructions. The time-current characteristics of fuses are specified and indicate the time required by the fuse to cut the power flow. Fuses for high voltage and high current applications are enclosed in ceramic tubes which are sand filled.

### Miniature Circuit Breaker (MCB)

A miniature circuit breaker is similar to a fuse, but re-settable. Most of the MCBs have two modes of operation—thermal and magnetic.

Thermal overload protection is used to protect cabling and equipment from long term over current damage.

Magnetic over load protection is used to protect cabling and equipment against very high fault currents caused by catastrophic component failure.

MCBs are rated by the normal carrying current, normal carrying voltage and by the maximum fault current and voltage that they can safely/repeatedly break.

### Residual Current Devices (RCD)—Earth Leakage Circuit Breaker (ELCB)

A residual current device is principally designed to offer personal protection against accidental electrical shock. Most types of RCDs work by monitoring the current normally flowing to and from a piece of equipment. Any difference (residual) from a current path to earth (e.g. caused by a person touching a live element) (or a current flowing to the body of an equipment which is in turn connected to the earth), is monitored and at a predetermined level, the device removes the power from the equipment. Double pole types of RCDs break both the phase and the neutral connections. RCD devices are rated by the time they take to achieve the break and by the nominal voltage and current they will carry. Common ratings are 30 mA and 30–40 millisecs.

### RCBO

An RCBO is a residual current circuit breaker with over current protection. It is a combination of an MCB and an RCD in one package. It offers both equipment protection and personal protection, generally used for space saving applications.

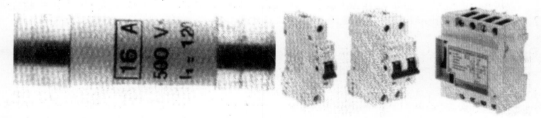

**Fig. A2.4:** HRC fuse, MCB and ELCB.

### Electric Motors

Electric motors are devices which convert electrical energy into mechanical energy. Electric motors work on the principle of electromagnetic induction which states that a current carrying conductor in a magnetic field experiences a mechanical force.

## Electric Motor Terminology

*Ambient:* The temperature of the space around the motor. Most motors are designed to operate in an ambient not over 40° C (104° F).

*Efficiency:* Efficiency is the ratio of output power divided by input power; usually expressed as a percentage. A measure of how well the electrical energy input to a motor is converted into mechanical energy at the output shaft. The higher the efficiency, the better the conversion process and lower the operating costs.

*Enclosure:* The motor's housing can be of following types;

- *Drip-proof (DP)*—Ventilation openings in end shields and shell so placed that drops of liquid falling within an angle of 15° from vertical will not affect the performance. DP motors are usually used indoors, in fairly clean locations.
- *Totally enclosed (TE)*—TE motors have no ventilation openings in the motor housing (but not airtight or waterproof). These are used in locations which are dirty, damp, oily, etc.
- *Totally enclosed, fan-cooled (TEFC)*—TEFC motors include an external fan, in a protective shroud, to blow cooling air over the motor.
- *Totally enclosed, non-ventilated (TENV)*—TENV motors are not equipped with an external cooling fan. Depends on convection air for cooling.
- *Totally enclosed, air-over (TEAO)*—In TENV motors, air flow from a driven or an external device provides the cooling air flow over the motor.
- *Hazardous location (explosion-proof)*—A totally enclosed motor designed to withstand an internal explosion of specified gases or vapors, and not allow the internal flames or explosion to escape.

*Full-load amperes:* Line current (amperage) drawn by a motor when operating at rated KW and voltage. This detail is usually shown on the motor nameplate. It is important for proper wire size selection and motor starter selection.

*Frame:* Usually refers to the system of standardized motor mounting dimensions, which facilitates replacement.

*Bearings:* Basic types: sleeve and ball.

- Sleeve bearing—preferred where low noise level is important, as on fan and blower motors. Sleeve bearing can be mounted in any position, including shaft-up or shaft-down (all-position mounting).
- Ball bearing—used where higher load capacity is required or periodic lubrication is impractical.

Two means are used to keep out dirt; shields and seals

- *Shields*—metal rings with close running clearance on one side (single-shielded) or both sides (double-shielded) of bearing.
- *Seals*—similar to shields, except have rubber lips that press against inner race, more effectively excluding dirt, etc.

*Hertz (Hz):* Frequency, in cycles per second, of AC power; usually 50 Hz, 60 Hz in USA.

*Insulation:* In motors, the insulation is usually classified by the maximum allowable operating temperature; class A—105° C (220° F), class B–130° C (266° F), class F—165° C (311° F), class H—180° C (356° F).

*Motor speeds:* Types of motors speeds–synchronous speed and full load speed.

- *Synchronous speed*—synchronous speed is the theoretical maximum speed at which an induction-type motor can operate. Synchronous speed is determined by the power line frequency and motor design (number of poles), and calculated by the formula.

$$\text{Synchronous RPM} = (\text{Frequency in Hz} \times 120)/\text{No. of poles}$$

$$\text{Speed (N)} = f \times 120/P$$

- *Full-load speed*—the nominal speed at which an induction motor operates under rated HP conditions. This will always be less than the synchronous speed and will vary depending on the rating and characteristics of the particular motor. For example, four pole, 50 Hz, fractional horse power motors have a synchronous speed of 1800 rpm, a nominal full load speed (which is usually shown on the name plate) of 1725 rpm, and an actual full load speed ranging from 1115 to 1745 rpm.

*Motor types:* Classified by operating characteristics and/or type of power required; induction, synchronous, etc.

Induction motors for AC operation—Induction motor is the most common type. Speed remains relatively constant as load changes. There are several types of induction motors. In fire service, induction motors are predominantly employed for pumps.

### Motors for Fire Pumps

Induction motors are predominantly used for driving centrifugal pumps for fire protection service. An induction motor is suitable for the constant speed requirements of centrifugal pumps used in fire protection systems. For the power needs, three phase induction motors are employed.

### Motor Starting

Three phase induction motors commonly employ starters like:

- Direct-on-line (DOL)
- Star-delta
- Autotransformer.

The choice of the starter depends on the voltage drop allowable in the supply system and the torque required to accelerate the pump load to normal running speed. It is preferable to use the star-delta starter for fire pump motors, since the initial starting current in a star-delta starter is only about 2.5 times the full load current, giving a soft start to the pump.

### Motor Size to Run a Pump

The motor rating required to operate a fire service pump is given by:

Motor HP $= (W \times Q \times H)/(75 \times \text{motor efficiency})$
Motor KW $= (W \times Q \times H)/(102 \times \text{motor efficiency})$

Where,  W = Specific weight of water (1000 kg/m$^3$)
        Q = Discharge capacity of the pump –m$^3$/sec.
        H = Pressure head of water from the pump–metres of water

The motor efficiency is the allowance for the efficiency of the motor and the power transmission–can be normally taken as 0.7 for small size pumps and 0.8 large size pumps, typically above 20 HP.

*Example:* The motor size required to drive a pump to deliver 2000 lpm (33.33 lps or 0.033 m$^3$/sec) at a pressure head of 7 bar (70 metres head)and a motor efficiency of 80% (0.8), is

Motor HP = (W × Q × H)/(75 × motor efficiency)
         = (1000 × 0.033 × 70)/(75 × 0.8)
         = (2310)/(60) = 38.5 HP

Motor KW = 38.5 × 0.735 = 28.3 KW.

**Fig. A2.5:** Extinguishing a transformer fire.

The types of hazards and equipment that can be protected using dry chemical extinguishing systems include the following:

1. Flammable or combustible liquids

2. Flammable or combustible gases

3. Combustible solids including plastics, which melt when involved in fire

4. Electrical hazards such as oil-filled transformers or circuit breakers

5. Textile operations subject to flash surface fires

6. Ordinary combustibles such as wood, paper, or cloth

7. Restaurant and commercial hoods, ducts, and associated cooking appliance hazards such as deep-fat fryers

## Electrical formulas

**Ohm's law**

Amperes = Volts/ohms

Ohms = Volts/amperes

Volts = Amperes × ohms

**Motor formulas**

$$\text{Torque kgm} = \frac{7875 \times \text{Horsepower}}{\text{rpm}}$$

$$\text{Synchronous rpm} = \frac{\text{Hertz} \times 120}{\text{Poles}}$$

**DC Circuit power formulas**

Watts = Volts × amperes

Amperes = Watts/volts

Volts = Watts/amperes

$$\text{Horsepower} = \frac{\text{Volts} \times \text{amperes} \times \text{effciency*}}{746}$$

### AC Circuit power formulas

| | **Single-phase** | **Three-phase** |
|---|---|---|
| Watts | = Volts × amps × power factor* | = 1.73 × volts × amps × power factor* |
| Amperes | = Watts volts × power factor* | = Watts/1.73 × volts × power factor* |
| | = KVA × 1000/volts | = KVA × 1000/1.73 × volts |
| | $= \dfrac{\text{Horsepower} \times 746}{\text{Volts} \times \text{efficiency*} \times \text{power factor*}}$ | $= \dfrac{\text{Horsepower} \times 746}{1.73 \times \text{volts} \times \text{effic*} \times \text{power factor*}}$ |
| Kilowatts | $= \dfrac{\text{Amps} \times \text{volts} \times \text{power facor*}}{1000}$ | $= \dfrac{1.73 \times \text{amps} \times \text{volts} \times \text{power factor*}}{1000}$ |
| KVA | $= \dfrac{\text{Amps} \times \text{volts}}{1000}$ | $= \dfrac{1.73 \times \text{amps} \times \text{volts}}{1000}$ |
| Horsepower | $= \dfrac{\text{Volts} \times \text{amps} \times \text{effic*} \times \text{pwr. factor*}}{746}$ | $= \dfrac{1.73 \times \text{volts} \times \text{amps} \times \text{effic.*} \times \text{pwr. fact.*}}{746}$ |

* Expressed as a decimal

**Fig. A2.6:** An electric motor for fire pump service.

| Electrical wire-dimensions and ratings (all data for solid copper wire) | | | | |
|---|---|---|---|---|
| AWG wire gauge | Diameter inches | Diameter mm | Resistance Ohms per km | Max allowable current capacity amps |
| 0 | 0.3249 | 8.252 | 0.3224 | 125–170 |
| 1 | 0.2893 | 7.348 | 0.4066 | 110–150 |
| 2 | 0.2576 | 6.544 | 0.5127 | 95–130 |
| 3 | 0.2294 | 5.827 | 0.6465 | 85–110 |
| 4 | 0.2043 | 5.189 | 0.8152 | 70–95 |
| 6 | 0.1620 | 4.115 | 1.296 | 55–75 |
| 8 | 0.1285 | 3.264 | 2.061 | 40–55 |
| 10 | 0.1019 | 2.588 | 3.277 | 30–40 |
| 12 | 0.0808 | 2.053 | 5.211 | 20–30 |
| 14 | 0.0641 | 1.628 | 8.286 | 15–25 |
| 16 | 0.0508 | 1.291 | 13.17 | 18 |
| 18 | 0.0403 | 1.024 | 20.95 | 14 |

**Fig. A2.7:** Cartridge fuses (BUSS make) (HRC fuse).

## Conductor Resistivity

Conductor resistivity, $R = \rho L / a$

where $\rho$ = specific resistance (or resistivity) (ohm metres, $\Omega{\cdot}m$)

$L$ = length (metres)

$a$ = area of cross-section (square metres)

Temperature correction

$$Rt = Ro (1 + \alpha t)$$

where,  Ro = resistance at 0° C (Ω)
         Rt = resistance at t° C (Ω)
         α = temperature coefficient which has an average value for copper of 0.00428 (Ω/Ω° C)

$$R2 = R1(1 + \alpha t1)/(1 + \alpha t2)$$

where,  R1 = resistance at t1 (Ω)
         R2 = resistance at t2 (Ω)

α values Ω/Ω °C

| | |
|---|---|
| Copper | 0.00428 |
| Platinum | 0.00385 |
| Nickel | 0.00672 |
| Tungsten | 0.0045 |
| Aluminum | 0.0040 |

| **Electric motors–full load current, amps** | | | |
|---|---|---|---|
| | | *3 phase AC induction motor* | |
| *HP* | *KW* | *220 V* | *440 V* |
| 1 | 0.745 | 3.6 | 1.8 |
| 2 | 1.5 | 6.8 | 3.4 |
| 3 | 2.235 | 9.6 | 4.8 |
| 5 | 3.725 | 15.2 | 7.6 |
| 7.5 | 5.587 | 22 | 11 |
| 10 | 7.45 | 28 | 14 |
| 15 | 11.175 | 42 | 21 |
| 20 | 14.90 | 54 | 27 |
| 25 | 18.625 | 68 | 34 |
| 30 | 22.35 | 80 | 40 |
| 50 | 37.25 | 130 | 65 |
| 100 | 74.50 | 24 | 124 |

The power source for an electric motor-driven fire pump must be reliable and have adequate capacity to carry the locked-rotor currents of the fire pump motor and accessory equipment. These two main requirements ensure that the fire pump will operate in the event of a fire without being accidentally disconnected and that the fire pump will continue to operate until the fire is extinguished, the fire pump is purposely shut down, or the pump itself is destroyed.

## Earth leakage protection recommended value of operating current of ELCB in consumer installation

| Circuit/equipment/apparatus | Rated operating current (mA) |
| --- | --- |
| 5A switched socket outlet | 30 |
| Water heaters/coolers | 30 |
| Refrigerator/washing machine | 30 |
| Domestic water pumps | 30 |
| 15 A switched socket outlets | 30 |
| General lighting | 30/100 |
| Flood lighting | 100/130 |
| Air-conditioner | 100 |
| Air handling units | 100 |
| Chiller | 100 |
| Electric cooker | 100 |
| Elevators/escalators | 300/500 |

## Classification of voltages (As per Indian Electricity Rules 2(1) (a))

| | | |
| --- | --- | --- |
| Low voltage | LV | Up to 250 V |
| Medium voltage | MV | Up to 650 V |
| High voltage | HV | 650 V to 33 KV |
| Extra-high voltage | EHV | Above 33 KV |
| Ultra-high voltage | UHV | Above 735 KV |

## Earthing current density
### (Bare conductor with no risk of fire for a 3 sec rating)

| | |
| --- | --- |
| Copper | 118 amps/mm$^2$ |
| Aluminium | 73 amps/mm$^2$ |
| Steel | 46 amps/mm$^2$ |

### Fuse ratings fuse for motors

| Motor rating HP | Back up fuse | | Earthing conductor | | |
|---|---|---|---|---|---|
| | DOL starting amps | Assisted starting amps | Aluminium cable size mm² | Copper SWG | Aluminium mm² cable |
| Upto 5 | 25 | 25 | 4 | 10 | 25 |
| 6 to 10 | 35 | 25 | 6 | 10 | 25 |
| 11 to 17.5 | 50 | 35 | 10 | 10 | 25 |
| 18 to 20 | 63 | 53 | 16 | 10 | 25 |
| 21 to 25 | 80 | 63 | 25 | 10 | 25 |
| 26 to 30 | 100 | 65 | 26 | 6 | 25 |
| 31 to 40 | 120 | 100 | 35 | 6 | 25 |
| 41 to 50 | 160 | 100 | 50 | 6 | 25 |
| 51 to 60 | 200 | 120 | 70 | 4 | 25 |
| 61 to 75 | 200 | 120 | 95 | 4 | 25 |
| 75 to 100 | 250 | 200 | 185 | 2 | 25 |

### Domestic electric supply

| Load in watts | Connecting wires SWG | Fuse rating amps | Fusing current amps |
|---|---|---|---|
| 500 | 39 | 2.5 | 4 |
| 1000 | 35 | 5.0 | 8 |
| 1500 | 32 | 7.0 | 11 |
| 2000 | 29 | 10.0 | 16 |
| 2500 | 28 | 12.0 | 1 |
| 3000 | 27 | 13.0 | 23 |
| 4000 | 23 | 20.0 | 38 |

### Power factor of some common loads

| Load | Power factor |
|---|---|
| Incandescent lamps | 1.00 |
| Fluorescent lamps | 0.6 to 0.8 |
| Induction motors | 0.8 |
| Neon signs | 0.4 to 0.5 |
| Arc lamps | 0.3 to 0.7 |
| Arc furnaces | 0.85 |
| Arc welding | 0.3 to 0.4 |
| Resistance welding | 0.65 |
| Induction furnace | 0.60 |
| Induction heating | 0.85 |

| Energy consumed by some common domestic appliances | Wattage rating | Time for one unit of consumption |
| --- | --- | --- |
| Incandescent bulb | 25 w | 40 hrs/unit |
| | 60 w | 25 hours/unit |
| | 100 w | 10 hours/unit |
| Fluorescent tube-light 2ft | 20 w | 50 hours/unit |
| Fluorescent tube 4ft | 40 w | 25 hours/unit |
| Night lamp | 15 w | 66 hours, 40 mins/unit |
| Mosquito repellent | 5 w | 200 hours/unit |
| Fan | 60 w | 16 hours, 40 mins/unit |
| Air cooler | 170 w | 5 hours, 50 mins/unit |
| Air conditioner 1.5 ton | 1500–2500 w | 40 to 30 mins/unit |
| Refrigerator 165 lits | 100 w | 10 hours/unit |
| Mixer | 450 w | 2 hours, 15 mins/unit |
| Toaster | 800 w | 1hour, 15 mins/unit |
| Hot plate | 1000–1500 w | 1 hour, 40 mins/unit |
| Oven | 1000 w | 1 hour/unit |
| Electric kettle | 1000–2000 w | 1 hour to 30 mins/unit |
| Iron | 450–700 w | 2 hours, 15 mins/unit |
| Water heater instant | 3000 w | 20 mins/unit |
| Water heater 10–20 lit | 2000 w | 30 mins/unit |
| Immersion heater | 1000 w | 1 hour/unit |
| Vacuum cleaner | 700–750 w | 1 hour, 20 mins/unit |
| Washing machine | 325 w | 3 hours/unit |
| Water pump | 750 w | 1 hour, 20 mins/unit |
| Television | 60–120 w | 16 to 8 hours/unit |
| Video | 20 w | 25 hours/unit |
| Computer | 120–500 w | 10 to 20 hours/unit |

| Resistance of human body parts | |
| --- | --- |
| Body part | Resistance, Ohm |
| Dry skin | 10 to 50 mega ohm |
| Wet skin | 1000 ohm |
| Hand-to-foot internal, excluding skin | 500 to 600 ohm |
| Ear to ear | 100 ohm |
| Internal, excluding skin | 100 ohm |

| Effect of electricity on human body | | | | | | |
|---|---|---|---|---|---|---|
| *Effect* | *Flow of current through part of the human body (milli-amperes)* | | | | | |
| | *DC* | *50 Hz, AC* | | *10 kHz* | | |
| | *Men/women* | *Men* | *Women* | *Men* | *Women* | |
| 1. Slight sensation on hand | 1 | 0.5 | 0.5 | 0.4 | 7 | 5 |
| 2. Perception threshold | 5 | 3 | 1.4 | 0.7 | 12 | 8 |
| 3. Shock not painful, muscular control not lost | 9 | 6 | 1.8 | 1.5 | 15 | 11 |
| 4. Shock-painful, muscular control not lost | 65 | 41 | 9 | 6 | 50 | 40 |
| 5. Shock-painful, let-go threshold | 75 | 50 | 15 | 10 | 70 | 50 |
| 6. Shock-painful, and severe muscular contractions, death certain | 90 3 s | 60 3 s | 23 3 s | 17 35 | 90 | 60 |
| 7. Shock-ventricular fibrillation, breathing stops, death certain | 500 (10s) | 500 (10s) | 100 (10s) | 100 (10s) | | |
| 8. Unsafe and dangerous high voltage surges (50 to 2500 micro-sec) | Current 5 mA | Energy 25 mJ | | | | |

| Electromagnetic radiation | | |
|---|---|---|
| *Radiation* | *Wavelength band- metres* | *Frequency band, hertz* |
| Radiowaves | Less than 0.1 m | More than $10^9$ |
| Microwaves | 0.4 mm to 10 cm | $10^{10}$ to $10^9$ |
| Infrared | 750 nm to 0.4 mm | $10^{14}$ to $10^{10}$ |
| Visible (light) | 400 to 750 nm | $10^{14}$ |
| Ultraviolet | 4 to 400 nm | $10^{16}$ to $10^{14}$ |
| X-rays | 0.1 Å to 100 Å | $10^{19}$ to $10^{16}$ |
| Gamma-rays | 0.01 Å to 0.1 Å | $10^{20}$ to $10^{19}$ |

Å–angstrom = $10^{-8}$

Dry chemical extinguishing systems shall not be considered satisfactory protection for the following:

1. Chemicals containing their own oxygen supply, such as cellulose nitrate.
2. Combustible metals such as sodium, potassium, magnesium, titanium, and zirconium.
3. Deep-seated or burrowing fires in ordinary combustibles where the dry chemical cannot reach the point of combustion.

Multipurpose dry chemical shall not be used on machinery such as carding equipment in textile operations and delicate electrical equipment.

Before dry chemical extinguishing equipment is considered for use in protecting electronic equipment or delicate electrical relays, the effect of residual deposits of dry chemical on the performance on electronic equipment shall be evaluated.

The following factors shall be considered in determining the amount of dry chemical required:

1. Minimum quantity of dry chemical
2. Minimum flow rate of dry chemical
3. Nozzle placement limitations including spacing, distribution, and obstructions
4. High ventilation rates, if applicable
5. Prevailing wind conditions, if applicable

## HYDRAULICS

Hydraulics is the science and engineering related to water–its characteristics, storage, flow and utilization.

### Water Flow

Water for utilization is generally conveyed through pipes and channels. Water flow in pipes is pressurized, whereas in channels, the water flow is non-pressurized and hence flows due to gravity.

### Types of Flow

*Uniform flow:* In an uniform flow, the velocity is equal across the flow cross-section as well as along the length of the flow. Uniform flow is generally seen in flows occurring in channels and in non-pressurized flow in pipes (unfilled flow in pipes).

*Non-uniform flow:* Non-uniform flow occurs in high pressure flows from pumps where the discharge pressure may not be constant and also phenomenon such as cavitation cause non-uniform flow.

*Stream line flow (laminar flow):* The fluid particles in laminar flow conform to definite path in a line.

*Turbulent flow:* The fluid particles in turbulent flow do not follow a definite path.

### Energy in a Fluid

*Potential energy:* Potential energy is the energy possessed by an object or body (water for example) by virtue of its position or height with respect to the ground level or a chosen reference level.

Potential energy (for a unit mass) is given by

$$P.E = g \times h,$$

Where, g is the acceleration due to gravity (9.8 m/sec$^2$)
And $h$ = height of the mass above the reference level (potential head)

*Kinetic energy:* Kinetic energy is the energy possessed by an object or body (water for example) by virtue of its velocity or motion.

Kinetic energy (for a unit mass) is given by

$$K.E = v^2/2\,g$$

Where      $v$ is the velocity of the body

$g$ is the acceleration due to gravity (9.8 m/sec²)

*Pressure energy:* Pressure energy is the energy possessed by an object or body (water) by virtue of its existing pressure.

Pressure energy (for a unit mass) is given by

$$Pr.\ Energy = p/w$$

Where      $p$ is the pressure of the fluid (kg/sq m)

and $w$ is the specific weight of the liquid (1000 kg/m³ for water)

## Total Energy in a Fluid

The total energy in a fluid is the sum total of all the above three energies, viz. potential energy, kinetic energy and pressure energy. Total energy is also termed as the total head.

$$Total\ energy = P.E + K.E + Pr.E$$

Therefore, total energy $= g\,h + v^2/2\,g + p/w$ (kg–m/kg.)

## Bernoulli's Principle

Bernoulli's principle states that the total energy or head in fluid in flow is constant at all points along the flow.

So,      Bernoulli's equation — $g\,h + v^2/2g + p/\ w.g = constant$

If loss of head (pressure) between two points A and B is $h_f$, then,

$$(P_A/w\ g) + (V_A{}^2/\ 2g) + Z_A = (P_B/w\ g) + (V_B{}^2/2g) + Z_B + h_f$$

## Application of Bernoulli's Principle

One of the main applications of the Bernoulli's principle is the venturimeter, which is used to measure the flow of fluids in a pipe-line. It is a very simple device with no moving parts and can be fitted in-line. The pressure drop at the venturi throat can be calibrated in terms of the flow.

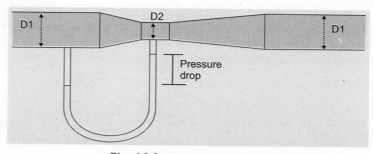

**Fig. A2.8:** The venturimeter.

## Hydrostatic Water Pressure

Water is a liquid that at any point exerts equal intensity of pressure in all directions. The intensity of pressure on a submerged surface is expressed in kgf/cm$^2$ or in metres of water above the atmospheric pressure at the point where the pressure is being read.

$$H = 10\ p \text{ or } p = H/10$$

where

$H$ = depth of submergence in metres, and
$P$ = intensity of pressure in bar.

The value of $p$ is also the gauge pressure since it is the pressure that would be shown on a pressure gauge. Absolute pressure is equal to the sum of atmospheric pressure and gauge pressure.

The total head at any point in a conduit flowing full

$$= \text{pressure energy} + \text{kinetic energy} + \text{potential energy}$$
$$= P/wg + V^2/2g + z$$

where

$P$ = pressure of the moving liquid (bar),
$w$ = density of the liquid (kg/m$^3$),
$g$ = acceleration due to gravity (m/s$^2$),
$V$ = velocity of flow (m/s), and
$z$ = height of the point above the datum line (m).

## Pressure of Atmosphere

The pressure of the atmosphere at the earth's surface is due to the weight of the column of air above. As the air is compressible and as a consequence the density varies, the atmospheric pressure is measured by the column of liquid it will support. This again varies with the amount of moisture in air and temperature. The average value is taken as 10332 kgf/m$^2$ or 1.0332 kgf/cm$^2$ (~ 1 bar) or 10.332 metres head of water.

## Gauge Pressure

The pressure of water in a pipeline or a vessel is measured by some type of gauge. The gauge registers the pressure above or below atmospheric pressure. To get the absolute pressure, the gauge pressure must be added or subtracted from the atmospheric pressure as the case may be. For pressures below the atmospheric pressure, the gauge pressures are observed on a vacuum gauge. If the pressures are above the atmospheric pressure, the gauge pressures are measured on a pressure gauge.

## Siphonage

Siphonage, aspiration, suction, negative pressure, partial vacuum or vacuum and other terms are used synonymously to indicate a pressure below atmospheric or below gauge pressure. The flow through a siphon is due to the difference in elevation of the free surfaces above and below the siphon.

The greatest height over which water could be apparently lifted by suction is equal to the atmospheric pressure which is 10.36 metres at sea level. This height is further reduced due to vapour pressure of water which varies with temperature and is equal to atmospheric pressure at boiling point of water (100° C). The atmospheric pressure also gets reduced with height above sea level. Table below gives the atmospheric pressure y in metres of water for different elevations × above sea level in metres.

The values can also be obtained from the equation

$$y = 10.366 - 0.00110857x.$$

| Atmospheric pressure for different elevations | | | | | | | | |
|---|---|---|---|---|---|---|---|---|
| X(m) | 0 | 500 | 1000 | 1500 | 2000 | 2500 | 3000 | 35000 | 4000 |
| Y(m) | 10.37 | 9.82 | 9.28 | 8.74 | 8.19 | 7.65 | 7.11 | 6.57 | 6.02 |

Vapour pressure is the pressure exerted by the tendency of a liquid to vapourize. This tendency varies with the temperature of the liquid as shown in Table below.

| Vapour pressure for different temperatures of water | | | | | | | | | | |
|---|---|---|---|---|---|---|---|---|---|---|
| Temp ° C | 0 | 5 | 10 | 15 | 20 | 25 | 30 | 35 | 40 | 45 | 50 |
| Vapour pr (mm Hg) | 4.57 | 6.54 | 9.2 | 12.8 | 17.5 | 23.7 | 31.8 | 42.2 | 55.3 | 71.8 | 92.51 |
| Temp ° C | 55 | 60 | 65 | 70 | 75 | 80 | 85 | 90 | 95 | 100 | |
| Vapour pr (mm Hg) | 118 | 149.4 | 187.6 | 234 | 289 | 355 | 434 | 526 | 634 | 760 | |

## Air and Gas Locks

A bend or hump extending upward above the regular line of the run of a pipe as shown in figure below or extending above the hydraulic grade line as in a siphon, may permit the accumulation of air or gas in the bend. It is also called air binding. The effect may be either to reduce or to cut-off flow in the pipe or to require pressure to force the trapped air through the pipe. The trapped air will diminish or stop flow through the pipe by reducing the cross-sectional area available for flow. It will act as a stoppage that no amount of prodding will remove and that is not to be found when the pipe is opened for examination. Air locks are likely to give trouble in pipes under low pressure and in siphons because of inadequate force to push the air along the pipe. The formation of an air lock can be prevented by avoiding' the creation of upward humps in a pipeline or by the installation of an air release valve at the highest point where air or gas is likely to accumulate.

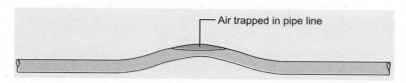

Air trapped in pipe line

Fig. A2.9

## Cavitation

Water vapourizes completely or boils at 100° C at atmospheric pressure. Water will boil or vapourize at a lower temperature if the pressure is reduced. This phenomenon may occur in plumbing pipes, equipment and pumps. It is called cavitation. It may be defined as a rupture of the continuity of a liquid as it turns to vapour owing to a sudden reduction of pressure. Low pressures are produced in conduits by a sudden increase of velocity. They are produced in equipment, e.g. in pumps when a moving object such as an impeller passes rapidly through the water. In other words, the pressure reduces as the velocity head increases in order that their sum may remain constant. The rapidity with which a high vacuum is made and broken, and water changing from a liquid to a vapour and back to liquid again may be so great as to create sounds varying from a rattle to a loud roar. A corrosive effect may appear on the surfaces of metal exposed to cavitation. The phenomenon is avoided by maintaining low velocities between liquids and surfaces in contact, and by avoiding sudden accelerations in velocities of flow in closed conduits.

## Water Hammer

If the velocity of water flowing in a pipe is suddenly diminished, the energy given up by the water will be divided between compressing the water itself, stretching the pipe walls and frictional resistance to wave propagation. This pressure rise or water hammer is manifest as a series of shocks sounding like hammer blows which may have sufficient magnitude to rupture the pipe or damage connected equipment. It may be caused by the nearly instantaneous or too rapid closing of a valve in the line or by an equivalent stoppage of flow such as would take place with the sudden failure of electricity supply to a motor driven pump. The shock pressure is not concentrated at the valve and, if rupture occurs, it may take place near the valve simply because it acts there first. The pressure wave due to water hammer travels back upstream to the inlet end of the pipe where it reverses and surges back and forth through the pipe getting weak on each successive reversal. The excess pressure due to water hammer is additive to the normal pressure in the pipe. Complete stoppage of flow is not necessary to produce water hammer as any sudden change in velocity will create it to a greater or lesser degree. The intensity of water hammer pressure varies directly with the velocity of flow in the pipe. If the velocity is kept below, the effect of water hammer could also be reduced.

Water hammer is held within bounds in small pipelines by operating them at moderate velocities because the pressure rise in bar (kg/cm$^2$) cannot exceed about 11.5 times the velocity expressed in metres per second. In larger lines, the pressure is held down by changing velocities at a sufficiently slow rate so that the relief wave returns to the point of control before excessive pressures are reached. If this is not practicable, pressure-relief or surge valves are used.

### Prevention of Water Hammer

The common causes in plumbing are the sudden closing of valves or taps, particularly of the automatic self-closing type and the quick-closing types. Water hammer may be caused also by displacing air from a closed tank or pipe from the top by the condensation of steam in water in a closed pipe, by reciprocating pumping machinery by the sudden stoppage of a pump and by other means. Water hammer can be prevented when a closed tank or pipe is being filled, by filling it from the bottom, allowing the air to escape from the top. Steam and water should not be allowed to come into contact in a closed pipe. To this end, downward dips in steam pipes should be avoided or suitable provision for drainage should be provided in case dips are unavoidable. The installation of an air chamber may control water hammer. Other methods of avoiding water hammer include the use of slow closing valves and taps such as the screw down types and of pressure reducing valves. Other types of control are by installing air chambers near the valve that is causing water hammer and is possible in a vertical position over the top of a riser pipe, which is used in a fire protection piping. The air chamber should have a capacity of at least 1% of the total capacity of the pipeline in which the water hammer is occurring. The purpose of placing the air chamber on the top of the riser pipe is twofold:

a. Air in the riser pipe will be compressed making way for excess water under pressure (trapped in the chamber, aiding in keeping air in the chamber); and

b. If it will receive the full thrust of the pressure from the vertical pipeline, it will be more effective in its operation. Provision should be made for renewing the air in the chamber. This can be done by the use of a stop-and-waste valve and a pet cock. If the water hammer is created in the water main in the street, the house plumbing may be protected by locating an air chamber on the service pipe as it enters the building. The installation of a pressure reducing valve on the supply line to the source of the water hammer will result in a reduction of the velocity of flow in the pipe.

## Flow of Water Under Pressure (Water Pipes)

The formula popularly used for the flow of water in conduits under pressure is the Hazen and William's formula given below:

(This is an alternate formula from the formula given earlier).

$$V = 0.849 \, C \, R^{0.63} \times S^{0.54}$$

where,

$V$ = velocity in metres per second,
$R$ = hydraulic radius in metres,
$S$ = slope of hydraulic gradient (metre per metre), and
$C$ = Hazen and William's coefficient.

For circular pipes, where $R = D/4$, the above formula becomes

$$V = 4.567 \times 10^{-3} \times C \times D^{0.63} \times S^{0.54}$$

Where, $D$ is the diameter of the pipe in mm.
The formula for $Q$, the discharge in kilolitres per day (kld) becomes

$$Q = 3.1 \times 10^{-4} \times C \times D^{0.63} \times S^{0.54}$$

The value of $C$ decreases with increasing surface roughness and the age of the pipe. The recommended values for new pipes and the values to be adopted for design purposes are given in following table.

| Conduit material | Recommended value of C | |
| --- | --- | --- |
| | For new pipe | For design purposes |
| Cast iron | 130 | 100 |
| Galvanized iron > 50 mm | 120 | 100 |
| Galvanized iron 50 mm and below | 120 | 55 |
| (used for house connections) | | |
| Steel, welded joints | 110 | 95 |
| Steel, welded joints, lined with | 140 | 110 |
| cement or bituminous enamel | | |
| Steel, welded joints | 140 | 100 |
| Concrete | 140 | 110 |
| Asbestos cement | 150 | 120 |
| Plastic pipes | 150 | 120 |

Hazen and William's formula is applicable to flow of water under pressure and at velocities normally used for fire protection piping.

## Flow of Water under Gravity (Drainage Pipes)

For free flow of water in conducts under gravity, the Manning's formula given below is recommended:

$$V = (1/n) \times R^{2/3} \times S^{1/2}$$

where,

$V$ = velocity in metres per second,
$R$ = hydraulic radius in metres,
$S$ = slope of hydraulic gradient (metre per metre), and
$n$ = Manning's coefficient.

For circular pipes where $R = D/4$, the formula for V becomes

$$V = (3.968/n) \times 10^{-3} \times D^{2/3} \times S^{1/2}$$

where D is the diameter of the pipe in mm.
The formula for Q, the discharge in litres per second, becomes

$$Q = (3.118/n) \times 10^{-6} \times D^{8/3} \times S^{1/2}$$

The value of $n$ varies directly with the roughness of the surface. The coefficients of roughness for different surface linings in clean straight channels, varies with the material. A value of n = 0.013 may be adopted for plastic pipes and 0.015 for all other pipes.

## Measuring the Level by Measuring the Pressure of a Liquid in Storage

This can be done by using the principle of Pascal, from which

$$P = h \times p \times g,$$

where     $P$ – pressure in Pascal's;
$h$ – height of the liquid in the storage tank;
$p$ – specific weight of the liquid in storage in $kg/m^3$;
$g$ – 9.8 $m/sec^2$.

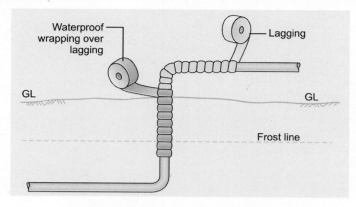

**Fig. A2.10:** Fire piping insulation.

For example, if the liquid density is 0.8 and the pressure in the pressure gauge is 160 m bar (16 kPa),

P = 16000, p = 800, h = 16000/(800 × 9.8) = 2.04 m.

### Relation between Volume Flow and Mass Flow

Volume flow: Volume/time = $m^3$/sec.
Mass flow: Mass/time = kg/sec.

To determine the volume from the weight, we utilize the specific weight of the fluid. Care has to be taken to use the same volume and the specific weight at the same temperature.

### *Example*

What is the mass flow of air from a compressor delivering a volume flow of 200 $m^3$/hour at 7 bar pressure and 20° C? P of air normal,(1.013 bar, 0° C) = 1.3 kg/ $m^3$.

$$P \text{ of air} = (7 + 1) \text{ bar, at } (273 + 20) °K = 8 \text{ bar at } 293 °K$$
$$\text{Specific weight of air} = 1.3 × (273/1.013) × (8/293) = 9.69 \text{ kg/m}^3$$
$$\text{Mass flow} = 9.69 \text{ kg/m}^3 × 200 \text{ m}^3/\text{hr.} = 1938 \text{ kg/hr} = 0.54 \text{ kg/s.}$$

### DESIGN OF PIPES

### Calculation of Pipe Diameter

Example–what should be the diameter of a pipe to carry water at the flow rate of 10 $m^3$/hour and a velocity of 1 m/s?

We know that flow rate Q = Cross-section area of pipe (A) × velocity of water in the pipe (V).

(See flow through pipes)

Q = A . V or A = Q/V
Q = 10 $m^3$/hr = 10/3600 m/s

V = 1 m/s.

A = (10/3600)/1 = 0.00277 m$^2$

$= 27.7/10000$ m$^2$ = 27.7 cm$^2$

A = Π D$^2$/4, therefore D (pipe diameter) = 6 cm.

## Pipe Design for Strength

Fire water supply pipelines are designed considering the maximum working pressure, corrosion effect in case of underground pipelines, surface protection provided and the quality of the pipe material and the installation.

| Material type | Pipe diameter/thickness | | |
|---|---|---|---|
| | Up to 150 mm | 180 mm | 250 mm and above |
| Welded | Schedule 10 | 0.34 mm (0.134 in) | 0.48 mm (0.188 in) |
| Roll grooved | Schedule 10 | 0.34 mm (0.134 in) | 0.48 mm (0.188 in) |
| Threaded | Schedule 30 | Schedule 30 | Schedule 40 |
| Cast grooved | Schedule 30 | Schedule 30 | Schedule 40 |

## Flow through Pipes

A pipe is a closed conduit generally of circular cross-section. When running full, the flow is under pressure. Flow under pressure only, is considered in flow definitions.

## Flow Equation

The flow through a pipe is given by

Q = A. V.

Where, Q is the flow in cubic meters per sec.

A is the inner cross-sectional area of the pipe, which is given by Π d$^2$/4, d being the inner diameter of the pipe.

V is the velocity head of the fluid flowing in the pipe and is given by

V = √ (2 × g × h),

g being the acceleration due to gravity = 9.8 m/sec$^2$ and

h being the pressure of the fluid in the pipe in metres of water.

## Loss of Head in Pipes

Any fluid flowing through a pipe suffers loss of head due to the frictional resistance to flow generated by the inner walls of the pipe and also due to losses suffered at various locations such as bends, elbows, valves, instruments. The fluid also loses energy at entry and exits.

## Loss due to Frictional Resistance

The frictional loss varies with the square of the velocity of the fluid, i.e. as the pressure of the flow increases, the head loss increases exponentially. The frictional resistance generated by the inner walls of the pipe varies with the nature of the inner surface of the pipe. A circular cross-section (pipe) offers the least frictional resistance.

## Darcy's Formula for Head Loss in Pipes

The head loss is given by

Head loss (in metres) = 4 f l $V^2$/2 g d

Where
- f– is the frictional coefficient (0.005 to 0.01)
- l– is the length of the pipe in metres
- v– is the fluid velocity in meters/second
- g– is the acceleration due to gravity—9.8 m/sec$^2$
- d– is the inner diameter of the pipe in meters.

## Entry and Exit Losses

The loss at entry to the pipe is given by

Head loss at entry (in meters) = 0.5 $V^2$/2 g

Where        V–is the fluid velocity in m/s

The loss at the exit is given by

Head loss at exit (in meters) = $V^2$/2 g

Where        V–is the fluid velocity in m/s

Therefore, total loss is given by loss in pipe + loss at entry + loss at exit.

Where l > 1000 d, i.e. in long pipes, where the length of the pipe is sufficiently long compared to the diameter of the pipe, the entry and exit losses can be neglected.

*Example:* Find the head loss due to friction in a pipe of 1 m diameter and 15 km long. Velocity of water in the pipe is 1 m/sec. Take frictional coefficient as 0.005.

D= 1 m, l= 15000 m. v = 1 m/sec, f = 0.005
Head loss is given by = 4flv$^2$/2 g d

= {4 . (0.005) . 15000 . (1)$^2$}/(2 . 9.81 . 1) = 15.29 m

If the initial head is 7 bar, what will be the pressure at the exit of 15 km?

Initial pressure = 7 bar = 70 m head
Head loss in 15 km = 15.29 m
Therefore, pressure at the exit of 15 km = 70–15.29
= 54.71 m.

What will be flow or discharge at the exit?

Discharge Q = A . V

V = $\sqrt{2 . 9.81 . 54.71}$ = 32.76 m/sec
Therefore Q = 3.14 . (1)$^2$/4 . 32.76= 6.425 m$^3$/sec
= 6425 litres/sec.

Dry chemical, when discharged, will drift from the immediate discharge area and settle on surrounding surfaces. Prompt clean up will minimize possible staining or corrosion of certain materials that can take place in the presence of moisture. Moisture can corrode metals such as steel, cast iron, and aluminum. Potassium bicarbonate, sodium bicarbonate, and urea-based potassium bicarbonate are slightly basic and, in the presence of moisture, can corrode metals such as aluminum, aluminium brass, aluminium bronze, and titanium. Such corrosion will vary from a dull or tarnished finish to mild surface corrosion. Corrosion should not be of concern when accompanied by prompt clean up. For the most part, these dry chemical agents can be readily cleaned up by wiping, vacuuming, or washing the exposed materials. A mono-ammonium phosphate-based agent will need some scraping and washing, if the exposed surfaces were hot when the agent was applied. Upon exposure to temperatures in excess of 121° C or relative humidity in excess of 50%, deposits will be formed that can be corrosive, conductive, and difficult to remove.

For flow through channels, the head loss is computed using Chezy's formula.
Time to empty a tank through a long pipe.

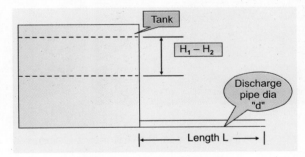

**Fig. A2.11**

If $H_1$ is the initial water level in the tank.
$H_2$ is the final level of water in the tank after a time ' t' seconds.
A is the cross-sectional area of the tank,
The volume of water discharged = A $(H_1–H_2)$
If l is the length of the pipe and d = diameter of the pipe
t = time for the discharge = 8 A $(\sqrt{(1 + (4 f l/d))}/\Pi d^2 (\sqrt{2 g})$

## PIPE AND HOSE FITTINGS

Water is the most widely used extinguishing medium in majority of fires. Since water is very cheap and easily available, fire services all over the world use hoses extensively. Water for fire fighting must be taken from a source, conveyed on to the scene of fire and delivered in the condition required. The mode of application depends upon the type of fire such as jet, spray, fog, etc. We may not get water on the scene of fire where it is required. Fire fighting pump draws water from the available source, imparts pressure and delivers it to the scene of fire. A fire hose is a high-pressure hose used to carry water or other fire retardant (such as foam) to a fire to extinguish it. Outdoors, it is attached either to a fire engine or a fire hydrant. Indoors, it can be permanently attached to a building's standpipe or plumbing system.

Suction hose draws water up to the pump and delivery hoses help to carry it to the required place. The fittings used to join or connect hoses and control the water flow are known as ' hose fittings'. The different types of hose fittings, which are commonly used in fire services are:

1. Couplings
2. Branches and nozzles
3. Collecting heads and suction hose fittings
4. Breechings
5. Adaptors
6. Miscellaneous hose fittings

**(a)**                    **(b)**

**Fig. A2.12:** (a) Fire hose; (b) Wall mounted hose reel.

## Couplings

Fittings used for joining two lengths of hose together or to an apparatus are called couplings. Couplings are fitted to either end of hose lengths used for suction or delivery purpose. The following couplings are normally used.

1. *Interlocking couplings*

    i. Instantaneous male
    ii. Instantaneous female coupling
    iii. Sure-lock coupling

2. *Hose reel couplings*

    i. Hermaphrodite coupling
    ii. MacDonald coupling

3. *Screw-type coupling*

    i. Round-threaded
    ii. 'V' threaded

All couplings are always a pair, one of which is named male and the other female. They differ in shape and function, except the hermaphrodite coupling, where each of the pair is similar in shape and function. Instantaneous couplings are used for delivery hoses and pump outlets, whereas screw type couplings are used on suction hose pump inlets.

### Interlocking Couplings

*Instantaneous type coupling:* Most widely used interlocking coupling is the instantaneous type. These couplings can be snapped on and off within an instant and therefore they are known as instantaneous couplings. They are provided with spring-loaded lugs and also with swiveling action for removing out of coils of hose. To make this coupling to be disconnected under pressure of water, pressure release type couplings have been developed. Pump and stand-pipe are fitted with a single cap twist-release type of coupling to permit instant disconnection without the necessity of reducing pressure. In these couplings, the plunger is raised by a cam action when the cap is twisted.

*Components of male coupling:*
  i. Serrated tail for fixing it into the hose.
 ii. Lip, which engages with the plunger in the female coupling.

*Components of female coupling:*
   i. Serrated tail for fixing it into the hose
  ii. Two hollow lugs
 iii. Spring-loaded plunger with wedged shape tooth.
  iv. Washer
   v. Metal disc
  vi. Cap
 vii. Self-locking nut

These may be of pull release type as on hose or twist-release type as on pump deliveries and hydrant stand-pipes.

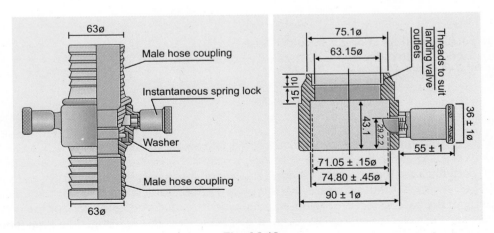

**Fig. A2.13**

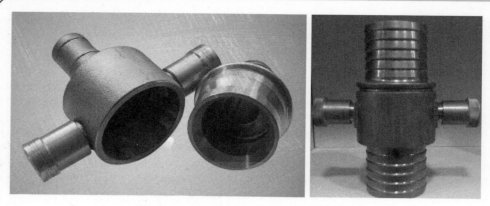

Fig. A2.14: Sure lock or bayonet type coupling.

*Components of male coupling:*

  i. Serrated tail for fixing it in to the hose
 ii. Wide flanges around which are three regularly spaced semi-circular indentations
iii. Three lugs lying centrally with indentations on the flanges.

*Components of female coupling:*

   i. Serrated tail for fixing it into the hose
  ii. A heavy collar slightly stepped back in two steps towards the tail.
 iii. Squared boss protruding from the collar
  iv. Spring-loaded pin running parallel with the coupling
   v. Holding nut with its head towards the shank end.
  vi. Three equidistant bosses inside orifice of the coupling
 vii. Rubber washer

## Hermaphrodite Couplings

*Components of male and female couplings*

  i. Serrated tail for fixing it in to the hose.
 ii. Two diametrically opposite lugs on a collar.
iii. Two tapering flanges
 iv. Two squared indentations

## BRANCHES AND NOZZLES

Branch pipe or branch as it is generally called is a tapering metal fitting, which is used at the delivery end of the line of hose. It is used for slowly narrowing down the area of the water-path to cause the velocity of water to increase. It consists of a brass or copper pipe, the length of which differs according to its type of manufacture. It has a male instantaneous coupling at the base to receive the hose and a male threaded end at the top for the nozzle to increase the velocity energy to provide a solid jet for fire fighting. The sizes of the nozzle and jet are known by the diameter of the nozzle orifice.

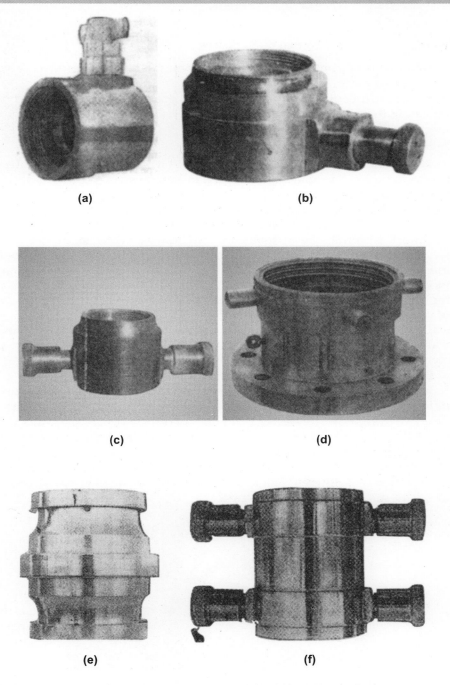

**Fig. A2.15:** (a) Female instantaneous coupling, (b) Female for fixing to hydrant, (c) Female double lug coupling, (d) Flange FRT, (e) Double male coupling, (f) Double female coupling.

## Classification of Branches

Branches may be divided into two main classes.

i. Branches without control features—these types of branches produce a stream of water in a jet form.

ii. Branches with control features—this type provides some type of control for regulating the flow.

### Branches without Control Features

i. *Standard branch/short branch/emergency branch:* Overall length of the branch is 20 cm for the threaded type and 21 cms for the instantaneous type. It consists of a solid brass casting weighing nearly about 2 kg for threaded type and 1.6 kg for the instantaneous type. It is able to withstand rough treatment.

ii. *Stream branch:* It is similar to the short branch in outward appearance but internally a cylindrical central tube is provided throughout its whole length. The tube, which is of 25 mm diameter, is open at both ends and is held in position by 3 longitudinal guide vanes. As the speed of water in hose is greater in the centre tube, it gives a more powerful jet. In a straight tube, the velocity of water, which travels through the centre tube, is more than that of water, which is in contact with the hose. Because of its tapering shape, the velocity of the slower moving water in contact with the wall of the hose is greatly increased inside the branch. On reaching the nozzle, the water actually travels faster than the central tube. The jet is composed of a central core held together by a smooth sheath to reduce the friction. This branch is usually of the same length as the short branch.

iii. *Twin or siamese branch:* This type of branch consists of a relatively long branch bifurcated at the lower end, each lug terminating in a coupling for attachment to a line of delivery hose. It is fed by twin lines of hose to deal with greater volume of water than a single line of hose. Jet from the twin branch can reach greater height than single branch.

iv. *Extended branch:* This type is sometimes used for fires involving grain, coal, hay stack, etc. It is of different length and size and is made of copper, brass or other metals to suit different requirements.

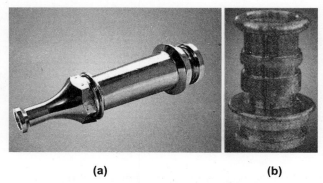

**(a)**            **(b)**

**Fig. A2.16:** (a) Branch pipe, (b) Diffuser.

A typical branch for dealing with coal stacks and bunkers consists of steel pipes usually of 3 meters in length and 63 mm diameter. At one end, it has a pointed head with a number of 3 mm diameter holes and the other end is threaded to suit a standard hose coupling or to extend the branch still further. A driving head may also be screwed on to enable the branch to be driven into the stack or bunker after which the head may be withdrawn and the coupling screwed in so that a hose may be connected. In some places, small branches of 600 to 900 mm long and less than 63 mm diameter are used specially for dealing with fire in hay stacks. A sliding adjustable collar is sometimes provided on such branches to vary the angle of discharge.

## Branches with Control Facilities

a. *Hand controlled branch:* This branch, which is also known as "London branch", has devices for jet, spray and off positions. They can provide a jet or spray independently or simultaneously and each is capable of independent control. By turning the rotating collar in anticlockwise direction, the spray can be varied from a conical angle of 30 to 40° to a flat of 180°.

b. *Diffuser branch:* This branch can give a jet or spray of variable sizes, which can be shut off as needed while the water is discharged from an intense mist over an angle of 180° to a complete shut off. However, this branch should not be mistaken for a hand control branch, which can give a jet and spray simultaneously or each independently.

c. *Fog nozzle branch:* It is also a hand-controlled type of branch but it is commonly known as "fog nozzle". It can be adjusted to give a water stream of straight jet, a fine or coarse spray and complete shut off. The fine spray is in the form of mist. A floating rotary valve connected with the handle facilitates the control operations. To prevent the entry of any foreign materials, a wire mesh filter is provided in the cylindrical tube. The spray head is detachable and held in position by a spring-loaded interlocking device. With this type of branch, the spray head or nozzle can be extended with an extension piece. Low-pressure application may be connected to extinguish small oil fires or discharge the fog straight to the burning surface. The fog nozzle is made of light alloy and the orifice diameter is usually 16 mm. This type is useful for indoor fire as well as small oil fires.

## Special Type of Controllable Branches

*Single revolving branch:* This is a gunmetal fitting having 63 mm male instantaneous or female threaded coupling at the lower end and a revolving round head supported on ball bearings on both sides with cylindrical cam and nut on top. The revolving round head, which appears like a dome, has 9 holes of 3 mm diameter. It is compact enough to be lowered down a ventilator or through a small hole cut on the floor.

*Double revolving branch:* This consists of a large gunmetal casting made in two parts. The upper part comprises a collecting breeching and carries four equidistant outlets. Two of these are in the form of branches diametrically opposite and fitted with standard nozzles. The other two are also located opposite to each other and are inclined to the center plane. It is rotated by the back-pressure of water passing through two outlets inclined to the casting. They maybe used in holds and basements.

*Cellar pipe branch:* It is a piece of equipment for use at smoky fires in basements and holds. It is fitted with a nozzle, which can be made to elevate, depress and rotate by remote control. The total length is 1.35 meters but can be folded when not in use.

*Gooseneck branch:* It is a "U" shaped metal fitting used in place of branch for filling tanks. The length is 600 mm and has a constant bore throughout. Diameter of the gooseneck may vary with the hose diameters.

*Radial branch:* It is a special type of metal branch provided with holder for use with jets up to 50 mm diameter. For the operation of this branch, only one man is needed. It consists of two double iron stays. To the front is a double stay pointed at the bottom and the back is fitted with steel studded plate both resting on the ground. The inclination of the first is controlled by a small winch gear carried at the head of the front stay. Pivoting the branch about the base plate can change the direction.

*Foam making branch:* This is applicable to any branch used for generating mechanical foam. It converts the foam solution into finished foam by entrainment of air.

*Monitor (fixed and mobile) branch:* It is a special form of branch with nozzles of bigger diameter. It can be used where large volume of water at high pressure is required. This can be used for extinguishing major fires.

*Low-pressure applicator:* This is an extra large form of ground monitor, which can deliver larger volume of water than ground monitors. The monitor can be rotated and swiveled. Low-pressure applicators of smaller size of fog head used to produce fog are used for small oil fires.

*Hose reel branch:* This is a first aid fire fighting appliance. The branch is connected to the hose reel and provides jet or spray operations. The diameter of the hose reel is normally 19 mm. Hose reels of 25 mm diameter are also in use.

## Nozzles

The nozzle is a small metal fitting made of gunmetal, brass or bronze, in a tapering shape. The purpose of the nozzle is to change pressure energy into velocity energy. As the velocity is increased, the stream of water can reach a longer distance with great striking power for efficient fire fighting. It is provided with a threaded female connection at the base for attachment to the branch. The base of the nozzle is hexagonally-shaped to permit the use of spanner when necessary, although a hand-tight joint is usually sufficient, if the washer is in good condition. Standard nozzles to fit any of the branches used with 63 and 70 mm hose are usually made of gunmetal or brass and vary in length from 100 to 115 mm. They are with female connection for attachment with branch and are of threaded types.

The term "working nozzle" is used for nozzle normally attached to the branch, which is first used on the scene of a fire by attachment to a branch and are of the threaded type. A working nozzle is usually of 19 mm or less. But it varies according to local preference and risks involved.

Besides the standard nozzles, special types of nozzles are also in use. Some of them are listed in Fig. A2.17.

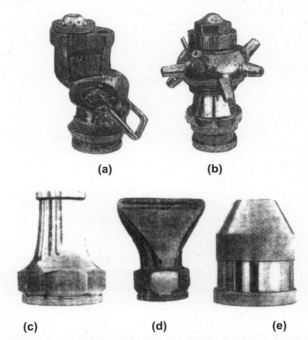

**Fig. A2.17:** (a) Fog nozzle, (b) Revolving nozzle, (c) Spray nozzle, (d) Flat spray nozzle, (e) Oil fire nozzle.

### First Aid Reel Nozzles

These nozzles are generally of a very simple pattern in which the jet can be turned ON/OFF by means of a rotating plug. In the majority of nozzles, the handle is placed in right angles in the OFF position and in line with nozzle in the ON position.

### Small Nozzles

Varieties of small nozzles are used for hand pumps, stirrup pumps and portable extinguishers. The nozzle designed for a jet are either with or without a control cock, or they have a double action nature like the thumb operated nozzles of the stirrup pump.

### Basement Spray

This is a gunmetal fitting with a suitable thread for attachment to a branch. On the lower end, it has a fork to which a reflector plate is bolted at right angle to branch orifice. It can spray water in all direction like a sprinkler head. This spray is used for dealing with fires in basement and holds where an all-round sealing effect is required.

### Miscellaneous Nozzles

Apart from those specified above, a great number of uncommon nozzles are in use. These are given on next page.

i. *Variable nozzle:* The nozzle gives a jet, which can be adjusted to any size from OFF position to 25 mm.

ii. *Triple purpose nozzle:* It is made of gunmetal. It is fitted at the outlet with teeth that can split the water apart diffusing into fine particles and completely saturating the core of the spray. The teeth are well protected with a large rim provided with heavy rubber bumper ring. The nozzle is capable of and can be used for providing a high-pressure jet, solid foam jet or a firewater curtain.

iii. *Oil-fire nozzle:* This type of nozzle is provided with an impeller inside the branch close to the outlet. The rotation of the impeller produces a fine mist. This is made of light alloy or gunmetal. It used for small oil fires. The approximate length of the nozzle is 60 mm.

## Collecting Head, Suction Hose Fittings and Standpipe

### Collecting Head

Collecting head consists of a metal casting usually of aluminium or gunmetal fitting with 2, 3, 4, 5 or 6 inlets for delivery hose and a single female suction coupling for attaching to the suction inlet of a pump. Each of the inlets is provided with a non-return valve, which consists of a metal casting and screwed into the body of the collecting head. It is fixed to a male instantaneous coupling. The valve is mushroom-shaped with either a synthetic rubber or a metal-to-metal round face. It is held against the seating with the help of a wire spring. The water pressure forces the valve off its seating and passes in. When the pressure is reduced, the valve returns to its seating and prevents the escape of the water. The valve lift should not be less than 16 mm. These fittings are primarily used in closed relays when they are fitted to the intermediate and delivery pumps. It is also used to augment the water supply to a pump from weak pressure fed sources. It can also help in priming a pump with a defective primer. It can serve the purpose of blank cap while a pump works from its mounted tank. It can also be used at the end of suction hose to avoid repeated priming if the pump is to be stopped from time-to-time.

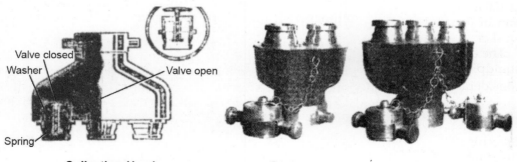

**Collecting Head**　　　　**2-way**　　　　**3-way**

Valve closed

Washer

Valve open

Spring

**Fig. A2.18:** Collecting head.

## Foot Valve

This comprises of a metal suction strainer fitted with an extension tube in the casting between the strain cover and the coupling. In the extension, a flap type valve is fitted in such a manner that water rising up into the suction, forces it to open. If the pump is shutdown for any reason, the flow ceases immediately in the suction and the flap valve falls back on to its seating. This prevents the water from escaping and keeps the suction fully charged. This helps to avoid re-priming of the pump when it is to be started again. A device is also provided to trip the valve for clearing the suction hose when necessary. It can also help in priming a pump with a defective primer.

## Deep Lift Suction Fitting

It is a very effective device for pumping out water from depths greater than maximum suction lift. It is extremely effective where it is not convenient to lay salvage hose from the pump and also where the pump is not self-priming type. It comprises of a light alloy body incorporating an inlet strainer plate and housing a two stage ejector unit. The inlet connection is 63 mm diameter, instantaneous male and outlet varies as per the applications. When used by fire services, the size usually is 100 mm but for marine uses the size normally is 125 mm diameter. Two eyes are provided for attaching ropes. It takes delivery of high-pressure water from the normal fire pump and can transfer the energy to a large volume of water at lower head. This fitting is capable of lifting from a depth of 25 meters.

## Standpipes

As it is almost impracticable to couple a hose directly to a water hydrant line, a piece of equipment known as a standpipe is used to extend the outlet of the hydrant above ground level. Use of a standpipe enables to overcome the access difficulties caused by the restricted size of hydrant pits and the depth below ground at which hydrant outlets are usually located. It also enables kinking to be avoided when connecting hose to hydrant supplies.

The term standpipe indicates the complete fitting including the head, although, with some types, the shaft and head are separate fittings which can be disconnected. The base casting is normally made of an alloy of copper or aluminum. It has an external diameter of 100 mm and internal diameter of about 60 mm. It is threaded from the bottom with 25 mm of female round thread, which terminates in a recess designed to take a leather washer.

The overall length of standpipe shafts often varies with different requirements. Most standpipes, in current use, are about 1 meter length. The standpipe heads differ considerably and can be made with either single or double outlet.

## Breeching

Breechings are generally used for the purpose of joining two lines of hose into one or for dividing one line into two. The breeching, used for joining two lines into one, is known as a collecting breeching. Other than these, there are special breechings like suction breeching, which brings two or more suctions into a common inlet.

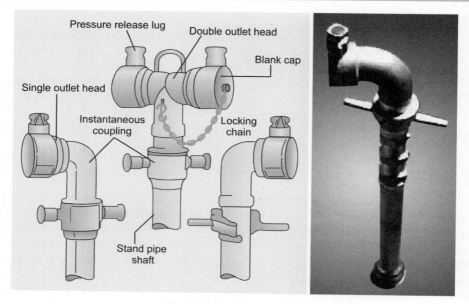

**Fig. A2.19:** Standpipes.

Breechings are manufactured in a light alloy for fresh water and gunmetal for salt water applications.

### Collecting Breeching

This is of two principal types. In the first, the two inlets and one outlet are designed to connect to a hose of the same diameter. In the second type, two inlets are designed to connect hoses of same diameter and the outlet of a larger diameter.

The simple form of collecting breeching is shaped like 'Y'. Each of the two lugs is fitted with a male coupling if the device is of instantaneous type and female coupling if the breeching is of threaded type. The coupling on the stem of reversed 'Y' is male if it is threaded type and female if of instantaneous type. A single lug pressure-release device is

| (a) | (b) | (c) |

**Fig. A2.20:** (a) Collecting head, (b) Dividing head, (c) Dividing breeches with control.

common for instantaneous female coupling. A control type breeching is fitted with a control valve to shut off either one or both of the inlets.

Collecting breechings for use with the large branches like a radial branch or a deluge branch, differs due to the use of a short length hose as feeder and for this reason breeching for that purpose has a larger diameter outlet coupling.

Suction breeching is now being replaced by collecting head. The breeching, in the case of a pump with one inlet, is used to gather water from two different sources. But in the case of a pump equipped with two inlets, the suction breeching is attached to one inlet and the second is connected to the suction hose or blanked off. In case of water relay, the absence of non-return valve in breeching may result in reverse flow when twin lines of hoses connected to it work at unequal pressure

### Dividing Breeching

The main purpose of dividing breeching is for dividing one line of hose into two. It is either made of gunmetal or bronze like the connecting breeching. In appearance, it is also shaped like 'Y'. But it consists of a male instantaneous coupling on the single inlet and female instantaneous couplings on the two outlets. The primary purpose of dividing one hose line into two is that when the pressure is sufficient, two jets can be provided from one, for purposes like damping down. The main disadvantage of this is that one line cannot be shut down without both having to be shut down. Dividing heads can also be provided with valves for control.

Breechings having the facility to control both the outlets at the same time by means of a valve are called control dividing breechings. By shutting off one outlet at a time, the water can be made available where it is most required, without wastage.

### Adaptors

These are metal fittings used for connecting hoses or couplings of different types or sizes, for connecting male-to-male or female-to-female parts of the same type of coupling.

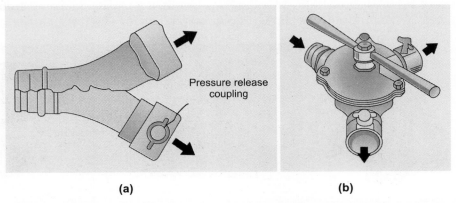

Pressure release coupling

(a)                                    (b)

**Fig. A2.21:** (a) Dividing breeches with pressure release, (b) Control dividing breeches.

Adaptors are usually cast as one fitting except in the case of screwed female ends, where nut and sleeve are separate. The average length is between 100 mm and 150 mm, the ends of the adaptors can be connected to hoses.

### Delivery Hose Adaptors

Delivery hose adaptors that are required are the male instantaneous to male round thread and female instantaneous to female round thread adaptors. The female is sometimes used to convert a hydrant outlet to enable delivery hose to be connected straight on to it.

### Hydrant Adaptors

These adaptors are used between the hose of the standpipe and the hydrant outlet. Normally, they are of the same pattern as delivery hose adaptors. Other combinations of threaded connections like female bayonet to female instantaneous, round thread to lugged, or lugged to male V thread are also used.

### First Aid Hose Reel Adaptors

These adaptors are used to enable hose reel couplings to be connected to hose reel equipment and to other types of coupling. Hermaphrodite hose reel couplings are connected to instantaneous female outlet. Similar adaptors with Mac Donald coupling to male instantaneous coupling are also used.

### Suction Adaptors

These adaptors are used for connecting delivery hose to suction or to the pump inlet or to join length of suction of different diameters. Suction adaptors usually consist of metal casting with either a male suction coupling and a female instantaneous or a male screw coupling to take the hose or a standard female screw coupling at the other end. Adaptors for connecting suction of different diameters usually are provided with female connection at one end and a male at the other end.

### Suction Reducing Piece

It consists of a short suction hose of 1 to 2 meters length. It is smaller in diameter than the main suction, with an appropriate hose coupling at one end for attachment to a standpipe

**Fig. A2.22:** Attaching a hose to a twin hydrant in a petroleum refinery.

head and a female suction coupling at the other for uniting the main suction hose or pump inlet.

## Miscellaneous Hose Fitting

### Nozzle Elbow

It is a metallic attachment for fixing between the branch and the nozzle to enable a jet to be delivered at right angle to the branch. It is used mainly for attacking fires in places where it is difficult to approach the fire due to thick smoke or some other reasons.

### Blank Cap

It is a metal cover, which is attached to close a delivery outlet or suction inlet while not in use. It is fixed to a length of chain and is connected to the standpipe.

---

### Jet reaction

When water is projected through a nozzle, a reaction equal and opposite to the force of the jet takes place at the nozzle and the nozzle tends to recoil in the direction opposite to the flow. This is known as jet reaction.

---

### Suction Coupling Wrenches

It is a pair of tools used for the purpose of tightening the suction hose coupling airtight. This is necessary, as a leak, however small, would allow air to enter in the suction and

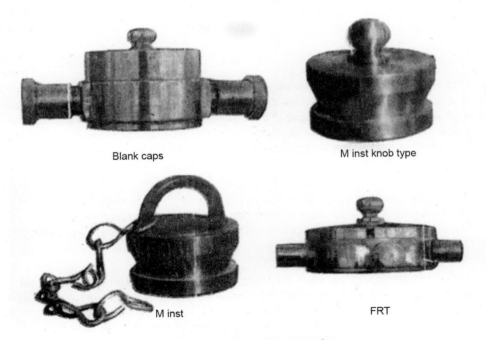

Blank caps                    M inst knob type

M inst                    FRT

**Fig. A2.23:** Tools and accessories.

make it impossible to obtain the necessary pressure drop to lift water. There are two types of suction wrench used. The latest type is designed to fit to any size of suction coupling. It has a tubular steel shank with one end flattened, slightly curved and shaped at the other end in a semicircular recess. A reinforced steel lever arm is pivoted to the bar of about 180 mm from the bottom end. The lever arm is also slightly curved and has an elliptical-shaped eye of about 25 cm from its end.

## Hose Ramp

One of the problems which often have to be faced when dealing with large fires is the obstruction on roads by lines of charged hose. If heavy vehicles are allowed to run over it, it will certainly cause damage to the hose. So a special equipment in the form of hose ramps is carried on most fire engines. It enables wheeled vehicles to cross-charged lines of hose without dragging or damaging the hose.

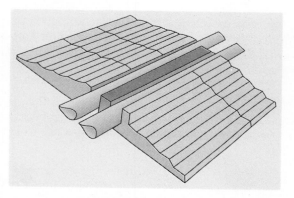

**Fig. A2.24:** Hose ramp.

Hose ramps of different patterns are used. They are all designed in such away as to provide a sloping up over the hose with sufficient clearance to prevent contact between vehicle wheels and the charged hose. Most ramps are designed to accommodate 70 mm and 90 mm diameter delivery hose. They are generally made of wood or steel. Steel ramps are used where heavy and continuous flow of traffic is confronted.

## Strainers

Strainers are designed and used to prevent solid objects, which might damage the pump, from being drawn-up the suction hose when working from open water sources. Two types of strainers, called metal strainers and basket strainers are mainly used.

### Metal Strainer

Metal strainer consists of an alloy strainer, cylindrical in shape and perforated with number of holes. The size of hole must be large enough for continuous flow of water and small enough to prevent entry of pieces of wood, stones or other solid objects. Metal strainers are fitted with a female suction hose coupling and made in sizes to fit the standard sizes of suction hose.

Sometimes it becomes necessary to pump from shallow water only a few centimeters deep and to facilitate this, specially designed low-level metal strainers of different patterns are produced.

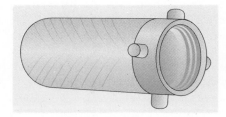

Fig. A2.25: Metal strainer.

## Basket Strainer

It is used in conjunction with a metal strainer but never alone. When resting on a soft surface like mud, the metal strainer may sink in and lower the percentage of its surface area. On such occasions, the basket strainer acting as an outer shell rests in the mud. As it has a largest surface area, it still leaves a considerable proportion exposed and thus allowing the metal strainer to function without any obstruction. The basket also protects the metal from damage. Besides, it helps in removing particles from the water entering the suction and protects the pump from damages.

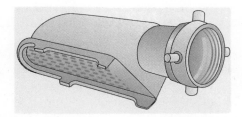

Fig. A2.26: Basket strainer.

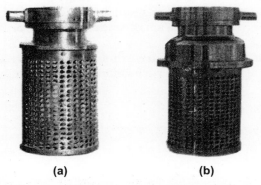

**(a)**          **(b)**

Fig. A2.27: (a) Strainer, (b) Foot valve strainer.

## Care and Maintenance of Hose Fittings

Hose fittings have to be maintained well, for ready and immediate use on fire scenes. After use, a fire hose is usually hung to dry as standing water that remains in a hose for an extended period of time can deteriorate the material and render it unreliable or unusable. As such, the typical fire station often has a high structure to accommodate the length of a hose for such preventative maintenance. Similarly, hose fittings should also be taken care of.

They must be periodically inspected, cleaned and kept in good condition. All moving parts should be slightly lubricated to ensure free operation. Condition of washers has to be checked and renewed if necessary. Rough edges on fittings like couplings, branches, etc. must be removed or smoothened. Particles of paint, tar, etc. should be removed with the help of suitable solvents. Threads of screw type couplings are to be checked and cleared of dirt. They are to be kept in the proper places designed and meant for each, so that they can be readily reached even in darkness.

> Fire hoses–tested to 14 bar pressure. Normal working pressure –7 bar.

## ADDITIONAL INFORMATION ON FIRE SERVICE EQUIPMENT

### Water Tender and Special Equipment

Water tender forms one of the first attendance appliances in most fire brigades and therefore it is very necessary that a fireman should have sufficient knowledge about its essential features.

Water tenders are designed for use by the fire service to deal with fires both in urban and rural areas. This appliance normally includes a fire pump, a water tank and an extension ladder. There are two types of water tenders available. Type 'A' which has no built- in pump and towed to a trailer pump instead. Type 'B' has a built-in pump and so does not need towing a trailer pump. Type 'A' appliance has become obsolete due to certain disadvantages like reduced speed of appliance and manpower required to handle trailer pump. Both types are designed for use in rural areas. But type 'B' with built-in pump has the added advantages of being suitable for use in urban areas. Small water tenders carry 1800 litres (typically in a small tender) of water, an extension ladder and a portable fire pump.

Both types have a normal two-wheel drive or a four-wheel drive chassis. A four-wheel drive is suitable for rural areas, where performance across the country may be needed.

### Features of a Small Water Tender

A water tender comprises the following essential features.

#### Water tank

This is tank with a capacity of 1800 litres and is mounted immediately behind cabin. The tank usually is elliptical in shape and is especially baffled to prevent water surge and to give the minimum center of gravity. It is fitted with an inspection manhole and a cover on the top, a cleaning hole and cover at the base. The tank is also fitted with overflow and filling pipes. To achieve a low center of gravity and facilitate body construction, there can be two tanks instead of one, each connected with the other at the bottom. In such a case, each tank will have separate inspection holes and drain holes. An over flow is also fitted in one of the tanks.

### Fire Pump

A water tender will have a main pump and a hose reel and may also carry a portable pump.

### Main Pump

The main pump is of centrifugal type and is powered by the engine which provides power for propelling the vehicle and forms an integral part of the appliance and cannot be removed for use elsewhere. The drive from the engine to the pump is generally through a gear box and is known as 'power take off' (PTO). The main pump may be mounted in any of the three positions, front, rear or middle.

### Hose Reel Pump

Many of the water tenders are fitted with hose reel pump for first aid purpose. The hose reel pump delivers water to the hose reel. It is generally a small rotary type of pump (gear type). The power for driving this pump is delivered from a subsidiary power take off. This pump can deliver 110 litres per minutes at 7 bar and can develop up to a maximum of 10.5 bar.

### Hose Reel and Hose

Hose reel is mounted on a hollow rotating shaft to the center of which water is fed. It is fitted with a stuffing box gland to enable water to be fed to the drum in which the reel is rotating. The hose is fitted to the outlet of the rotating shaft and is generally kept wound on the reel and is not less than 60 meters of length with 19 mm internal diameter. Hose reel can be fitted with a straight shut-off branch giving jet and spray as needed.

### Function of Hose Reel

The function of the hose reel system is to ensure an immediate supply of water for extinguishing small fires and for attacking larger fires with a minimum delay possible, in order to exercise a measure of control until the main pump is ready to work. As a shut-off nozzle is fitted, its use prevents excessive water damage in extinguishing operations.

### Operation of a Five Way Valve

*T-P-R (tank to pump to reel):* A T-P-R is required on arrival at a fire. Water is fed by gravity from the tank to the pump and delivered under pressure to the hose reel.

*H-R (hydrant to reel):* An H-R is required when the hydrant has good pressure. In this case, the pump is disengaged from the hose reel.

## Fire Services Equipment Specifications

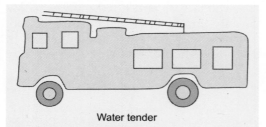

Water tender

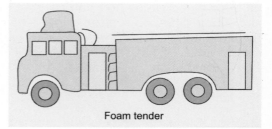

Foam tender

### Water tender

Chassis–Tata- LPT 608, 1210, 1612, 1616
   –Ashok Leyland- Comet, Beaver
Pump–Single/double stage
   1800 lpm/2250 lpm/3200 lpm
PTO–Full torque drive line PTO
Water tank–1800 to 6000 litres
Optional-water monitor, foam system, ATP.
Specifications–IS: 950; 6067, TAC approved

### Foam tender

Chassis–Tata- LPT 608, 1210, 1612, 1616; 2416
   –Ashok Leyland-Comet, Beaver;
   Tusker; Taurus
Pump–Single/double stage
   1800 lpm/4000 lpm
PTO–Full torque heavy duty PTO
Water tank –1000 to 4500 litres
Foam tank –450–3000 litres
Foam equipment–Foam monitor; foam
   system; automatic/manual ATP
Optional–DCP; $CO_2$; BCF
Specifications–IS; 951; 10460

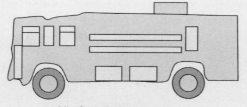

Mix-O-matic-foam tender

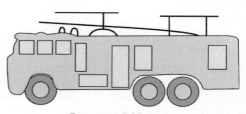

Foam cum DCP tender

### Mix-o-matic foam tender

Chassis–Tata- LPT 1210/42 1612/48
   – Ashok Leyland- Beaver
Foam tank –4000–7000 lts (SS)
Foam monitor –2000–700 lpm
Range—60 to 75 mtrs
Foam System –8 × 63 mm
   foam/water outlets.
   Mixomatic vent
   proportioners with foam
Hydrant–    4 × 63 mm
Manifold    inlets both sides
Optional–Fire pump 2250–4000 lpm

### Foam cum DCP tender

(Optional high pressure type pump with fog
   nozzle)
Chassis–Tata–LPT  1616/48,  2416/48
   leyland–Taurus, tusker, beaver
Pump–3200–4000 LPM(normal pressure);
   High pressure –(400 LPM/40 bar)
PTO–heavy duty full torque PTO
Water tank –2000–4000 ltrs.
Foam tank –400–800 ltrs
Foam monitor –2400 lpm 60 mtrs range
DCP –500/1000 kg, DCP monitor
Hose reel –2 × 30 mtrs water fog, hose reel
   with foam guns.

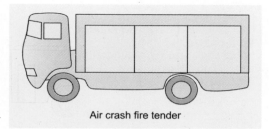

Air crash fire tender

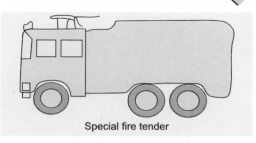

Special fire tender

### Air crash fire tender

Chassis–leyland –4 × 4, 6 × 6(F-23) or imported Iveco Mogirus
Pump –4500 LPM 8.5 ksc
PTO–Engine dependent
Water tank –4500–6000 ltrs.
Foam tank –550–800 ltrs
Foam system–manually cabin operated, remote controlled capacity –2400 lpm
Foam headlines –4 × 63 mm FB 10 foam branch
BCF system –2 × 50 kg BCF cylinders with applicators
Acceleration –0 to 80 kmph in 40 secs
Spec-IS: 951/ICAO defence/NFPA

### Special fire tenders

Foam nurser truck
Jumbo water
Lighting van
Hose van/towing van
LPG recovery vehicles
Riot control vehicles

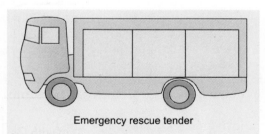

Emergency rescue tender

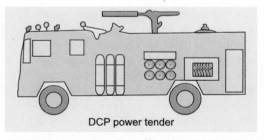

DCP power tender

### Emergency rescue tender

Chassis–Tata LPT 1210 Ashok Leyland-comet
Generator–12 KVA/15 KVA PTO
Lights –5 mtrs telescopic mast with 3 × 1000 watts halogen floods lights(rotating and tilting type)
Cable winch –5 ton pulling capacity with 60 mtr rope.

### DCP powder tender

Chassis–Tata-608/1210, 1612; Leyland–comet tusker
DCP vessels –500–1000 kg
2000m–4000 kg(single/double vessel)
Monitor–variable output –15–40 gr per sec, range –45 mtrs
DCP hose –2 × 30 mtrs hose reel
Expellent system –Nitrogen cylinders.
Special–expansion shock device
Spec- IS: 10993

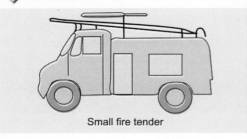

Small fire tender

Small fire tender
Chasis–LCV Tata-407 Mahindra
Pump –1350 lpm; 7 bar
PIO–Heavy duty drive line PIO
Water tank –2000 ltrs
Spec- IS: 2096; IS: 938

## Important Features and Specifications

### Small Fire Tender

The water tank has a capacity of 1800 litres. It should be mounted immediately behind the cab so as to allow the full contents to flow to the pump. It should be suitably baffled and its shape and mounting should be such as to bring the center of gravity as low as possible in the chassis. It should be fitted with a cleaning hole of 250 mm diameter at the base and an inspection manhole on top. The tank should be fitted with filling and over flow pipes. The diameter of the over flow pipe should not be less than 50 mm.

The pump should be of centrifugal type and preferably of single stage design. It must be as light as possible. The impeller rings have to be manufactured from high quality bronze. The gland should be self adjusting type.

A drain plug should be provided at the bottom of the pump casing. The suction inlet should be provided with a standard 100 mm suction hose connection with removable strainer and blank cap. There should not be less than two delivery valves. The pump should be mounted either at the front or at the rear of the water tender appliances. Preference may be given to rear mounting.

A speed of 72 kmph on level ground should be readily obtainable with the appliance fully laden, without trailer. The acceleration should be such that with a warm running engine, the fully laden appliance shall attain a speed of 64 kmph from a standing start, through the gears in a maximum time of 40 seconds.

The minimum capacity of the chassis should not be less than 5 tonnes. The wheel base should not exceed 2.3 metres and overall width should not exceed 3.3 metres. The turning circle should be as small as possible.

Control valves shall be accessible from the driver's seat. The body must provide enclosed accommodation for 7 men with or without partition. It should be fitted with scaffolds for carrying 10.5 m extension ladder. All windows should have safety glasses. Lockers or other suitable accommodation shall be provided for all the equipment. The lockers rack must be painted and marked for ready identification of missing equipment. Adequate lighting should be provided in both compartments. A master switch for isolating the locker lighting circuit should be provided. Handrails and non-slippery steps with sufficient grip shall be provided where required.

### Portable Pumps

A portable pump is sometimes mounted on a water tender. It is useful for working in places inaccessible to a wheeled vehicle and also for providing additional capacity to that of the appliance on which it is carried. The portable pump is to be operated, if hydrant pressure is too low. The portable pump is also used for refilling the tank. When the pump is not needed, it should be disengaged.

### Ladder

The appliance carries an extension ladder mounted on it, in the normal way, with ladder scaffolds. The ladder is usually 10.5 m in length and may be either of aluminium or wood. Aluminium ladder is the usually trusted type.

### Sparge Pipe

A sparge pipe and road-watering pipe may be provided in the front and rear of the appliances to protect the tire from ground fires to water the road. For the sake of protection of tyres, the appliance is usually fitted with a sparge pipe at the front side of the vehicle. It is fitted with a control valve, which regulates the water spray. Sparge pipe is particularly useful in case of rural areas where vehicle may have to travel over ground fires.

### General

The appliance provides seating accommodation for seven men including the driver. In addition, it is provided:

a. A searchlight fitted on a tripod with approximately 90 meters range.
b. Spot lights of adjustable types
c. Fire bell

### Need for a Fire Protection System

It is wrong to assume that if a public fire brigade service is available, then fire precautions and fire protection systems are not required in a facility. No matter how fast and efficient the outside fire service, there is always a lapse of time before the fire service team can arrive at a scene of fire. Therefore, to attack and suppress a fire incident, it is worthwhile to invest in fire protection installations to prevent loss of life and property.

## TAC GUIDELINES FOR SPRINKLER FIRE PROTECTION SYSTEMS

The Tariff Advisory Committee (TAC) adopted in 1978, recommends guidelines for automatic sprinkler installations with regard to design, maintenance, extension, etc. Installations which satisfy TAC rules are allowed substantial reduction in the fire insurance premium.

For the purpose of providing automatic sprinkler protection, the TAC rules have classified occupancies as:

- Extra-light hazard: Buildings with low fire hazards such as schools, hospitals and hotels.
- Ordinary hazard: The majority of commercial and industrial buildings.
- Extra-high hazard: Process risks such as paint, foam plastic and foam rubber; high piled storage risks.

These are primarily based on the fire load of the occupancies and the likely heat release rates in case of fire occurrence.

The TAC guideline stipulate adequate rate/density of discharge of water with:
- Evaluation of the area over which discharge of water must be made
- Period for which the water should be available.

| | Discharge density (lit/min/m²) | Area of discharge M² | Time of water availability Minutes |
|---|---|---|---|
| Extra-light | 2.25 | 21 | 30 |
| Ordinary | 5 | 12 | 60 |
| Extra-high | 7.5 | 260 to 300 | 90 |

## Important Features of NFPA 13 on Sprinkler Systems

- The design of the sprinkler system is to be based on the occupancy classification and the hazard classification.
- The water supply requirement should be calculated.
- The sprinkler spacing and location is to be designed.
- The appropriate piping material should be selected.
- Sprinkler systems should be subjected to acceptance tests.

## Sources of Water Supply for Sprinkler Systems

The sources of water for a sprinkler system can be: (1) Gravity tank, where elevation provides the pressure to the water; (2) Pressure tank, where the water is kept under pressure with compressed air; (3) Automatic pumps actuated by a fall in water pressure which occurs when a sprinkler head opens.

The pumps may take suction from gravity tanks, reservoirs or perennial sources of water such as lakes, rivers, etc. In some cases, water supply from secondary sources such as storage tanks may be necessary. The water must be free from solid suspended matter which can accumulate in pipe-work and block the passage.

## Water Quality for Sprinkler Systems

Natural fresh waters contain dissolved calcium and magnesium salts in varying concentrations, depending on the sources and location of the water.

If the concentration of these salts is high, the water is considered hard. A thin film composed largely of calcium carbonate, $CaCO_3$, affords some protection against corrosion where hard water flows through the pipes. However, hardness is not the only factor to determine whether a film forms. The ability of $CaCO_3$ to precipitate on the metal pipe surface also depends on the water's total acidity or alkalinity, the concentration of dissolved solids in the water, and its pH. In soft water, no such film can form.

## Pipe Work

Water from the source of supply is conveyed to the sprinklers through a network of pipelines. The network may of the branched type or grid type.

There are three basic types of piping layouts for sprinkler systems—tree, loop and grid. The tree configuration is the traditional piping layout. The loop and the grid configurations are used where the installation is hydraulically designed.

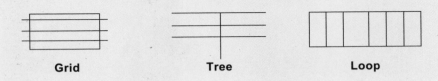

| Grid | Tree | Loop |

Black or galvanized MS pipes and fittings as per IS: 1239 can be used for sprinkler installations.

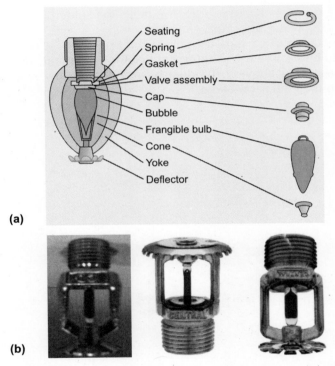

**(a)**

**(b)**

**Fig. A2.28:** (a) Components of frangible bulb type sprinkler head; (b) Frangible bulb type sprinkler head.

A sprinkler head is a thermally actuated valve which, when its heat sensitive element reaches a specific temperature, opens up and releases water as a spray. The pattern of the spray depends on the design of the deflector.

Sprinklers are essentially of two types–frangible bulb type and fusible link type.

In the frangible bulb type, a sealed glass bulb contains liquid and a small air bubble. At high temperatures, the liquid expands to absorb the bubble and the resulting increase in pressure ruptures the bulb and this allows the water to flow from the sprinkler.

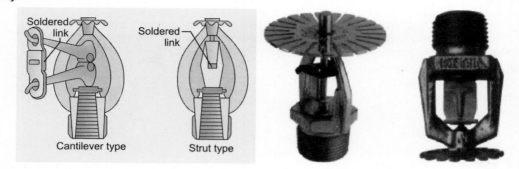

**Fig. A2.29:** Fusible link type sprinkler heads.

In the fusible link type, the heat from the fire melts the solder allowing the levers to separate and letting the water to escape.

Sprinklers are available to various temperature ratings. TAC rules specify that the chosen rating should be as close to the highest anticipated temperature conditions in the area, but not less than 30° C above it. To identify sprinklers easily, colour coding of the yoke or the liquid in the bulb is observed.

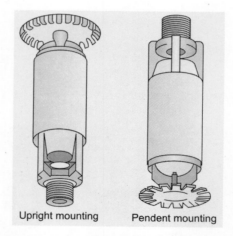

**Fig. A2.30:** Different sprinkler deflector types.

Maximum protection area for each standard sprinkler

| Hazard class | Max protection area (sq.m) |
| --- | --- |
| Light | 20–25 |
| Ordinary | 10–15 |
| Extra-hazard | 8–10 |

## Other Components of the Sprinkler System

*Main stop valve:* The main valve is always kept open so that water flow to the sprinkler system is always available. The valve is closed only to carry out maintenance activities such as changing the sprinklers, etc.

*Alarm valve:* The alarm valve is a check valve which is normally in a closed position, when the system is activated, the clack lifts up and allows water to flow to the sprinklers.

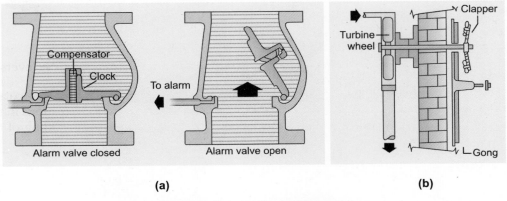

**(a)**                                                              **(b)**

Fig. A2.31: (a) Alarm valve, (b) Water alarm.

The water-flow device should be field adjusted so that an alarm is initiated no more than 90 seconds after a sustained flow of at least 10 gpm (40 litres/min).

Features that should be investigated to minimize alarm response time include:

1. Elimination of trapped air in the sprinkler system piping
2. Use of an excess pressure pump,
3. Use of pressure drop alarm-initiating devices
4. A combination thereof of the above.

| Periodic testing and maintenance chart for sprinkler installations | | |
|---|---|---|
| *Subject* | *Activity* | *Duration* |
| Reservoir | Level checking | Weekly |
|  | Cleaning | Once a year |
| Pump | Running test | Daily 5 mins |
|  | Flow test | Annually |
|  | Lubrication | Quarterly |
|  | Gland packing | Weekly |
|  | Overhaul | Once in 2 years |

*(Contd.)*

| Periodic testing and maintenance chart for sprinkler installations | | |
|---|---|---|
| *Subject* | *Activity* | *Duration* |
| Engine | Running test | 5 mins daily |
| | Lubrication | Quarterly |
| | Battery | Weekly status |
| | Load test | Annually |
| | Overhaul | Once in 2 days |
| | Fuel tank check | Daily |
| Motor | Lubrication | Weekly |
| | Starter contact checking | Weekly |
| | Insulation resistance | Half yearly |
| Main piping | Flushing | Once in 2 years |
| | Gauge pressure | Daily |
| Valves | Operation | Monthly |
| | Alarm check | Weekly |
| | Lubrication | Quarterly |
| Sprinklers | Cleaning | Quarterly |
| | Flow test | Quarterly |
| Detector element | Performance | Half yearly |
| Sprinkler installation | Performance | Quarterly |
| | Physical check | Monthly |
| Pressure gauges | Calibration | Annually |
| Painting | | Every 2 years |

Sprinklers are available in the following sizes:
    10 mm–code K 57 used for light hazard
    15 mm–code K 80 used for ordinary hazard
Both size sprinklers are available with following spray patterns:
    Upright
    Pendent
    Conventional
    Side wall
Operating temperature colour code for sprinklers (in° C)
    Orange – 57        Green  – 93
    Red     – 68        Blue    – 141
    Yellow  – 79        Mauve – 182
                              Black    – > 183

*Alarm gong:* When the clack is lifted, a small portion of the water is bypassed to a turbine wheel connected to a gong to sound an alarm. Alarms can be electric type also, actuated by a pressure switch or flow switch.

*Pressure gauges:* One gauge immediately above the alarm valve and one immediately below it is required to be present in a sprinkler system to indicate the water pressure in the installation and the other to indicate the water supply pressure.

## ADDITIONAL INFORMATION OF FOAM SYSTEMS

### Foam Making Equipment

Mainly there are two types of foam making equipment:
1. Foam making branches or foam generators
2. Foam concentrates induction and injection equipment.

The primary aspirated foam making equipment may be divided into following main categories;
1. Low-expansion (LX) hand held foam-making branches (FMB)
2. LX hand-held hose reel foam unit
3. LX hand-held water branch
4. LX foam generators
5. LX foam monitors
6. Medium-expansion (MX) hand held foam-making branches (FMB)
7. MX hand-held water branch
8. MX foam pourers
9. High-expansion (HX) foam generators

A few types of foam making equipment are fitted with means of picking up foam concentrate at the equipment, through a length of tube. These are known as "self-inducing types". Certain types of these operate at fixed induction rates while others have control valves. It is also possible to turn off the induction facility completely and to use premix solutions.

With all other types of foam making equipment, the concentrate should be introduced to the water stream by some form of induction or injection equipment.

### LX Hand Held FMB

#### Working Principle

In the diagram shown below, there are two orifice plates. The upstream orifice is larger of the two. Its function is to create turbulence in the space between the two orifice plates so that when the jet issues from the downstream orifice, it rapidly breaks up into a dense spray. The spray fills the narrow inlet section of the foam making tube and entrains large quantities of air through the air inlet holes. The downstream orifice is smaller and it is calibrated to give the designed foam solution flow rate at the recommended operating pressure (225 lpm at 7 bar pressure).

Most foam making branches have a narrow section at the inlet end in which the air entrainment occurs and then a wider section in which the foam forms. The wider section of the foam making tube some times contains "improvers" (e.g. semicircular baffles, gauze cones, etc.).

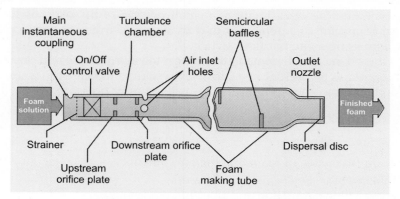

Fig. A2.32: Cross-section of a foam making branch.

Improvers are designed to work the foam solution in order to produce longer draining finished foam. At the outlet, the branch is reduced in diameter to increase the exit velocity thus helping the finished foam to be thrown to an effective distance.

Too narrow an outlet generates back-pressure. This results in less air entrainment and low expansion ratio. If the outlet is large, the expansion is higher, but the throw gets reduced. For LX-FMBs, the longer the foam making tube, the better the working and mixing of foam solution with air. This results in a more stable finished foam with drainage times that are longer than those produced by shorter FMBs.

Some FMBs are designed particularly for use with —film forming foam concentrates in crash fire situations. These branches have adjustable jaws at the outlet providing the option of a jet or spray. They also are equipped with on/off trigger mechanism.

A 225 lpm adjustable jaw type FMB when operated at 7 bar pressure is capable of throwing up to 7 meters in spray mode and up to 13 meters in jet mode.

Fig. A2.33: FB-225 lpm foam makers @ 7 bar.

## LX Hand-held Hose Reel Foam Unit

This consists of a portable hand held unit, very much like an extinguisher (Fig. A2.34). It can contain up to 11 litres of foam concentrate, which can be P, FP or AFFF. An appliance hose reel is connected to an adaptor at the top of the unit. Water is applied at between 2 and 10.5 bar pressure.

*Function:* A small proportion of the water from a hose reel is diverted to fill a completely deflated flexible bag within the container. Inflation of the bag helps displacing the foam concentrate through a siphon tube. The foam concentrate enters the main water stream and passes to an integral LX-FMB to give a jet of primary aspirate foam. The unit can be controlled by an on/off valve on the adaptor.

When operated at 3.5 bar with a flow of 45 lpm, the unit produces foam with an expansion ratio of nearly eight or 360 lpm of finished foam.

## "Fog Foam" Hose Reel Unit

The "fog foam" is a small inline inductor, which can be clipped in between the hose, reel tubing and the branch and has a third coupling set at right angles (Fig. A2.35). A cylindrical magazine pre-fitted with 1% of AFFF concentrate can be connected to the coupling or a drum fed pickup may be connected.

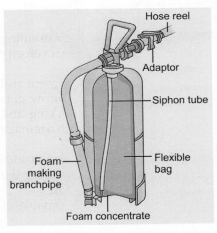

Fig. A2.34: Hand-held foam unit.

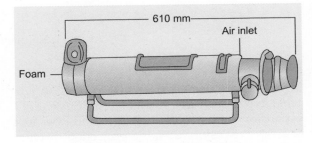

Fig. A2.35: Fog–foam unit.

The fog foam is used in conjunction with a conventional hose reel branch, which produces non-aspirated foam.

As an alternative to FMB, an LX foam generator can be made use of. This, when inserted in a line of hose, induces appropriate amounts of foam concentrate and air into the water stream to generate finished foam. This is delivered through the hose to a water type branch and nozzle. The foam concentrate is induced on the same principle as that of a inline inductor. The air is drawn in through the inlets adjacent to the water-head. Normal water requirement of such generator is 255 lpm at 10.5 bar. These are generally used with 70 mm hose with a water branch of 38 mm nozzle.

## LX Foam Monitors

Primary aspirating LX foam monitors are larger versions of FMBs which cannot be held by hand. They are free standing and portable, mounted on trailers, etc. They generally have multiple water connections and are often self-inducing or fixed with inductors.

There are various types of LX foam monitors in use coming in a wide range of nominal flow and inlet pressure requirements.

For example:

1. 7 bar, 1800 lpm, horizontal range 50 m, max, height of throw –18 m.
2. 10 bars, 43000 lpm, horizontal — 60 m, height 24 m.

These LX monitor can be found in fixed installation at oil tanker jetties, refineries, aircraft hangers etc.

**Fig. A2.36:** Trailer mounted FMB.

### MX Hand-held FMBs

MX FMBs are usually designed to be used with synthetic foam concentrates, although other types such as FP, AFFF, AFFF-AR, FFFP, FFFP-AR, etc. may also be used. The expansion ratio usually is 25:1 to 150:1 because of large expansion ratio, the projection distance is much less than LX foam.

With MX FMBs, an inline inductor is usually used to introduce the foam concentrate as a premix. The branch then diffuses and aerates the stream of foam solution and projects it through a gauze mesh to produce bubbles of a uniform size.

Normal flow requirements of MX FMBs are from 225 to 450 lpm with inlet pressure from 1.5 to 8 bar. The expansion ratio of the foam produced is usually in the region of 65:1 with throws ranging from 3 to 12 metres.

### MX Foam Pourers

As the name suggests, it signifies that the foam pours out of them instead of projecting. These pourers are much larger than the hand-held

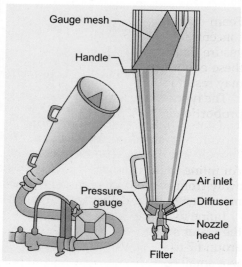

**Fig. A2.37:** Foam monitor.

models. They have higher flow requirements and hence produce greater volume of foam.

They are designed to stand on their integral legs for the unattended delivery of MX foam into bunded areas, such as those surrounding fuel storage tanks.

They have a nominal flow requirement of 600 to 800 lpm at 2.5 bar inlet pressure with an expansion ratio of 40:1. They operate in the similar way as the hand-held MX FMBs.

## HX Foam Generators

High expansion foam generators are designed to be used with synthetic foam concentrates only. They generally produce finished foam at expansion ratio from 200: 1 to 1200: 1.

Air is blown through the generator with the help of a fan. Foam solution is sprayed into the air stream and this is directed on to the surface of a fine net screen. This produces foam with a mass of bubbles of uniform size, which is poured rather than projected.

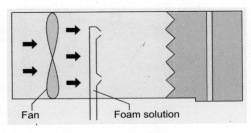

**Fig. A2.38:** HX foam generator.

## Principle of an HX Foam Generator

Some HX foam generators are self-inducing, some use separate inline inductors and some can be operated either way. Because the foam cannot be projected, it is generally fed through a large diameter flexible tube. It may be used without dueling also, e.g. placed on the side of a ship's hold or in a fixed installation in an aircraft hanger.

## Foam Concentrate Induction and Injection Equipment

Foam concentrate induction and injection equipment are used to introduce the foam concentrate (FC) into the water supply in order to produce foam solution (premix).To ensure that the finished foam is of optimum quantity and also to avoid wastage of FC, these equipment must work with precision. Even slight variations in the FC, the flow may result in foam solutions much weaker or stronger than required, being produced.

The types of equipment most widely used are: (i) Inline inductors, (ii) round the pump proportioners.

## Inline Inductors

An inline inductor is kept in a line of delivery hose, usually not more than 60 meters away from the foam making equipment. This enables the foam maker equipment to be moved around freely without moving the FC containers.

Inline inductors make use of the venturi principle to induce the foam concentrate into the water stream. Water is fed into the inlet of the inductor generally at a pressure of around 7 to 10 bar. This passes through the smaller diameter nozzle within the inductor to a small induction chamber and then to the inductor's large diameter outlet through flow improvers.

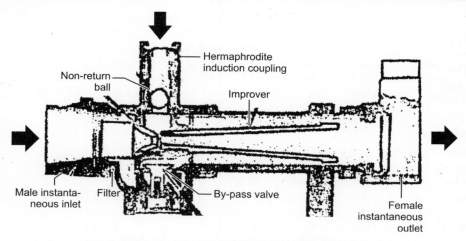

Fig. A2.39: Cross-section of an inline foam inductor.

*Venturi Principle*

As the water enters the small nozzle, its velocity goes up dramatically causing its pressure to drop. This causes the pressure in the induction chamber to fall below the atmosphere pressure. The partial vacuum sucks the foam concentrate through the pick-up tube and into the low pressure induction chamber. A non-return valve is fixed in the FC pick-up line to prevent water from flowing back to the FC container.

*Advantages of the Use of Inline Inductors*

1. Simple, robust and moving parts are less.
2. Cheapest induction system.
3. Quick deployment/re-deployment on fire ground.
4. Foam solution does not pass through the pump/appliances.

Fig. A2.40: Inline foam inductor drawing FC.

*Round the Pump Proportioners (Continuous Proportioners)*

This type of inductor is connected across a pump and can either be a permanent fixture in the appliances or independent with adaptor and connecting hoses. It has a nominal induction flow range of 0–45 lpm or 0–90 lpm, which can be regulated by a rotating grip on the handle.

## *Disadvantages*

1. The inductor should be matching to the foam making equipment for optimum performance.
2. The inductor should be matching to the type and concentration of the foam solution used.
3. Accuracy of proportioning will vary with the pressure of the flow.
4. Pressure losses through the inductor in excess of 30% can be expected at the normal working pressure range.

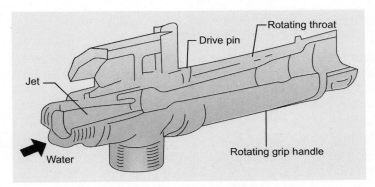

**Fig. A2.41:** Proportioning inductor.

When pumping begins, some quantity of water flows to the deliveries and some to the proportioner. The proportioner induces foam concentrate, to produce a rich foam solution, which passes back to the suction side of the pump. Before re-entering the pump, the foam solution mixes with a fresh intake of water. Most of it passes to the deliveries, while a small amount returns to the proportioner, where more foam concentrate is induced and the process is repeated.

The proportioning inductor has an operating pressure range of between 3 and 14 bar. The recommended pressure is 7 bar with a water requirement of 192 lpm.

*Pressure control valve:* Continuous proportioner can only function correctly, if the pressure on the suction side of the pump is less than one-third of the pressure on the delivery side. The pressure control valve brings down the pressure in the pump inlet line to one-fifth of the hydrant pressure. The valve may be inserted into the pump inlet line at any convenient position. It can be fitted as an integral part of the pipe work system on a water tender appliance.

Water under pressure from the hydrant, passes through the valve over a moveable butterfly regulator. This butterfly regulator is connected to a hydraulic piston, which receives pressure from both the sides of the butterfly regulator. The area of the piston which is subjected to pressure on the upstream side is one fifth of the area of the piston on the downstream side. When the downstream pressure is one-fifth of the upstream pressure, the force acting on the piston will balance. When the upstream (hydrant) pressure rises, the downstream will experience a proportionally greater pressure increase.

This makes the piston move, closing the butterfly regulator and thereby reducing the downstream pressure until 5:1 ratio is restored. If the hydrant pressure falls, the reverse process will take place.

## Foam Systems for Petroleum Storage Installations

Petroleum products (other than LPG) are classified according to their closed-cup flash points as:

Class A–liquids which have a flash point below 23° C
Class B–liquids which have a flash point between 23 and 65° C
Class C–Liquids with flash point between 65 and 93° C
Class D–liquids with flash point above 93° C.

For petroleum tanks (floating roof or fixed roof) with diameter exceeding 18 m and storing class A and class B petroleum liquids, fixed foam system or semifixed foam system is provided for fire protection.

### Fixed Foam System

A fixed foam system comprises of fixed piping for water supply at adequate pressure, foam concentrate tank, inductor, suitable proportioning equipment for drawing foam concentrate and making foam solution, fixed piping system for onward conveying to foam makers for making foam, vapour seal box and foam pourer. Suitable detection system is provided to activate the foam system.

### Semifixed Foam System

A semifixed foam system gets supply of foam solution through a mobile foam tender. A fixed piping connected to the foam maker and vapour seal box conveys foam to the surface of the tank.

### Protection System for a Fixed Roof Tank

System should be designed to create foam blanket on the burning surface in a short period. Foam should be applied at a high rate to overcome the fire; foam makers should be located not more than 24 m away from the periphery of the tank; the foam system should be designed to provide 600 mm foam dam height; a minimum of 2 foam pourers per tank should be provided.

### Foam Application Rate

The minimum foam flow rate should be between 5 to 18 lpm/m$^2$ of liquid surface area of the tank, depending on the type of tank.

### Duration of Foam Discharge

The duration of foam application for tanks containing class A and class B is 65 minutes. For only spill fire protection, it can be 30 minutes.

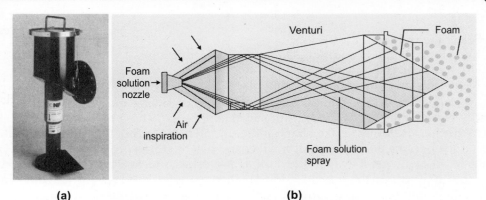

| (a) | (b) |
|---|---|

**Fig. A2.42:** (a) Foam pourer, (b) Foam making venturi principle.

## SMALL GEARS AND HAND TOOLS

Tools are devices which enhance human capability. Tools help us to perform tasks which are not possible to perform by our hands alone.

It is important that the fire operator must possess the requisite knowledge of the application and utility of various fire service equipment such as hose, hose fittings, ladders, fire protective clothes, breathing apparatuses and a variety of tools carried on the appliances. These varieties of tools and equipment are referred to by the term "small gears". To perform tasks efficiently, not only tools should be available, we should also know how to use them.

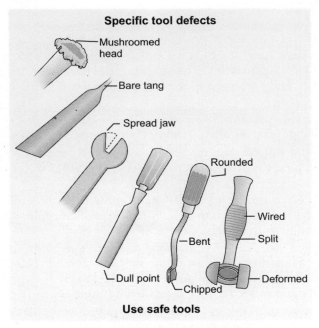

**Fig. A2.43:** Tool-use guide.

## Tool Rules

- Select the right tool for the job
- Keep tools in good condition. Do not use defective tools.
- Use tools correctly
- Keep tools in a safe and designated place.

Items under "small gears" appear to be composed of diverse tools but they have been carefully chosen after many years of practical experience. These tools may be different from each other, or similar in size, shape, material and function, but they are of great utility in actual fire fighting and rescue. All small gears shall be stored properly, so that the crew members can easily get it even in the dark. Hooks, slots, brackets and straps are used to secure the small gears in the appliance. Most of the small gears need very little maintenance.

It is highly important, in the interest of the safety of the personnel, to remember that the equipment with insulated handles like the fireman's axe, pliers, etc. should be used only after observing necessary precautions on live electrical installations.

Small gears are classified into three groups:

  i. General purpose tools and equipment
  ii. Special tools and rescue equipment
  iii. Lamps and lighting sets

## General-purpose Tools and Equipment

They are further divided into two groups

  a. Conventional tools
  b. Non-conventional tools

### Conventional Tools

*Spades and shovels:* They have steel blades and wooden handles. A spade has rounded sides and a shovel has square sides. The length of the handle varies depending on the nature of application. They are used to dig, scrape, scoop-up a small dam, to cut shrubs and bushes, etc.

*Pickaxe:* A crescent-shaped steel head having two sharp ends, one sharp and other one pointed with a well balanced wooden handle. They are used to dig or turnover an object.

*Axe:* It consists of a forged steel head with a fine cutting edge and a well-balanced wooden handle. It is used to cut through wooden partitions like doors, windows, screens, etc.

*Sledge hammer:* It has a steel head of heavy weight with round ends at both sides. It is usually fitted with steel or wooden handles. It helps to break through various obstructions. It is used in conjunction with chisels and wedges.

*Steel crowbar:* It is a steel bar of nearly one-metre length with one or both the ends shaped at an angle. Some times one end may be claw-shaped. The claw-shaped end is used to pull off nails from boards. The other end is used to lift sliding doors, knock down unsafe walls, break open locks, remove doors from hinges, etc.

*Steel wedges:* They are steel pieces of different lengths, rectangular at one end, tapering in shape and sharp at the other. Wedges are used to tighten or secure, one or other part of structures. They are used with the help of hammers.

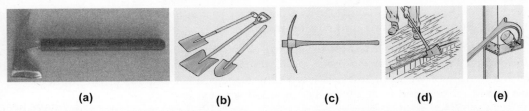

**Fig. A2.44:** (a) Fire man's axe, (b) Spades and shovels, (c) Pick axe, (d) Using the axe and (e) Using the hook.

*Hammer:* A steel head, flat at one end and clawed at the other, it may be ball peen type to get better hammering effect. The handle is made of wood. They are made out of forged steel of different weights and sizes. Hammers of different shapes are available to suit variable applications. They are used to strike a job or to break an object while engaged in rescue operations Sledge hammers and other types of heavy hammers are used to break walls or any such obstructions during rescue operations. It is used to hammer, break, flatten sharp edges, pull off nails, etc.

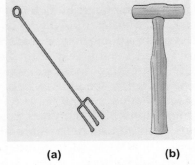

**Fig. A2.45:** (a) Fire fork, (b) Sledge hammer.

*Saws and hacksaws:* Saws are wood cutting tool and hacksaws are metal cutting tools. They are available in different sizes and thickness and are operated by hand. Saws of different size are used for cutting wood and hacksaws are effectively used for cutting metals of various sizes and thickness during rescue operations. A saw is primarily a wood cutting tool. It is made out of hardened and tempered steel strip. It has sharp teeth at one edge. The size and shape of the teeth varies from saw to saw. There are different types of saws for various applications. Some of them are:

a. Cross-cut saw        b. Rip saw
c. Hand saw           d. Floor saw
e. Keyhole saw       f. Tenon saw, etc.

Chain saws, circular saws, etc. are operated by electrical power.

*Hacksaw:* It is a metal cutting hand operated tool. It has a steel frame, which can easily be adjusted to take different length of blades. Blades are made of carbon steel or alloy steel and have fine teeth cut on one side. They are used to cut metals like padlocks, window frames, railings, etc.

*Pliers:* They have forged steel cutting bits operated by long handle for better leverage. The handle may or may not be insulated. They are used for gripping and cutting purposes. They are usually classified depending on the type and overall length. They are used for cutting wires and cables, for gripping, bending, holding rods, pipes, etc. They are used to cut wires, to bend metal bars, etc. Pliers with insulated handles can be used for cutting live electrical wires with necessary safety precautions.

*Chisels:* They are used for cuffing and chipping of wood and metals. They are made of carbon steel and are available in different sizes and shapes to suit different applications. Chisels for chipping and cutting wood and metals are available in various sizes. They have forged steel blades with varying width for different applications. Wood chisels are used for cutting wooden boards; ply woods, partitions, etc. whereas metal chisels are used for chipping, scraping and cuffing metals.

*Screw driver:* They are used for tightening or loosening. Screw drivers are available in wide ranges depending upon their applications. It is a long steel rod of varying length with a handle of wood or fibre. The working edge could be flat or cross-pointed. They are used to screw and unscrew bolts, threaded screws, etc.

*Spanners:* They are available in different sizees, shapes and types for the dismantling and assembling process. They are made of high carbon steel and are classified on the basis of type, size of the bolt or nut. They are forged steel or cast iron pieces with recess, ring, sockets at one or both ends. The spanners, which are socket shaped, are pieces used in conjunction with Tommy bars. Spanners with adjustable size are also used. They are used to tighten or loosen bolts or nuts.

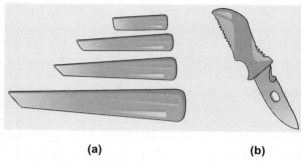

**(a)**                               **(b)**

**Fig. A2.46:** (a) Chisels, (b) Quick release knife.

*Wrenches:* They are similar to spanners but bigger in size. They are made out of forged steel. Examples are pipe wrench and stilson wrench. They are generally used for tightening or loosening cylindrical objects such as pipes.

*Ratchet braces, wheel braces and bits:* These are used for drilling holes through wood or metal. Different sizes of drill bits in varying length and materials such as high speed steel, carbon steel, alloy steel, etc. are used for making drill bits. They are also known as breast drills.

*Quick release knife:* It is a special type of knife, with a slightly curved and sharp steel blade. The upper and outer edges are rounded. They are used to set free persons from moulded harness or seat belts and used for other rescue work.

*Chimney rod:* The equipment comprises of a number of flexible canes, which are screwed together. A chimney nozzle is attached to the head of the first rod. A hose reel is connected to this nozzle and the rods are pushed up the chimney to required height. They are generally used to extinguish chimney fires.

*Persuader:* It is a steel piece-shaped like a cigar tapering at one end. It cannot be used to cut metals or other objects as the points is rounded. It is inserted into the loop of the hasp and the head struck with a sledge hammer. It breaks the hasp by its expanding effect. A wire handle is attached to hold it in position safely. They are use to break open hasp, pad locks, etc.

*Spreader:* This is a special type of tool with a nut, which carries two 'Y' shaped lugs with threaded stems. At the center of the nut, a hole is provided for inserting Tommy bar. The 'Y' shaped lugs can spread outwards by rotating the nut. They are used to rescue trapped persons, animals, etc.

## Field Equipment

Some specialist tools like beaters, stack drags, pitchforks and thatch knives are still required, carried and used. They can be very useful in outdoor fires.

### Beaters or Fire Bats

i. *Green branches:* A handle l to 1.25 m long with green branches tied securely at one end.
ii. *Wire net:* A piece of wire-netting is attached to a frame of 30 cm width and 35 cm length. The frame is attached to a wooden handle.
iii. *Gunny sacks:* A small single gunny sack is fitted in side of a frame and a handle is attached to the frame.

All the above types of beaters are used to extinguish bush fires, small fires, etc. by beating.

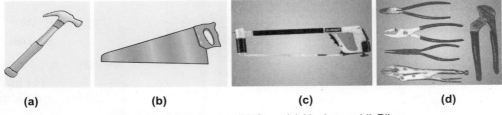

| (a) | (b) | (c) | (d) |

**Fig. A2.47:** (a) Hammer, (b) Saw, (c) Hacksaw, (d) Pliers.

### Forks

i. *Drag hook:* A wooden handle fitted to an iron fork at right angles. The fork has four prongs.
ii. *Pitch fork:* A wooden handle fitted to an iron fork, which has two slightly curved prongs.

### Hay Knife

It is a special type of knife, which is used to cut burned portion of a haystack.

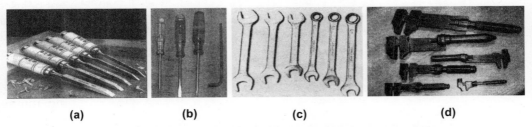

| (a) | (b) | (c) | (d) |

**Fig. A2.48:** (a) Chisels, (b) Screw driver, (c) Spanners, (d) Wrench.

## OTHER GEARS

### Fire Blankets

Fire blankets are used to prevent or resist penetration by molten metals during cutting or welding operations, to protect from radiant heat, to extinguish small fires by smothering, etc. The materials used to make fire blankets are asbestos, fiber glass and leather. Presently fiber glass and leather are widely used.

### Fire Protective Clothing

This consists a complete suit with helmet and visor, tunic with breathable fabrics, reinforced boots, leather gloves, etc. They are used to tackle fires of high temperature. A recent development is made of heat reflecting 'aluminized asbestos fabrics' or aluminized Kevlar. This is used as an approach or entry suit for rescue efforts against a very high temperature. It needs careful maintenance to retain the reflectivity of aluminum. To enter into an atmosphere of ammonia, it requires further protection. A special suit known as 'ammonia suit' is used with breathing apparatus.

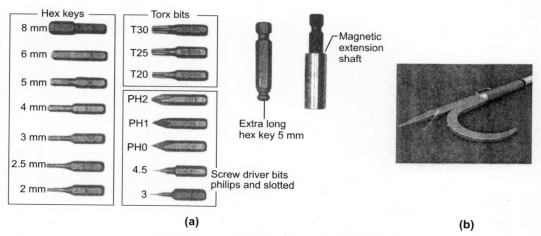

**(a)**                                                                                      **(b)**

**Fig. 5.49:** (a) Tool bits, (b) Fire man hook.

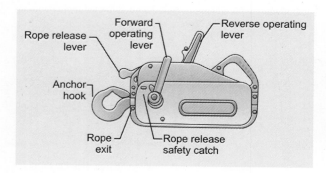

Fig. A2.50

**Medical first aid equipment**
The first aid box containing sterilized bandages, medicines, etc. to give medical aid to injured persons.

## SPECIAL SERVICE

### Rescue Equipment

This type of equipment can be divided into two categories, namely powered and non-powered. Powered equipment are driven by petrol or diesel engine or electricity or by motor driven hydraulic and pneumatic drives. Hand-pumped equipment, whether pneumatic or hydraulic are considered non-powered equipment.

### Non-powered Equipment

*Hydraulic jack:* This type of jack is very much like a car jack. The length of their stroke and their capacity for lifting usually identifies them. A lever fitted to the side of the jack, pumps the hydraulic oil and load can be released by operating a valve, which permits the hydraulic oil to pass back to the reservoir. Jacks of different capacities ranging from 5 to 50 tons are available.

*Tirfor:* It is a band-operated pulling and lifting machine. A standard tirfor has a safe working load of 1600 kg lifting and 2500 kg pulling. It is operated by pulling directly on a wire rope. The pull is applied by means of two pairs of smooth jaws, which exert a grip on the rope in proportion to the load being pulled or lifted.

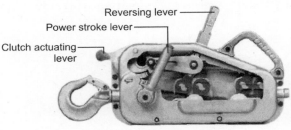

**Fig. A2.51:** Tirfor–cross-section.

### Flame Cutting Equipment

Flame cutting equipment work with the combination of oxygen and acetylene gases or oxygen and propane gases. The gases are led to a cutting torch through control valves, gauges and specially designed hoses. When oxygen and a flammable gas (acetylene or propane) are delivered through the torch and a spark is applied, the gases will ignite producing a flame of high temperature. This flame is used for cutting metals for rescue operations.

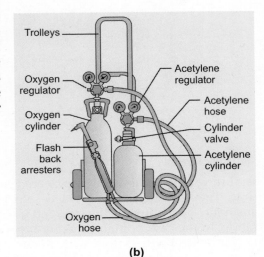

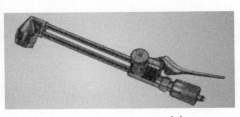

**(a)**            **(b)**

**Fig. A2.52:** (a) Cutting torch, (b) Cutting set.

## Thermal Cutting Equipment

These equipment are operated with high-pressure oxygen. High-pressure oxygen is fed from the cylinder to the box, where it is reduced to a pressure of 27 bar. A release valve is incorporated in the box, which is set to 30 bar. Two gauges in the control box show the cylinder pressure and the running pressure. The cable itself consists of high tensile steel galvanized with wire laid up to from a wire rope. It is covered in a plastic sheath to contain the oxygen flow. The center of the cable is hollow to enable a good, fast passage of oxygen to feed the combustion at the tip.

## Hand Pumped Hydraulic Equipment

There is a wide variety of fittings which may be used along with this equipment such as alligator jaw spreader, wedges, jacks, etc. A hand pump sucks the oil from the tank and the delivery of the pump may be connected through hoses to rams, wedges, etc. The hydraulic pressure lifts the ram when the pump is operated.

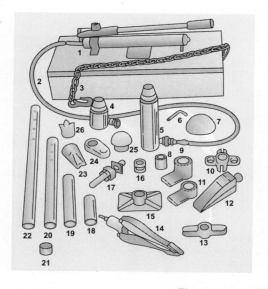

| | |
|---|---|
| 1. Pumps and handle | 14. Alligator jaw spreader |
| 2. Lengths of hose | 15. Flat base |
| 3. Chain | 16. Tube coupler |
| 4. Ram (short stroke) | 17. Slip lock extension |
| 5. Ram | 18. Extension tube |
| 6. Lock pin | 19. Extension tube |
| 7. Flexible rubber base | 20. Extension tube |
| 8. Short male connector | 21. Saddle |
| 9. Spreader ram toe | 22. Extension tube |
| 10. Chain plate | 23. Wedge head |
| 11. Spreader plunger toe | 24. Clamp head |
| 12. Wedgies | 25. Flexible rubber head |
| 13. Pull plate | 26. 90° vic base |

**Fig. A2.53:** Hydraulic equipment set.

## Air Lifting Units

Air lifting units are inflated by powered compressors or by air cylinder through a control unit. They are generally divided into low-pressure air bags and high-pressure air cushions. Depending on the type and use, bags can be made of neoprene coated or nylon fabric. In the case of high-pressure type, high quality rubber reinforced with layers of kevlar or steel wire is used with an outer layer of neoprene. Operating pressures range from 0.5 bar for low-pressure bags to 8.5 bar for the high-pressure bags. Lifting capacity ranges from 5 to 70 tonnes.

## Powered Equipment

### Power Supply Units

Because of the increased use of the hydraulic equipment, many fire brigades are turning to small petrol driven compressors. They are usually of 2 KW power and the compressors can supply hydraulic oil up to 720 bar to operate a variety of tools. A few tools have their own power such as petrol driven chain saws and some are driven electrically from a 11 KV generator on a rescue unit or emergency tender. Many electrical appliances are driven by mobile generators.

*Hydraulic cutter:* They are a variety of these tools carried on appliances. They can exert a cutting pressure of any thing up to 29 tonnes. They are very useful for removal of vehicle roofs, doors, etc. Even solid steel bars up to 20 mm thick can be cut with a hydraulic cutter.

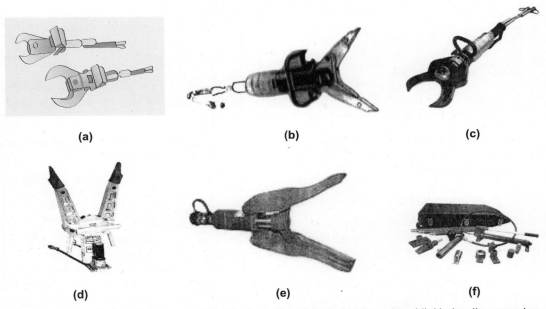

(a)      (b)      (c)

(d)      (e)      (f)

**Fig. A2.54:** (a) Hydraulic cutter, (b) Hydraulic cutter, (c) Hydraulic cutter (d) Hydraulic spreader, (e) Hydraulic spreader, (f) Hydraulic ram.

*Hydraulic spreader:* This tool can be used to force apart or pull together where necessary. Only the tips are used and the spreading distance varies with types, as does the spreading force. The working pressure is 720 bars and this equipment can also be used in conjunction with chains attached to the tips. This type of equipment should not be used for lifting purposes.

*Hydraulic rams:* They are available in various sizes and capacities with single or double acting rams. A set having a lifting capacity of 10 tonnes can use three sizes of rams ranging from 680 to 1660 mm. They can be used in practically any mode and have safety control check valves which block the load in the event of an oil supply failure.

*Electric power tools:* Various types of power tools and appliances are used for the rescue operations by fire brigades. Some of the examples are disc grinder, disc cutter, drill, chainsaw, blower, etc.

## Lamps and Lighting Sets

The need for enough illumination at an incident is very essential. Whether it is flood lighting of the whole area or just focusing on a point, illumination certainly helps prevent accidents and helps fire fighters do their work efficiently even in unfavourable conditions. Fire brigades have a variety of lighting appliances ranging from powerful halogen floodlights on powered masts to personnel hand-held torches.

### Hand Lamps

a. *Electrical hand lamps:* Operating on dry cells. It has a bulb with a reflector.
b. *Portable electric box lamp:* Operating on accumulators. They are able to provide better light for a long period.
c. *Hurricane lamps:* It bums by paraffin. It is used in deep wells, sewage, etc. where light is insufficient.
d. *Safety lamps:* They are made with gas tight joints so that they can be used in flammable atmosphere. They are operated on accumulators and they give safe and sustained light. They are used in conjunction with BA sets.

*Precautions:* The accumulators must be maintained properly. The glass funnel of hurricane lamps must be cleaned, wicks properly trimmed and the tank of paraffin should be kept filled up always.

### Flood Lights and Search Lights

These are used for flooding the required area with sufficient light for rescue and fire fighting operations. The powerful full beam can be focused across some distance. Most flood lighting system are electrical and powered either by portable generators or by vehicle's 24 V batteries. Certain appliances have electro/air operated flood lighting masts, which can be raised to a height of 5.4 m and rotated through 340°. The mast is equipped with two 500 W halogen flood lamps and a blue flashing warning beacon or 4 to 6 smaller flood lights. Different types of portable flood lights which can be mounted on tripods, are also used.

## Maintenance of Small Gears

1. Material made of iron or steel must be checked periodically for rust, stain, etc. If found rusty, the same should be cleaned thoroughly and suitable antirust compounds should be applied.
2. Storage and stowage must be in such a way that there is no dampness.
3. They must be cleaned and polished carefully.
4. Care should be taken to see that wooden equipment are not damaged by white ants.

5. Care should also be taken to prevent damage caused by fungi.
6. A coat of varnish or paint should be given for preserving wooden parts.
7. Rubber equipment should be kept under ideal conditions and temperature for storage. They should not be subjected to very extreme (low or excessive) temperatures.
8. They should not be allowed to come into contact with oils, chemicals, solvents, etc.
9. Liberal use of chalk powder will prevent hardening and cracking of rubber equipment.
10. They should be kept in the appropriate places provided for each, for easy access.

## PORTABLE FIRE PUMPS

Portable fire pumps provide pumping facility to bring fire water from its source to the scene of fire. Centrifugal fire pump supplements the fire fighting equipment in forest service, fire departments, industrial parks, and timber yards. They are constructed to be light weight for portability. The portable pumps are usually engine driven, so that they can be used in areas remote from sources of electrical power supply. Time consuming activities like wiring, etc. are not required. The portable pumps can be stationed wherever necessary such as close to water sources and this can reduce the length of piping hoses required. The engine-starter should be of the quick starting type.

**Fig. A2.55:** An engine driven portable pump mounted on a skid.

### Pipe Data

| Standard pipe dimensions | | | | | |
|---|---|---|---|---|---|
| *ANSI* | | | *ISO* | | |
| *Nominal pipe size* | *Actual OD* | | *Nominal pipe size* | *Actual OD* | |
| *Inches* | *Inches* | *mm* | *Inches* | *mm* | *Inches* |
| 1 | 1.315 | 25 | 0.984 | 32 | 1.260 |
| 1¼ | 1.660 | 32 | 1.260 | 40 | 1.575 |
| 1½ | 1.900 | 40 | 1.575 | 50 | 1.959 |
| 2 | 2.375 | 50 | 1.969 | 63 | 2.480 |
| 2½ | 2.875 | 65 | 2.559 | 75 | 2.953 |
| 3 | 3.500 | 80 | 3.150 | 90 | 3.543 |
| 4 | 4.600 | 100 | 3.937 | 110 | 4.331 |
| 5 | 5.563 | 100 | 4.921 | 140 | 5.512 |
| 6 | 6.625 | 150 | 5.906 | 160 | 6.299 |

## Branch Pipe Sizing

The method given below is helpful in determining the branch pipe sizes whenever a pipe or hose is to be branched to provide water at different points. The equal area method of sizing pipe manifolds for branching is based on maintaining constant total cross-sectional area in all portions of a piping train, regardless of the number of branches in each portion.

In the sketch below, the equal area method requires that area of X = 2 times area of Y = 6 times area of Z. The advantage of this method is that once the size of the smallest branch has been determined, by velocity pressures or any other valid method, the remainder of the piping system can be correctly sized without any additional calculations. This method helps in preventing excessive pressure losses at the transfer points. But, if the calculation of the smallest branches is in error, the entire system will be incorrectly sized.

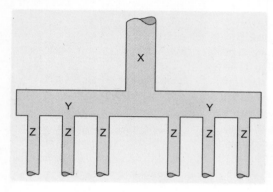

For example,

If the size of Z–the final branches is 1 inch (area of 3 sq in × 6 pipes = 4.7 sq in), the manifold Y should of 6 inches and the supply pipe X should be 12 inches.

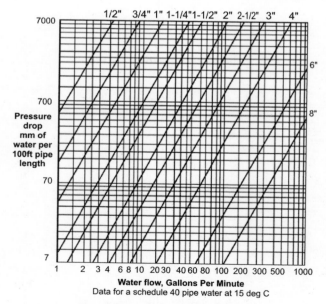

**Fig. A2.56:** Sizing water piping.

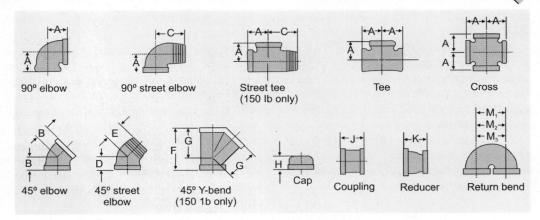

Fig. A2.57

## Some Common Pipe Fittings

*Elbow:* An angular or jointed part, steel elbow are used during the installation of pipes that allow for easy connection between pipes and smoothen the flow disturbances, reduce swirl and develop highly symmetrical flow with minimal pressure loss.

*Bend:* Bend is the curved portion of tube or pipe, bend helps in a smooth change of direction water by balancing the reaction force, downstream pressure, and velocities. There are various angle in degrees to which the pipe bends are formed, i.e. 45°, 90°, 180°, which lessens the flow resistance and possible leakage.

*Stub-end:* They provide joints between plain end pipes and pipe fittings.

*Tee:* A tee is used for connecting pipes of different diameters or for changing the direction of pipe runs. A common type of pipe tee is the straight tee, which has a straight-through portion and a 90° take-off on one side.

*Reducer:* A reducer is used to connect two pipes of different diameters.

*Flanges:* Flanges are metal-discs used for joining two lengths of pipe to form a flange coupling. Couplings also help to relieve any minor misalignment between the two lengths.

*Nipple coupling:* A nipple coupling is a short length of pipe with a male thread on each end. It is used for extension from a fitting.

*Union:* There are two types of pipe unions. The ground joint union consists of three pieces, and the flange union is made in two parts. Both types are used for joining two pipes together and are designed so that they can be disconnected easily.

*Socket:* An opening or a cavity into which an inserted part is designed to fit, socket is used in a pipe with an expansion at one end to receive the end of a connecting pipe.

*Valves:* A movable control element, valves regulate the flow of gases, liquids or loose materials through piping by opening, closing or obstructing ports or passageways.

## Different Types of Valves

*Check valves:* Check valves are self-actuated. These valves are opened and sustained in the open position by the force of the liquid velocity pressure. They are closed by the force of gravity or backflow. The seating load and tightness is dependent upon the amount of back pressure. Typical check valves include swing valves, foot valves lift check and stop check valves. Swing check valves are used to prevent flow reversal in vertical pipelines.

*Ball valves:* Ball valves are low cost, compact, light weight, easy to install and easy to operate valves. They offer full flow with minimum turbulence and can balance or throttle fluids. Typically ball valves move from closed to full open in a quarter of a turn of the shaft and are, therefore, referred to as quarter turn ball valves. Low torque requirements permit the ball valves to be used in quick and automatic operations and ball valves have a long operating life.

*Gate valves:* The gate valve is one of the most common valves used in liquid piping. A gate valve is an isolation valve used to turn on and shut off the flow, isolating either a piece of equipment or pipeline as opposed to actually regulating flow. The gate valve has a gate-like disc which operates at a right angle to the flow path. As such it has a straight through port that results in minimum turbulence erosion and resistance to flow. However, because the gate or the seating is perpendicular to the flow, gate valves are impractical for throttling purposes and are not used for frequent operation applications.

*Globe and angle valves:* Liquid flow does not pass straight through globe valves. Therefore, it causes an increased resistance to flow and considerable pressure drop. Angle valves are similar to globe valves, but the inlet and outlet ports are at 90° angles to each other, rather than at 180°. Because of this difference, angle valves have slightly less resistance to flow than globe valves.

*Butterfly valves:* Butterfly valves provide a high capacity flow with low pressure loss and are durable, efficient and reliable. The chief advantage of a butterfly valve is its seating surface. The reason for this advantage is that the disc impinges against a resilient liner and provides bubble tightness with very low operating torque. Butterfly valves exhibit an approximately equal percentage of flow characteristics and can be used for throttling or for on/off control.

*Pinch valves:* Pinch valves, pinch an elastomeric sleeve shut in order to throttle the flow through the pipeline. Because of the streamlined flow path, the pinch valve has very good fluid capacity. Pinch valves typically have a fairly linear characteristic. Pinch valves can be used for turbid water also.

*Plug valves:* Plug valves are another type of isolation valve designed for uses similar to those of gate valves, where quick shutoff is required. They are not generally designed for flow regulation. Plug valves are sometimes also called as cock valves. They are typically a quarter turn open and close. Plug valves have the capability of having multiple outlet ports. This is advantageous in that it can simplify the piping. Plug valves are available with inlet and outlet ports with four-way multi-port valves which can be used in place of two three or four straight valves.

## Other Flow Control Devices

### Pressure Relief Devices

*Pressure relief valves:* Pressure relief valves are automatic pressure relieving devices that protect piping systems and process equipment. The valves protect systems by releasing excess pressure. During normal operation, the valve disc is held against the valve seat by a spring. The spring is adjustable to the pressure at which the disc lifts. The valve disc lift is proportional to the system pressure so that, as the system pressure increases, the force exerted by the liquid on the disc forces the disc up and relieves the pressure. The valve will reseat when the pressure is reduced below the set spring pressure.

*Rupture discs:* A rupture disc is another form of a pressure relief device. Rupture discs are designed to rupture automatically at a predetermined pressure and will not reclose. These discs can relieve very large volumes of liquid in a rapid manner. Materials of construction include metals, graphite or plastic materials held between special flanges and of such thickness, diameter and shape, and material, that it will rupture at a pre-determined pressure.

*Backflow prevention:* Backflow prevention is often handled by three main methods, one of which is check valves. Another method is the use of pressure and vacuum breakers. The third method is use of reduced pressure backflow prevention assembly.

## Pipeline Protection

### Corrosion Protection of Fire Protection Piping

The integrity and life of a fire protection piping system is dependent upon corrosion control. Internal corrosion of piping is controlled by the selection of appropriate piping materials, wall thickness, and linings and by regulating the quality of the water by addition of treatment chemicals. External corrosion can be addressed through materials of construction, coatings and linings and by providing cathodic protection.

### Cathodic Protection

Cathodic protection is employed to provide corrosion protection to buried (underground) piping. The two main types of cathodic protection are galvanic protection and impressed current. Cathodic protection reduces corrosion by minimizing the difference in potential between the anode and the cathode.

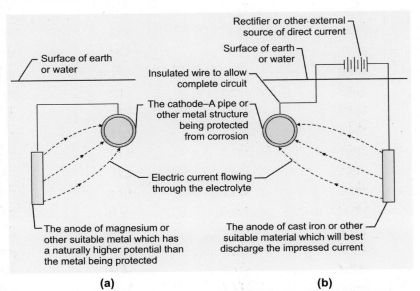

**(a)**                    **(b)**

Fig. A2.58: (a) Galvanic anode system; (b Impressed current system.

## Protective Coatings

Since corrosion of metallic piping is electrochemical in action, if a protective coating which is continuous, impervious and insulating, is applied to the pipe exterior, the electrical circuit cannot be completed and therefore corrosion will not occur. The basis for the selection of an exterior coating is chemical inertness, adhesiveness, electrical resistance, impervious and flexibility to adjust to the pipe contour and any pipe deformation and environmentally induced stress. The coating must be applied without any gaps, pinholes or cracks.

Pipelines which are above ground can be painted after suitable surface preparation.

| Actual Pipe Dimensions | | | | |
|---|---|---|---|---|
| Schedule 40 (SI units) | | | | |
| Nominal pipe Size (in) | Outside diameter (mm) | Inside diameter (mm) | Wall thickness (mm) | Flow area (m²) |
| 1/8 | 10.3 | 6.8 | 1.73 | $3.660 \times 10^{-5}$ |
| 1/4 | 13.7 | 9.2 | 2.24 | $6.717 \times 10^{-5}$ |
| 3/8 | 17.1 | 12.5 | 2.31 | $1.236 \times 10^{-4}$ |
| 1/2 | 21.3 | 15.8 | 2.77 | $1.960 \times 10^{-4}$ |
| 3/4 | 26.7 | 20.9 | 2.87 | $3.437 \times 10^{-4}$ |
| 1 | 33.4 | 26.6 | 3.38 | $5.574 \times 10^{-4}$ |
| 1 1/4 | 42.2 | 35.1 | 3.56 | $9.653 \times 10^{-4}$ |
| 1 1/2 | 48.3 | 40.9 | 3.68 | $1.314 \times 10^{-3}$ |
| 2 | 60.3 | 52.5 | 3.91 | $2.168 \times 10^{-3}$ |
| 2 1/2 | 73.0 | 62.7 | 5.16 | $3.090 \times 10^{-3}$ |
| 3 | 88.9 | 77.9 | 5.49 | $4.768 \times 10^{-3}$ |
| 3 1/2 | 101.6 | 90.1 | 5.74 | $6.381 \times 10^{-3}$ |
| 4 | 114.3 | 102.3 | 6.02 | $8.213 \times 10^{-3}$ |
| 5 | 141.3 | 128.2 | 6.55 | $1.291 \times 10^{-2}$ |
| 6 | 168.3 | 154.1 | 7.11 | $1.864 \times 10^{-2}$ |
| 8 | 219.1 | 202.7 | 8.18 | $3.226 \times 10^{-2}$ |
| 10 | 273.1 | 254.5 | 9.27 | $5.090 \times 10^{-2}$ |
| 12 | 323.9 | 303.2 | 10.31 | $7.219 \times 10^{-2}$ |
| 14 | 355.6 | 333.4 | 11.10 | $8.729 \times 10^{-2}$ |
| 16 | 406.4 | 381.0 | 12.70 | 0.1140 |
| 18 | 457.2 | 428.7 | 14.27 | 0.1443 |
| 20 | 508.0 | 477.9 | 15.06 | 0.1794 |
| 24 | 609.6 | 574.7 | 17.45 | 0.2594 |

## Discharge from an Orifice

Let A = cross-sectional area of the orifice = $(\pi/4)d^2$

and $A_c$ = cross-sectional area of the jet at the vena contracta = $(\pi/4)d_c^2$

then $A_c = C_c A$

or $C_c = \dfrac{Ac}{A} = \left(\dfrac{d_c}{d}\right)^2$

where $C_c$ is the coefficient of contraction

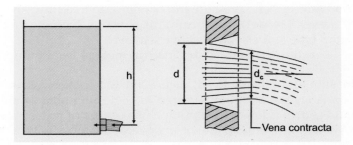

At the vena contracta, the volumetric flow rate Q of the fluid is given by

Q = area of the jet at the vena contracta × actual velocity

$= A_c v$

or $Q = C_c AC_v \sqrt{2gh}$

The coefficients of contraction and velocity are combined to give the coefficient of discharge, $C_d$

i.e. $C_d = C_c C_v$

and $Q = C_d A \sqrt{2gh}$

Typically, values for $C_d$ vary between 0.6 and 0.65

Circular orifice: $Q = 0.62 A \sqrt{2gh}$

Where Q = flow $(m^3/s)$ A = area $(m^2)$ h = head (m)

Rectangular notch: $Q = 0.62 (B \times H)^{2}/_{3} \sqrt{2gh}$

Where B = breadth (m) H - head (m above sill)

Triangular right angled notch: $Q = 2.635 H^{5/2}$

Where H = head (m above sill)

## PUMPS

A pump is a mechanical device which transfers energy to a fluid.

### Pump Terms

*Head:* Head is another measure of pressure; expressed in meters or bar. Usually applies to centrifugal pumps, indicates the height of a column of water being lifted by the pump, neglecting friction losses in piping.

*Flooded suction:* In a flooded suction setup, the liquid source is higher than the pump and liquid flows to the pump by gravity. Flooded suction is preferable for centrifugal pump installations.

*Lift (suction lift):* The liquid source is lower than the pump. Pumping action creates a partial vacuum and atmospheric pressure forces liquid up to pump. Theoretical limit of suction lift is 10 metres (34 feet); practical limit is about 7.5 metres (25 feet) or less, depending on pump type and elevation above sea level.

*Static discharge head:* It is the vertical distance (in metres) from pump to the point of discharge.

*Total head:* The sum of discharge head, suction lift and friction losses.

Prime/priming: A charge of liquid required to begin pumping action of centrifugal pumps when liquid source is lower than pump. This charge of water may be held in the pump by a foot valve on the intake line or a valve or chamber within the pump. This action of filling the suction line with water before starting the pump is called priming.

Methods to avoid priming of pumps:
- By using a foot valve.
- By designing flooded suction .
- By submerging the pump in the water.

*Specific gravity:* Specific gravity is the ratio of the weight of a given volume of a liquid to the same volume of pure water. Pumping heavier liquids (specific gravity greater than 1.0) will require more drive power.

*Viscosity:* Viscosity is the 'thickness" of a liquid or its ability to flow. Temperature must be stated when specifying viscosity, since most liquids flow more easily as they get warmer. The more viscous the liquid, the slower the pump speed required.

*Seal:* A seal is a device mounted in the pump housing and/or on the pump shaft, to prevent leakage of liquid from the pump.

*Flow:* Flow is the measure of the liquid volume capacity of a pump. Given in litres per minute (lpm) or gallons per minute (GPM).

*Pressure:* Pressure is the force exerted on the walls of a container (tank, pipe, etc.) by the liquid. Measured in bar or Pascals.

*Lip:* A flexible ring (usually rubber or similar material), with the inner edge held closely against the rotating shaft by a spring. A lip is used together with a seal.

*Seal-less (magnetic drive):* Here no seal is used; power is transmitted from motor to pump impeller by magnetic force, through a wall that completely separates motor from impeller.

*Relief valve:* A relief valve is usually used at the discharge of a positive displacement pump. An adjustable, spring loaded valve opens or "relieves," when a preset pressure is reached. A relief valve is used to prevent excessive pressure and pump or motor damage, if the discharge line is closed.

*Unloader valve:* Similar to relief valve.

*Check valve:* A check valve allows liquid to flow in one direction only. It is generally used in the discharge line to prevent reverse flow.

*Foot valve:* A foot-valve is a type of check valve with a built-in strainer. Used at the point of liquid intake to retain liquid in the system, preventing loss of prime when liquid source is lower than pump.

## Centrifugal Pump

A fan shaped impeller rotating in a circular housing, pushing liquid towards a discharge opening. A centrifugal pump has a simple design where the only wearing parts are the shaft seal and bearings. Centrifugal pumps are usually used where large flow of liquid at relatively low pressure (head, lift) is desired. Self-priming centrifugal pumps have the same features as straight centrifugal pumps, but will self-prime without a foot valve after an initial filling of the pump casing. Non-self-priming centrifugal pumps work best with the liquid source higher than the pump (flooded suction/gravity feed). As the discharge pressure (head) increases, flow and drive power requirements increase. Maximum flow and motor loading usually occur at minimum head.

## Positive Displacement Pump

In a positive displacement pump, the pumping action is created by moving chambers or pistons. They are also called as reciprocating pumps. The flow rate of this pump is almost the same at any pressure level. They are generally self-priming. They should never be operated dry, because of internal wearing of rubbing parts. If the discharge flow is restricted (higher pressure or head), the drive power requirement increases. A relief device should be provided on the discharge line to prevent over pressure and damage to pump or motor if discharge line is closed off or severely restricted.

*Mechanical seal:* A mechanical seal has a rotating part and a stationary part with highly polished touching surfaces. It has excellent sealing capability and life, but can be damaged by dirt or grit in the liquid.

The most common positive displacement pump types are given below.

*Diaphragm pump:* It consists of a flexible diaphragm which moves up and down in a chamber, creating suction and pressure. As the diaphragm is moved up, it creates a vacuum which opens the suction valve and draws fluid into the chamber. When the diaphragm is forced down, fluid is forced out through the discharge valve. Diaphragm pumps handle fluid mixtures with a much greater percentage of solids (e.g. silt, mud, sludge and waste).

*Gear pump:* A gear pump consists of two meshed spur gears in a housing. As the gears rotate, fluid is carried in the space between the teeth. Gear pumps cannot handle any solid material, because of close running tolerances of gears. Gear pumps are suited for pumping more viscous liquids, at slower speeds.

*Flexible impeller pump:* A flexible impeller pump consists of a flexible vaned member, usually rubber, rotating in an eccentric housing. The volume of the spaces between the vanes changes as the pump rotates, creating pumping action.

*Rotary screw pump:* A rotary screw pump consists of a screw-shaped rotor, turning within a flexible stator, usually of rubber. Progressing cavities between screw and stator carry the fluid. These pumps can handle abrasive mixtures or slurries, at slower speeds.

*Roller or vane pump:* In a vane pump, rollers or vanes in a rotor are rotating in an eccentric housing like a flexible impeller pump.

*Piston pump:* In a piston pump, the fluid is drawn in and forced out by pistons moving within cylinders.

*Jet pump:* A jet pump utilizes water flow through a narrow opening or nozzle (jet, ejector) to bring water from a well. As water is forced through the nozzle, an area of low pressure is created and atmospheric pressure forces additional water from the well into the system. In shallow well systems (up to 7.5 m or 25 feet lift), the jet is located at the pump. In deep well systems, it is located at the bottom of the well.

## Deep Well Submersible Pump

A submersible pump is a centrifugal pump in which a number of impeller assemblies in a housing, are mounted on a shaft directly coupled to a submersible motor. The entire assembly is located at the bottom of the well. The required electric power is brought to the motor by a waterproof cable.

## MEASUREMENTS

The physical state or condition of a material, object or system can be described by its characteristics known as physical quantities or variables. Physical quantities such as length, mass, time, need to be described to be able to understand and compare two or more entities. The quantities should be measured and expressed quantitatively. Only when numbers represent units of measurement, they convey meaning. Measurement is the expression of physical quantities in terms of numbers and prescribed units by comparison to a standard. Measurement helps in data collection for study, control and performance of human and technological operations. Measurement provides a meaningful approach to observe relationships.

*Fundamental physical quantities and their measurement units:* The fundamental physical quantities are—*mass, length, time, temperature, electric current, luminous intensity and amount of substance*—All other physical quantities can be expressed in terms of the fundamental units.

## Units of Measurement

Measurement variables are expressed in terms of numbers or units. The units of measurement are based on the international system of units (SI system). The SI system consists of seven base units, two supplementary units and a series of multiples and sub-multiples of the various units.

The base units are metre, kilogram, second, ampere, kelvin, mole and candela. The supplementary units are radian and steradian.

The units mass (or force), length and time (M, L, T) are considered the fundamental units of measurement or units of fundamental quantities.

Measurement may be direct or indirect.

## Direct Measurement

In direct measurement, a comparison is made with a standard. Direct comparison, e.g. using a scale or micrometer to measure a linear dimension.

## Indirect Measurement

Indirect measurement is based on the effect or relationship with the parameter being measured, e.g. measuring the temperature by the expansion of a liquid column. This is a measurement based on the effect.

| SI base units | | |
|---|---|---|
| *Quantity* | *Name of SI unit* | *Symbol* |
| Length | metre | m |
| Mass | kilogram | kg |
| Time | second | S |
| Electric current | ampere | A |
| Thermodynamic temperature | kelvin | K |
| Luminous intensity | candela | Cd |
| Amount of substance | mole | mol |

| Derived units | | | |
|---|---|---|---|
| *Quantity* | *Name of SI unit* | *Symbol* | *In base units* |
| Frequency | Hertz | Hz | $1\,Hz = 1\,s^{-1}$ |
| Force | Newton | N | $1\,N = 1\,kg\,m\,s^{-2}$ |
| Pressure, stress | Pascal | Pa | $1\,Pa = 1\,N\,m^{-2}$ |
| Work, energy, quantity of heat | Joule | J | $1\,J = 1\,N\,m$ |
| Power | Watt | W | $1\,W = 1\,j\,s^{-1}$ |
| Quantity of electricity | Coulomb | C | $1\,C = 1\,A\,s$ |
| Electric potential | Volt | V | $1\,V = 1\,W\,A^{-1}$ |
| Electric resistance | Ohm | $\Omega$ | $1\,\Omega = 1\,V\,A^{-1}$ |
| Radioactivity | Becquerel | Bq | $1\,Bq = 1\,s^{-1}$ |

| Other physical quantities | | |
|---|---|---|
| *Quantity* | *Symbol* | *Definition* |
| Volumetric mass, mass-density, density | D | Mass divided by volume |
| Relative volumic mass, relative density | d | Ratio of the density of a substance to the density of a reference substance under conditions that should be specified for both substances |
| Massic volume, specific volume | $v$ | Volume divided by mass. $v = 1/D$ |
| Lineic mass, linear density | m/l | Mass divided by length |

*(Contd.)*

| Quantity | Symbol | Definition |
|---|---|---|
| Surface density | $D_s$ | Mass divided by area |
| Moment of inertia | I | The moment of inertia of a body about an axis is the sum (integral) of the products of its elements of mass and the squares of their distances from the axis |
| Momentum | $p$ | Product of mass and velocity |
| Force | F | The resultant force acting on a body is equal to the derivative with respect to time of the momentum of the body |
| Weight | W | The weight of a body in a specified reference system is that force which, when applied to the body, would give it an acceleration equal to the local acceleration of free fall in that reference system |
| Moment of force | M | The moment of a force about a point is equal to the vector product of any radius vector (r) from this point to a point on the line of action of the force, and the force. $M = r \times F$ |
| Moment of a couple | M | Sum of the moments of two forces of equal magnitude and opposite direction not acting along the same line |
| Torque | T, M | Generalization of the moment of a couple |
| Gravitational constant | G | The gravitational force between two particles is given by $F = G (m_1.m_2)/r^2$ where $r$ is the distance between the particles, and $m_1$ and $m_2$ are their masses |
| Pressure | P | Force divided by area |
| Normal stresss | ó | |
| Shear stress | s | |
| Linear strain, (relative elongation) | t, e | $e = dl/l_0$ where dl is the increase in length and $l_0$ is the length in a reference state to be specified |
| Shear strain | t | $t = dx/x$, where dx is the parallel displacement of the upper surface with respect to the lower surface of a layer of thickness $d$ |
| Volume strain (bulk strain) | $v$ | $õ = dv/v_o$, where dv is the increase in volume and $v_o$ is the volume in a reference state to be specified |
| Poisson ratio, poisson number | ì | Lateral contraction divided by elongation |
| Modulus of elasticity | $E$ | $E = ó/e$, $E$ is also called the young modulus. |

*(Contd.)*

| Quantity | Symbol | Definition |
|---|---|---|
| Shear modulus, modulus of rigidity | G | $G = s/t$ |
| Surface tension | ó | Force perpendicular to a line element in a surface divided by the length of the line element. |

## Force

Force is a vector quantity, a push or pull which changes the shape and/or motion of an object.

In SI, the unit of force is the newton N, defined as a $kg.m/s^2$.

## Weight

Weight is the gravitational force of attraction between a mass, m, and the mass of the earth.

In SI, weight can be calculated from

Weight $= F = mg$, where $g = 9.81$ m/s$^2$.

## Newton's Second Law of Motion

An unbalanced force F will cause an object of mass $m$ to accelerate $a$, according to:

$$F = ma$$

## Torque Equation

$$T = I \alpha$$

where, $T$ is the acceleration torque in Nm, $I$ is the moment of inertia in kg m$^2$ and $\alpha$ is the angular acceleration in radians/s$^2$.

## Momentum

Momentum is a vector quantity, symbol $p$,

$p = m\,v$ expressed in SI unit as kg.m/s.

## Work

Work is a scalar quantity, equal to the (vector) product of a force and the displacement of an object. In simple systems, where $W$ is work, $F$ force and $s$ distance.

$$W = F\,s$$

In SI, the unit of work is the joule, *J*, or kilo joule, *kJ*

$$1 J = 1 Nm$$

## Energy

Energy is the ability to do work, the units are the same as for work; *J*, *kJ*.

### Kinetic Energy

Kinetic energy is the energy due to motion.

$$E_k = 1/2mv^2$$

### Potential Energy

Potential energy is the energy due to position in a force field, such as gravity

$$E_p = m g h$$

### Thermal Energy

In SI, the common units of thermal energy are *J*, and *kJ*, (and *kJ*/kg for specific quantities)

### Electrical Energy

In SI, the units of electrical energy are *J*, *kJ* and kilowatt hours *kWh*. In imperial, the unit of electrical energy is *kWh*.

## Power

Power is a scalar quantity, equal to the rate of doing work.
   In SI, the unit is the watt *W* (or *kW*)

$$1W = 1 J/s$$

## Pressure

When a force is exerted on the surface of a body, the pressure is defined as the force divided by the surface area. Pressure is the force per unit area. Pressure is a vector quantity.
   In SI, the basic units of pressure are pascals *Pa* and *kPa*.

$$1 Pa = 1 Nm^2$$

Pressure = Force/area, i.e. $P = F/A$

Where $P$ is expressed in pascals, $F$ is expressed in Newtons, and $A$ is expressed in sq metres.
   Since the Pascal (SI unit) is a very small unit, we use the bar more often and which corresponds to $10^5$ pascal.

## Pressure Definitions

### Atmospheric Pressure

It is the pressure exerted by the atmospheric layer (about 11 km in height) surrounding the earth on all bodies which are in it.

The normal atmospheric pressure is equal to 1.01325 bar (at 0° C under normal gravity acceleration, g = 9.8066 m/s²) at sea level.

The atmospheric pressure decreases as we go up. It decreases by about 130 Pa for a 10 metre height variation.

The device used to measure the atmospheric pressure is called–barometer.

> 1 kPa = 0.294 in. mercury = 7.5 mm Hg
> 1 kPa = 4.02 in. water = 102 mm water
> 1 bar = 14.5 psi = 100 kPa
> 1 kg/cm² = 98.1 kPa = 14.2 psi = 0.981 bar
> 1 atmosphere (atm) = 101.3 kPa = 14.7 psi

### Relative or Effective Pressure

It is the pressure which is measured in relation to the atmospheric pressure. It is generally measured using manometers. Relative pressure is given by the difference between the absolute pressure and the atmospheric pressure.

### Absolute Pressure

It is the pressure measured in relation to complete vacuum.

Relationship between the three pressures:

Absolute pressure = Atmospheric pressure + Relative pressure

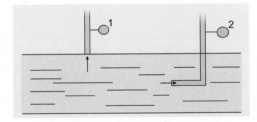

**Fig. A2.59:** Pressure in moving fluids.

On manometer no 1 velocity of the fluid has no effect. It measures the static pressure. Manometer 2 shows a higher pressure than 1. It also measures the pressure exerted by the velocity, called the dynamic pressure. Therefore manometer 2 measures the total pressure.

Total pressure = Static pressure + Dynamic pressure

## Static Pressure

It is the pressure exerted on the pipe walls by the fluid. It acts perpendicularly to the fluid flow direction.

## Dynamic Pressure

It is the pressure caused by the fluid velocity in the pipe. It can be felt by obstructing the flow.

It is expressed in relation to the density, in $kg/m^3$ to the velocity $v$ in $m/s$, by the expression

$$P_d = \text{density} \times \text{velocity}$$

## Total Pressure

It is the sum of the static and dynamic pressures.

## Differential Pressure

We know that the measure of the relative pressure is measuring of the pressure difference between the absolute pressure and the atmospheric pressure. If the atmospheric pressure is prevented from acting on one of the u-branch of the manometer and if the atmospheric pressure is substituted by any other pressure P2, we get another relative pressure called differential pressure (Pi and P2 are the pressures at either end of the U-tube).

$$\text{Differential pressure} = P1–P2$$

## Pressure Reduction or Loss due to Flow

Let us consider a long horizontal pipe in which a fluid is flowing. When the pressure is measured at different locations, we notice that the pressure is gradually decreasing in the direction of the flow. The fluid is losing its pressure as it flows–this is called pressure loss or head loss.

This is due to the reason that the fluid particles are rubbing one on to another and also against the walls of the pipe. This is also known as frictional loss.

The value of the head loss ($\Delta$ p) per linear metre of pipe is given by

$$\Delta P = \lambda \, (V^2/2g) \, (L/D)$$

Where  $\lambda$ = Flow coefficient
L = Length of the flow
$\Delta P$ = Head loss in metres of column of the fluid per linear metre
V = Average velocity of the fluid in m/s
D = Inner diameter of the pipe in metres.

We note that the head loss varies as the square of the fluid velocity and is inversely proportional to the pipe diameter. In a moving fluid, in a horizontal pipe, the pressure decreases in the direction of the flow.

## Measurement of Pressure

The u-tube manometer: The u-tube is generally a glass tube made with a single piece, curved to form a u-shape or two glass tubes connected together at their ends with a rubber tube. There are variations of the basic u-tube manometer such as the inclined tube, etc.

The tube is filled with a liquid of known density (water has a density of 1.0; water + 50% glycerine has density of 1.125, mercury has a density of 13.56).

The lower the liquid density, the greater the level difference between the two branches for the same pressure difference.

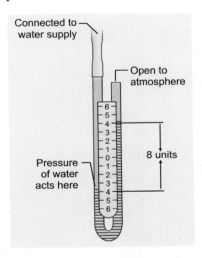

Fig. A2.60: A typical u-tube manometer (pressure gauge). Here the manometer indicates a pressure of 4 inches (100 mm) in water column.

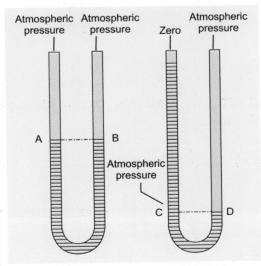

Fig. A2.61: U-tube showing the effect of atmospheric pressure in terms of height of water column with vacuum on the other side.

*Pressure Calculation Example*

What is the pressure exerted by a water column of 10 m?

$$P = p \times g \times h$$
$$p = 1000 \text{ kg/cu.m}; g = 9.8 \text{ m/s}^2$$

therefore $P = 10 \times 1000 \times 9.8 = 98000$ pascal

{10.2 m of water column (W.C)} = 1 bar,
10200 mm W.C = 100,000 Pascal or 1 mm W.C = 10 Pa.

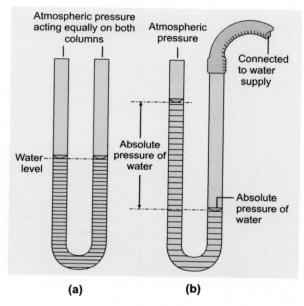

**Fig. A2.62**

*Gauge Pressure*

The pressure indicated on a pressure gauge is actually the difference between atmospheric pressure and the absolute pressure.

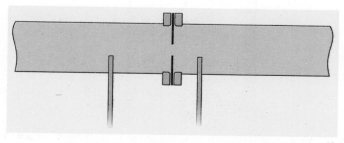

**Fig. A2.63:** Principle of an orifice plate.

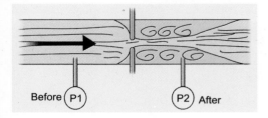 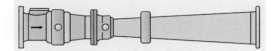

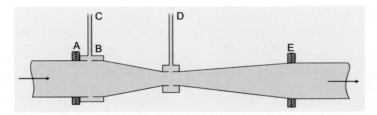

**Fig. A2.64:** Principle of a venturimeter.

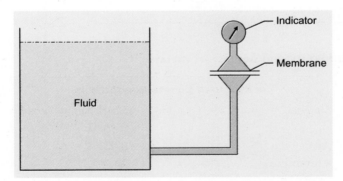

**Fig. A2.65:** Diaphragm type pressure gauge.

| Conversion of pressure units | | | | | | | | | |
|---|---|---|---|---|---|---|---|---|---|
| *Units of pressure* | *Pascal (Pa)* | *Bar* | *kgf/ cm²* | *Atm* | *cm H₂O* | *mm Hg* | *mbar* | *In Hg* | *Psi* |
| 1 pascal | 1 | $10^{-5}$ | $1.02 \times 10^{-5}$ | $0.9868 \times 10^{-5}$ | $1.02 \times 10^{-2}$ | $0.7 \times 10^{-2}$ | $10^{-2}$ | $0.2953 \times 10^{-3}$ | $0.1551 \times 10^{-3}$ |
| 1 bar | $10^{5}$ | 1 | 1.02 | 0.9869 | 1.020 | 750 | 1000 | 29.53 | 14.51 |
| 1 kgf/ cm² | $98 \times 10^{3}$ | 0.980 | 1 | 0.986 | 1.00 | 735 | 980 | 28.96 | 14.22 |

*(Contd.)*

| Units of pressure | Pascal (Pa) | Bar | kgf/cm² | Atm | cm H₂O | mm Hg | mbar | In Hg | Psi |
|---|---|---|---|---|---|---|---|---|---|
| **Conversion of pressure units** | | | | | | | | | |
| 1 atm | 101325 | 1.013 | 1.033 | 1 | 1.033 | 760 | 1013 | 29.92 | 14.70 |
| cm H₂O | 98 | 98 × 10⁻⁵ | 10-3 | 0.968 × 10⁻³ | 1 | 0.735 | 0.98 | 0.02896 | 0.01422 |
| mm Hg | 133.3 | 13.33 × 10⁻⁴ | 1.36 × 10⁻³ | 1.315 × 10⁻³ | 1.36 | 1 | 1.333 | 0.03937 | 0.01934 |
| 1 mbar | 100 | 1 × 10⁻³ | 1.02 × 10⁻² | 0.9869 × 10⁻³ | 1.02 | 0.750 | 1 | 0.02953 | 0.01451 |
| 1 in Hg | 3.386 | 33.86 × 10⁻³ | 0.03453 | 0.03346 | 34.53 | 25.4 | 33.85 | 1 | 0.4910 |
| 1 Psi | 6.890 | 6.89 × 10⁻² | 0.07031 | 0.068 | 70.3 | 51.75 | 68.947 | 2.041 | 1 |

## Thermal conductivity and density of metals

| Coefficient of thermal conductivity | |
|---|---|
| Material | Coefficient of thermal conductivity $W/m° C$ |
| Air | 0.025 |
| Aluminium | 206 |
| Brass | 104 |
| Brick | 0.6 |
| Concrete | 0.85 |
| Copper | 380 |
| Cork | 0.043 |
| Felt | 0.038 |
| Glass | 1.0 |
| Glass, fiber | 0.04 |
| Iron, cast | 70 |
| Plastic, cellular | 0.04 |
| Steel | 60 |
| Wood | 0.15 |
| Wallboard, paper | 0.076 |

## Stress Strain and Modulus of Elasticity

$$\text{Direct stress} = \frac{\text{load}}{\text{area}} = \frac{P}{A}$$

$$\text{Direct strain} = \frac{\text{extension}}{\text{original length}} = \frac{\Delta l}{L}$$

Modulus of elasticity

$$E = \frac{\text{direct stress}}{\text{direct strain}} = \frac{P/A}{\Delta l/L} = \frac{PL}{A\Delta l}$$

$$\text{Shear stress } \tau = \frac{\text{force}}{\text{area under shear}}$$

$$\text{Shear strain} = \frac{x}{L}$$

Modulus of rigidity

$$G = \frac{\text{shear stress}}{\text{shear strain}}$$

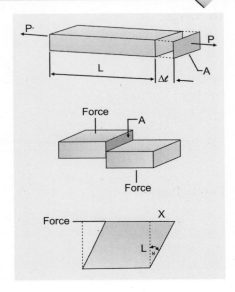

**Movement of smoke:** Smoke like all fluids moves wherever there is difference in pressure between two points. The most obvious force which acts on the smoke is the buoyancy created by heat which makes it travel upwards. The smoke within the building also travels to other parts of the building due to air currents caused by air conditioner systems, operating fans or due to temperature difference between the inside and outside areas of the building.

## Mensuration

### PLANES

### Square

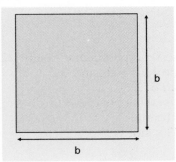

Area = b × b = b²
Length of diagonal = √2 × b
Perimeter = 4 × b

## Rectangle

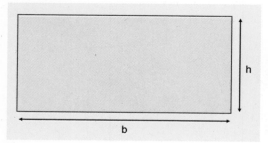

Area = b × h
Length of diagonal = $\sqrt{(b^2 + h^2)}$
Perimeter = 2 (b + h)

## Parallelogram

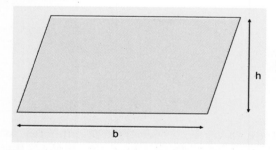

Area = b × h
Perimeter = 2 (b + h)

## Trapezium

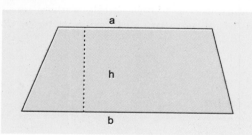

Area = (a + b) × h/2

## Triangle

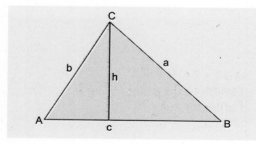

Area = (base × perpendicular height)/2
Area = bc sin A/2 = ab sin C/2
        = ca sin B/2
Area = $\sqrt{(s \, (s-a) \, (s-b) \, (s-c))}$; s = (a + b + c)/2
Equilateral triangle
Area = 0.433 × side$^2$

# Circle

*Sector of a Circle*

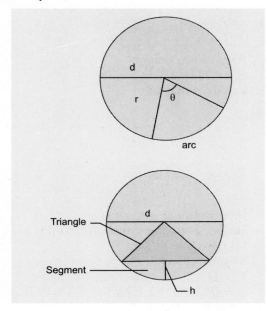

Circumference of a circle
C = Πd
Area of a circle
A = Πr² = (circumference × r)/2
   = (Πd²)/4 = 0.7854 d²
Area of a sector of a circle
A = (arc × r)/2

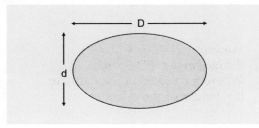

Area of a segment of a circle
A = area of sector–area of triangle
Also approximate area =
4/3 h² √ ((d/h)–0.608)

# Ellipse

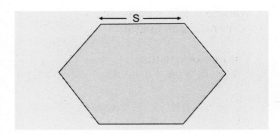

Area = Π/4 D × d
Circumference = Π (D + d)/2

# Hexagon

Area of a hexagon
A = 2.6S² where S is the
length of one side

## Octagon

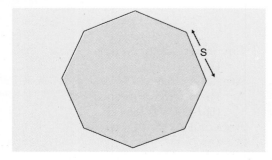

Area of an octagon
$A = 4.83 S^2$ where S is the
length of one side

## SOLIDS

### Sphere

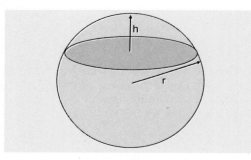

Total surface area $A = 4 \Pi r^2$
Surface area of segment $A_s = \Pi d h$
Volume $V = 4/3 \Pi r^3$
Volume of segment
$V_s = \Pi h 2/3 (3r–h)$
$\quad = (\Pi h/6) (h2 + 3 a^2)$
a = area of the segment

## Cylinder

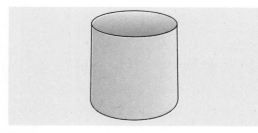

Volume $= \Pi d^2 h/4$
Surface area $= 2\Pi r (r + h)$
Where r = radius of the cylinder
h = height of the cylinder

## Pyramid

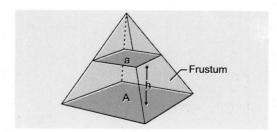

Volume $= 1/3 \times$ base area $\times$
perpendicular height

Volume of frustum
$\quad V_F = h/3 (A + a + \sqrt{Aa})$

# Cone

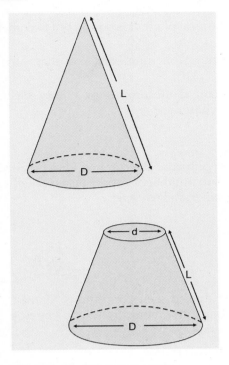

Area of curved surface of cone = $\Pi$ D L/2
Area of curved surface of frustum
  = $\Pi$ (D + d) L/2
Volume of cone =
base area × perpendicular height/3
Volume of frustum =
perpendicular height × $\Pi$ ($R^2 + r^2 + Rr$)/3

## ROPES, KNOTS AND LADDERS

### Ropes

Ropes and lines form one of the very important item of equipment of fire services. They are needed for various requirements in fire fighting and rescue operations. Therefore every fireman must know details about the construction, use and maintenance of ropes and lines. The sizes of ropes are measured in terms of their diameter. Rope can be of either fibre or metal. When we say–rope, it means fibre ropes and metal wire ropes are termed "wire ropes."

### Materials

Following materials are used in single or combined form to make ropes:

1. Natural fibre
2. Man made or synthetic fibre
3. Steel

### Natural Fibre Ropes

Following natural fibers are used to manufacture ropes:

*Hemp:* Raw material for this fibre is the stalk of certain hemp plants which is quite soft but strong.

*Manila:* This fibre is made from the leaf sheath of a palm growing in Manila. It is also strong like hemp, and very cheap.

*Sisal:* This fibre is prepared out of the leaves of the sisal plant. It is strong but becomes slippery and swells when it gets wet.

*Coir:* This fibre is made from shell-husk of coconut. Coir rope is weaker than other ropes. It floats on water, as they are lighter.

*Cotton:* This fibre is obtained from the cotton plant. Cotton ropes are highly elastic and soft to handle, but they get easily damaged by chaffing.

## Method of Manufacturing

The fibre is arranged in a parallel order and number of turns. The combed fibre is called sliver. The sliver is twisted together to form "yarns" and number of yarns twisted together to form 'strands'. A number of strands twisted together makes rope.

## Synthetic Ropes

The following man made fibers are used to make ropes:

a. Nylon     b. Terylene

Synthetic ropes are smooth, flexible and strong due to their continuous filaments. They do not require rot-proofing or water repellant treatment. Synthetic ropes are manufacture by:

(i) Twisting yarns, (ii) Forming strands, (iii) Laying strands

## Steel Wire Ropes

Steel wires are arranged around a center fibre core to form wire rope. These ropes are stronger than other ropes but less flexible and hence difficult to handle.

## Causes of Deterioration of Ropes

Different causes of mechanical deterioration of ropes are:
a. Abrasion and chaffing
b. Breakage of internal fibers
c. Kinking or sharp bends
d. Dragging on ground
e. Overloading
f. Entry of grits into the fibre

Various causes of chemical deterioration of ropes are:
a. Exposure to chemicals such as acids, alkalis, etc.
b. When the ropes are submerged for a long time in water.
c. Exposure to excessive heat and sunlight.
d. Mildew.

## Storage of Ropes

To maintain the existing strength of any rope that is properly prepared, it should be stored safe from deleterious fumes, heat, chemicals, moisture, sunlight, rodents and biological attack.

Rope should be stored in a dry place where adequate air-circulation is present. The air should not be extremely dry. However, small ropes can be hung-up and larger ropes can be laid on gratings, so that air can get underneath and around them.

Rope should not be stored or used in an atmosphere containing acid or acid fumes as it will quickly deteriorate. Signs of deterioration from this cause are dark brown or black spots on the rope.

Ropes should not be stored unless it has been cleaned. Dirty rope can be hung in loops over a bar or beam and then sprayed with water to remove the dirt. The spray should not be so powerful that it forces the dirt into the fibers. After washing, the rope should be allowed to dry and then, be shaken to remove the rest of the dirt.

The uses of ropes in fire service are:
- Ropes are used as a safety line.
- Ropes are used for rescue works of all types.
- Ropes are used for hauling up and securing lines of hose.
- Ropes are used for securing suction hose, hauling up or lowering equipment.
- Ropes are used for hauling and lowering branches.
- Used for hauling up hose, hose reel punch line or light gears and a guide back line to open air.
- Ropes are also used for assisting a man to return to open air when entering thick smoke, and for securing ladders.

## Different Causes of Deterioration of Steel Ropes

Wire ropes also get weakened or damaged in the same way as fibre ropes. It is very important that kinked wire rope should not be subjected to load because the kink cannot be removed. Once a kink has been formed, the strength of the rope is permanently reduced.

## Inspection and Test

Periodical visual inspection of ropes must be carefully done. Load test of lines used for rescue purposes must be necessarily carried out. Tie one end of the rope to a strong point and six men must take position at a distance of 2 meters and pull steadily without any jerk. After the pull, the line must be examined for any visual damage. Ropes should be tested:
  i. Once in a month
 ii. After every use at fire or any incident
iii. Before every drill

New rope should be inspected along its entire length thoroughly for any damage or defect. In-service ropes should be inspected every 30 days under ordinary conditions and more often if used in critical condition. Inspection consists of an examination of the entire length of the rope, for wear, abrasions, powdered fibre between strands, broken or cut fibres, displacement of yarns or strands, variations in size or roundness of strands, discoloration and rotting. To inspect the inner fibres, the rope should be untwisted in several places to see whether the inner yarns are bright, clear and unspotted.

If exposed to acids, corrosive atmosphere, natural fibre ropes such as manila should be scrapped or retired from critical operation, as visual inspection will not always reveal acid damage. A rope, like a chain, is only as strong as its weakest link (or in a rope, its cross-section).

The factors of safety variations among different types of rope are decided from factors such as chaffing, cutting, elasticity, diameter-strength ratio and general anticipated mishandling. If possible, a rope should not be dragged as this abrades the outer fibers. If the rope picks-up dirt and sand, abrasion within the lay of the rope will rapidly wear it out. Precaution should be taken to keep rope in good condition. Kinking, e.g. strains the rope and may overstress the fibers. It may be difficult to detect a weak spot made by a kink. To prevent a new rope from kinking while it is being uncoiled, lay the rope coil on the floor with the bottom end down. Pull the bottom end up through the coil and unwind the rope counter clockwise. If it uncoils in the other direction, turn the coil of rope over and pull the end-out on the other side.

Twisted rope should be handled so as to retain the amount of twist (called balance) that the rope seeks when free and relaxed. If rotating loads and improper coiling and uncoiling change the balance, it can be restored by proper twisting of either end. Severe unbalance can cause permanent damage, localized over-twisting causes kinking or hocking.

Sharp bends over an unyielding surface cause extreme tension on the fibers. To make a rope fast, an object with a smooth round surface of sufficient diameter should be selected. If the object does have sharp corners, pads should be used. To avoid excessive bending, sheaves or surface curvatures should be of suitable size for the diameter of the rope.

When lengths of rope must be joined, they should be spliced and not knotted. A well made slice will retain up to 103% of the strength of the rope, but a knot only half.

Use of wet rope, or of rope reinforced with metallic strands, near power lines and other electrical equipment is extremely dangerous. Rope must be thoroughly dried out before reuse.

| Efficiency of ropes with splices, hitches and knots | |
|---|---|
| *Jointure* | *Percent efficiency* |
| Full strength of dry rope | 100 |
| Eye splice over metal thimble | 90 |
| Short splice in rope | 80 |
| Timber hitch. Round hitch, half hitch | 65 |
| Bowline slip knot, clove hitch | 60 |
| Square knot, weaver's knot, sheet bend | 50 |
| Flemish eye over-hand knot | 45 |

| Different types of lines | | | |
|---|---|---|---|
| *Name of rope-line* | *Length metres* | *Circumference cm* | *Purpose* |
| 1. Extra long | 50 to 75 | 50 | Used as a safety line |
| 2. Lowering line | 30 | 50 | Used for rescue works |
| 3. Long line | 30 | 50 | Used for hauling and securing |
| 4. Short line | 12–15 | 12.5 | Used for securing hoses |
| 5. Line | 3 | 25 | Used to haul up hoses |
| 6. Escape line | 5 to 7 | 25 | Used to assist men to come out or for securing ladders |

## Rope Knots

A knot is a twisted rope design, used for various purposes such as joining two ropes, tying a rope to an object, or for securing objects together.

Every fireman should be fully acquainted with the several types of knots and they should also be able to make the knots which are required in emergency rescue work.

Knots are used in:

- The moving part of a rope-line, which is loose and used to hoist or lower.
- The free end of a rope-line.
- The binding together of two or more ropes.
- The part of the rope-line, which is fixed.
- The hiding of the end of a rope-line with twine to prevent in un-laying

## Details of Use

- Used as a simple stopper and tied at each end in a length or burst of fire-hose, when laid out.
- Used to join ropes lines.
- Used as sling to lower insensible person.
- A number of knots are used for securing suction hose, hoisting up a branch etc.

## Bends and Hitches

The different bends and hitches are useful in fire fighting and rescue work and also in various other work areas. The terms used in rope-skills are:

- Bend–To fasten a rope-line to another object.
- Bite–The looped or loose part of a line between two ends.
- Hitch–A simple fastening of a line to some object by passing the line around the object and crossing one part over the other.
- Running part–The moving part of a line, which is loose and used to hoist or lower.
- Running end–The free end of a line
- Seizing–The binding together of two or more ropes
- Standing part–The part of a rope line, which is fixed.
- Whipping–The hiding of the end of a line with twine to prevent untying.

| Different types of knots | |
|---|---|
| *Knot list* | *Use* |
| Bowline | A non-slipping eye used for a variety of purposes. |
| Over hand/thumb knot | Used as a simple stopper and tied at each end of a rope. |
| Reef knot | Used where frequent and easy untying is necessary |
| Chair knot | Used as sling to lower unconscious/disabled persons |
| Half hitch | Used for securing suction hose, hoisting up a branch nozzle, etc. |
| Clove hitch | Used to secure a rope-line to any round object. |
| Rolling hitch | Same as clove hitch but the knot will not slip along the object when a side-wise pull is applied |
| Timber hitch | A simple knot to tie and hoist any irregular shaped object |

## Some Important Knots

*Bowline:* The bowline knot is one of the most useful knots and hence should be memorized. The bowline forms a secure knot which will not jam and is also easy to untie. It can be tied around objects, can be tied into any size loop and even after being under load, can be untied. It is truly a very versatile and useful knot to learn.

To make a bowline knot, form an eye in the rope with the standing part underneath. Run the free end up through the eye then take a turn around the standing straight part. Feed the free end back down into the eye and hold down there, while pulling on the standing straight part to tighten the knot.

*Square knot:* The square knot is also known as reef knot. Square knot is secure and easy to untie and is used by sailors and scouts.

*Clove hitch:* A useful and easy to tie knot, it is a good binding knot. However, as a hitch, it should be used with caution, because it can slip and loosen, if the object it is tied to rotates or if constant pressure is not maintained in the line.

*Half-hitch:* Although the half-hitch is a knot in its own right, it is rarely used alone. Two half-hitches can be used to tie a rope to a tree, a boat or other object. It is often used in a supporting role, e.g. to increase the security of a primary knot.

*Double fisherman's knot:* Although rarely used in fishing, it is a good knot for tying two ropes together. It is essentially two knots that slide together when tightened to form the finished knot. Tying just one side of the knot is also used by mountain climbers to tie a backup knot with the tag (free) end of the primary knot.

*Monkey's fist (Paw knot):* The key to tying this knot successfully is to make a small ball or core to insert into the knot before tightening it up. The core must match the size of the knot–which is dependent on the size of the rope being used for the knot to finish right.

*Heaving line knot:* This knot adds bulk and weight to the end of a rope making it easier to throw the line.

*Butterfly knot:* The butterfly knot is used by climbers to tie in the middle climber when travelling three to a rope. Also useful for making non-slip loops in the middle of a rope to attach any object. It can take load in any three directions, independently or together. The knot can also be used to isolate any damaged section of a rope.

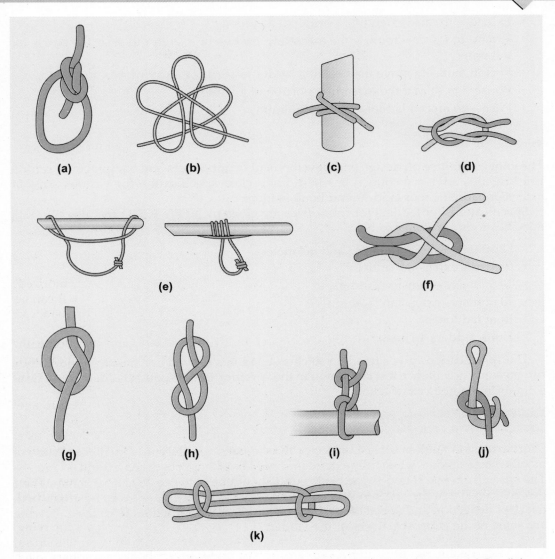

**Fig. A2.66:** (a) Bow line knot, (b) Butterfly knot, (c) Clove hitch knot, (d) Square knot, (e) Prusik knot, (f) Sheet bend knot, (g) Overhand knot, (h) Figure-eight knot, (i) Two half hitches, (j) Half hitch, (k) Sheep shank.

## Ladders

### Terminology

Firemen should first make sure that they are familiar with the proper terms for the various parts of a ladder. Firemen should get themselves well acquainted with the following terms used in operations with the ladders.

- Extend—to raise extending portion of a ladder.
- Extend to lower—to raise the extending portion of a ladder to clear the pawls for lowering.
- Reel in, out—to move the heel of a ladder towards or away from a building.
- Rouse—to retract the extending portion of a ladder.
- Pitch—to erect a ladder against a building.

## Specifications

The joint committee on design and development of appliances and equipment of central fire brigades advisory council has laid down clear specification for various sizes of extension ladders, roof ladders and hook ladders.

Specifications for ladders for fire service are given by NFPA, JCDD and also BIS (ISI). Specifications are available for:

- Short extension ladders 5.5 and 6.7 meters.
- 9 meters extension ladders
- 10.5 meters extension ladders
- 13.5 meters extension ladders
- Roof ladders
- Hook ladders 4.1 meters

The specifications gives freedom to choose ladders of wood or metal provided they can pass any acceptance test laid down in the specifications. Wooden ladders could be of solid or laminated construction.

## Short Extension Ladders

Short extension ladders consist of two or three extensions. Recent ladders mostly are of metals. These ladders when fully extended should be of length between 5.5 and 6.5 meters. The space from one round to the next round should be between 280 and 305 mm. Their design is such that they fit into various criteria, including degree of sway when pitched, overlap, deflection and speed of separation for use as individual section. The following are some of the general features of the ladder:

i. Reasonable rigidity and freedom from sway under prescribed conditions of pitch and load.
ii. One main and one extending section of the ladder having a width inside the string of not less than 300 mm.
iii. Easy extension by one man with an endless rope over a pulley or other approved means.
iv. Safety devices including pawls of an approved design and nonskid shoes on the heels of the main ladder.
v. The weight should not go above 56 kg.
vi. Operating gear should be so designed that it is easy to remove the extension.
vii. The ability to pass stringent deflection and round test.

### Nine Meter Ladder

*Wooden ladders:* Most wooden ladders are trussed. The trussing should be on the underside of the ladder when it is in operational use. Pawls seldom operate automatically, except when the ladder is being housed. They are usually cleared initially by extending to lower.

*Metal ladders:* Typically these are of a riveted trussed construction with high tensile aluminum alloy extrusions and square section rounds ribbed on the tread surface to make them less slippery. The extending section runs on nylon rollers in guide channels integral with the stills and double action automatic pawls are fitted. The line is 16 mm in diameter and rot-proofed. The head of the ladder usually has wheels for easy running on wall surfaces.

### 13.5 Meter Ladder

This type of ladders differs in several areas from those of 9 and 10.5 meters ladders. The main difference is that the 13.5 meter ladder consists of one main section and two extensions, the width of the narrowest being not less than 305 mm. The ladder also has suitable props with universal joints. These props are attached to the main section of the ladder and have non- skid attachments on their feet. They must stow comfortably along side the strings when not in use.

The ladder should also have quadrants or spikes to help in elevation and continuously adjustable jacks. Their position has to be such that they allow room for firemen's steps between the jack and the main ladder string. The extra extension is operated by a cable, which should meet various requirements for approved fixing, minimum breaking load (1.3 tonnes), sheaves, etc. Exact weight limits for the ladder are maximum 113 kg, preferable 102 kg.

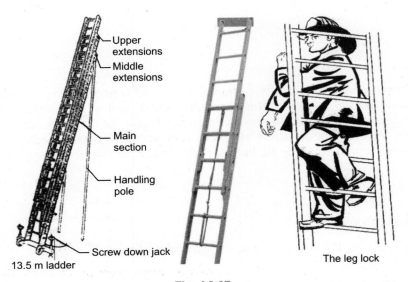

Fig. A2.67

## Ladders as per Indian Standards

Aluminium extension ladders for fire brigade use, IS: 4571-1977.

## Specification

Aluminium ladders being lighter are now being preferred to timber ladders by the fire fighting personnel.

Extension ladders—The aluminium extension ladder consists of one main and one extending section. The-design shall be such as to ensure easy sliding of the extending section without excessive clearance in the guide and over extension of the ladder. The extending section shall be guided throughout the full range of extension in a manner such that the sections cannot separate, retaining clips being on the main section.

The width of the extending section inside its strings shall be not less than 30 cm. The rounds shall be with non-slip serrations running the full length. These shall be fixed by expanding and flaring and shall be spaced at 25 cm centre to centre.

The aluminium alloy section for stiles or strings shall conform to Grade HE-20 (in WP temper) of IS: 3921-1966 or IS: 733-1975.

The aluminium alloy section used for the rounds of the ladder shall conform to Grade HV (in W temper) of IS: 1285-1975.

Aluminium alloy latch shall be produced from aluminium alloy conforming to IS Designation A-6 of IS: 617-19759 by chill casting.

Rubber feet of aluminium ladder shall have shear hardness of 50 to 60.

Extension ladders when fully extended shall be in the three sizes, namely, 4.5 m, 7.5 m and 10.5 m.

The ladders shall be as light as possible, their total mass shall not exceed as given below:

| | |
|---|---|
| 4.5 m | 20 kg |
| 7.5 m | 30 kg |
| 10.5 m | 48 kg |

## Acceptance Tests

The specification has laid down acceptance test for all size of extension ladders. These consist of a deflection test and a round test and a slide test for a 13.5 meter ladder. The other ladders are also subject to acceptance tests. Those for the hook ladder, especially on its hook, anchorage and safety ring, are very stringent. Other acceptance tests for hook ladders comprise test for the ladder deflection, rounds and balance, and a general test.

## Constituent Parts of Extension Ladder

Main ladder with rounds.

*The lowest section:* It consists of two strings and a number of rounds.

*Extending ladder with rounds:* It is known as upper section or extending section, consisting of two strings and a number of rounds.

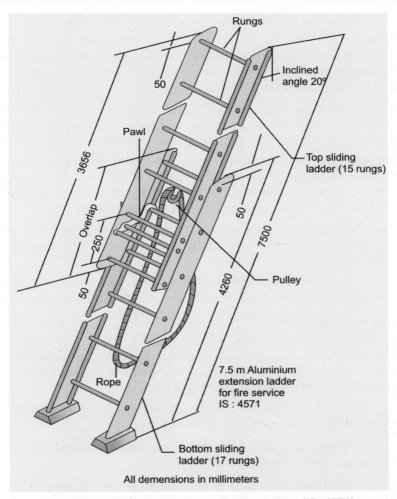

Rungs

50

Inclined angle 20°

3656

Pawl

Top sliding ladder (15 rungs)

Overlap

250

50

50

7500

4260

Pulley

Rope

7.5 m Aluminium extension ladder for fire service IS : 4571

Bottom sliding ladder (17 rungs)

All demensions in millimeters

**Fig. A2.68:** A 7.5 m aluminium extension ladder (IS: 4571).

*Extending line :* The extending line is needed for extending or lowering the extending section of a ladder. It is also useful for operating the movement of the pawls. During the movement of the extension or lowering, it takes the weight of the extending section. The extending line is made of a three strand, manila rope of 4 cm size. The extending line passes between the sections over a head pulley on the main section. The ends are attached to a lever in the middle of the rotating shaft, and thus making the endless line. Plenty of slack is provided on this line to allow for shrinkage when wet. It should never be shortened nor tied in a knot when the ladder is stowed. The loose portion has to be folded around the bottom rounds of the ladder in such a way that as soon as the ladder is erected and the rope is pulled, the slack releases itself automatically.

*Head pulley:* It is fitted to the top most part of the main ladder. The head pulley must be secured rigidly to the round, otherwise the rope may slip off the groove and then extension or lowering may become difficult. If the head pulley is loose it may foul with the round while extending.

*Rotating shaft:* The rotating shaft is fitted in between the bottom and second round of the extending section. This shaft is also known as pawl shaft as two pawls of the claw pendulum type are mounded on this shaft.

*Lever:* Lever is a specially designed steel plate, which is rigidly secured to the rotating shaft in the center. Both the ends of the extending rope are secured to the lever by means of a spring loaded sharp hook. The operating level controls movement of the pawls.

*Pawls:* Two pawls are rigidly secured to the rotating shaft. These are of claw pendulum type and when engaged or tripped, the weight of the extending ladder is released from the extending rope and taken by the main ladder. In some ladders, spring loaded pawls are fitted underneath the string of the extending section.

*Guide plates:* They are fitted at the bottom of the top section. These plates fit into grooves, which are cut into inside of the strings of the main ladder and thus guide the top section when extending.

*Strengthening wires:* The strings of the ladders are reinforced by a 5-gauge, 7 ply galvanized wire which is stretched tight along the wider side of the spring of both sections.

*Staple and screw:* Staples and screws at the head and heel of the ladder secure the strengthening wire.

*Tie rods:* The rounds are strengthened by wrought iron tie rods under the 1st, 5th, 9th, and 13th rounds of the main and extending sections. In some ladders, the tie rods are secured underneath all the rounds of both sections.

*Grooves along the strings:* Grooves are cut into the inside of the strings of the main ladder and thus guide the top section while extending.

*Head iron plates:* Iron plates are tied to the strings near to the side of both sections to protect the strings against wear and tear.

*Heel iron plates:* The heels of the strings of the main ladder are protected against wear by rubber heal boots. The heels of both the strings are fitted with non-skid shoes to protect the string and give good ground grip. In some ladders, the heel iron plates are corrugated to reduce the tendency of the ladder to slip when on a hard surface.

*Metal roller:* Metal rollers are fitted near to the side of the main ladder. They make smooth and easy extension of the extending ladder.

*Safety stops:* Provision is made for safety in order to prevent over extension of the ladder. They are fitted near to the head side of the main ladder and near to the heel side of the extending ladder.

### Uses of Extension Ladders

- Extension ladder is usually employed by fire service to have access to upper floor, roofs of the buildings or such other elevated places for performing rescue and fire fighting operations.

- Extension ladder can be used as substitute for stairs.
- Extension ladder maybe used for bridging purposes over unsafe floor, two tall buildings, etc.
- Extension ladder can be used as an improvised stretcher.
- Extension ladder is more helpful in fires where a wheeled fire escape is not necessary or cannot be used due to requirements of space, such as narrow passages, lanes, small compounds or gardens, etc.
- Two extension ladders may be placed in an angular shape supporting each other and can be used as a stepladder to reach the middle of a high ceiling for want or support.
- Extension ladder can be used for facilitating an entry in to a basement or a hold.

### Pitching and Climbing Hints

Pitching and climbing hints are to be observed strictly by fire officers so as to enable them to pitch the ladders properly and to climb safely and efficiently both for rescue and fire fighting purposes.

### Pitching

a. The trussing must be on the under side in order to provide a clear working space. The trussing on the upper side may interfere while carrying down.
b. Ladder should always be pitched in such a way that a little space is left over on the right side and more space on left side.
c. Head of the ladder must be well above the sill of a window. It is then easier to step "ON" or to step "DOWN" the ladder even when bringing down a casualty.
d. Where a window is narrow, ladder should be kept in such a way that it rests on the wall preferably to right side and not on the windowsill, so that the maximum space is left for working and entering the building.
e. Ladder should be pitched at a safe working angle, in which case, the distance of the head of the ladder from the wall should be between one-fourth and one-fifth of the extended span of the ladder.

Fire rescue team with SCBA on a rescue mission going up a fire-ladder

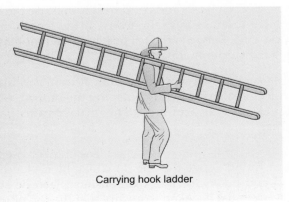

Carrying hook ladder

Fig. A2.69

## Climbing

a. A ladder has to be climbed steadily by keeping the body erect, the arms straight and slightly looking upwards. The climber should not watch his feet while ascending or descending the ladder.

b. The feet must be kept towards the center of the rounds.

c. The hands must move gently on the rounds at a level between the waist and shoulders. (Operating above shoulder level gives less control and requires more energy.)

d. The rounds must be gripped by the hands, keeping the palms down and the thumb under the rounds.

e. Both hands should never be on the same rounds at the same time. When ascending or descending, the hands and foot on the same side of the body, should move in uniform (left hand and left foot, right hand and right foot).

f. Ladder must not be left unattended when a man is ascending or descending.

It is to be remembered that "ATTENTION TO DETAIL" is important as the basic effective ladder work. Neglect of details leads to slowness of movement, mind and habit, causes accidents and lowers the standard of performance and morale of the crew.

## Safety Holds

It is necessary to perform work on a ladder, which requires the use of both hands. The following safety holds allows free use of both hands. They are referred to as leg-lock, arm-hold, body-hold, and crotch-hold. Because there are several types of ladders and physical difference between individuals, it may be necessary to occasionally vary some of the direction given to assume a comfortable working position to a man on a ladder.

*Leg-lock:* While working on a ladder, sometimes it becomes necessary to have both hands free to direct a jet, to handle a small ladder or some small gear, etc. This can be accomplished by taking "leg-lock" on the ladder. To do this, the ladder man must stand on one round and keep the other leg over the second round above the one on which he is standing. Then he should bend the leg and use the inside of the knee and back of the calf of the leg to make himself steady. If the length of the leg so permits, the foot may be brought back between the rounds, and the foot maybe placed under the above round on which the ladder man is standing.

*Arm-hold:* When working a branch from a ladder, it is at times advisable to put one hand round a string before grasping the branch, while the other directly holds the base. This is known as arm-hold. This procedure is particularly helpful in extremely cold weather.

*Leg-lock facing away from ladder:* The climber keeps his leg through the ladder over the second round above the one on which he is standing, then brings it back over the first round and hooks his foot over strings. Then he steps down one round with his other foot. Care must be taken to ensure that the top of the ladder is secure so that the pulling action would not pull the ladder away from the building.

*Body hold:* The body-hold gives more freedom to the man being secured and is particularly helpful in holding a man, who is operating tools from a ladder.

*Crotch-hold:* One man to secure another man on a ladder uses the crotch-hold. The man being secured may face either toward or away from the ladder depending on the work to be performed.

Take a position on the round below the one on which the other man stands and knee clamp one of his legs. Then place an arm between his legs and grasp the second or third round above the one the man is standing on, pass the other arm around waist and take hold of the round nearest to this level, and pass closely against him.

### Standard Tests and Inspection *(Ladders can be tested to NFPA 1932)*

*Tests:* Ladders should be tested and examined quarterly, after operational use and on such occasions considered as necessary.

1. *Testing of strings:* The ladder must be pitched to its full working height with the head leaning against a building. A rope-line must be made fast to each string between two adjacent rounds at the center of the middle extension in case of a 13.5 m ladder and at the center of the overlap of the two sections of 10.5 m and 9 m ladders and must be so arranged that the weight of three men in all, can be applied equally to the two strings and then released.

2. *Testing of rounds:* The ladder has to be extended fully and pitched at a safe working angle. Each round is then to be jumped, keeping firm grip with both hands, and feet well apart, to test the rounds and sockets (it is not necessary to jump the rounds of metal ladders).

   The proper method of testing the rounds by jumping is to transfer the weight sharply downward from each round to the next. The height of the jump must not be increased nor should a jump be used to deliver a violent blow to the round. The head of the ladder can be secured, if necessary, to get stability and two men must be on the heels during this test.

   In a 13.5 m ladder, for the top rounds which cannot be jumped, a line should be made fast at the center of the top round of the upper extension with the help of a round turn and two half hitches. The weight of two men should be applied to the line and then should be released. This test should be repeated to other rounds that cannot be tested by jumping.

3. *Testing of extending line:* The ladder must be pitched without being extended against a wall; the weight of two men must then be applied to the line as for extending the ladder as low as practicable, while two other men apply their weight to the extension to stop it moving. The remaining part of the line must be tested by applying the weight of two men, but with the ladder fully extended.

*Inspection:* A ladder has to be thoroughly checked for any movement of the timber, for looseness of bolts or rivets, in the fitting for loose wedges and to see that the shoulders of the rounds fit closely up to the strings. Riveted and screwed joints of metal ladders must be examined to make sure that they are tight and welded joints of metal ladder should be examined to ensure that there are no cracks.

Special attention should also be paid to the following points:

1. The moving parts are clean and sufficiently lubricated.
2. Extending lines and cables are correctly moving and run freely through the various pulleys.
3. The ladder pawls are operating correctly.
4. The anchorages for lines and cables are secured in 13.5 m ladders with extending wing.
5. The positive pawls and section breaks are functioning correctly.
6. Extending cables show no sign of excessive kinking.
7. When ladders are placed on appliances, the mounting and securing gear are functioning correctly.

### Annual Examination *(13.5 m ladders only)*

Every 12 months the following additional examination must be done unless the ladder is subjected to an annual examination and test in a brigade workshop.

The ladder has to be kept horizontally on two trestles; gears, shaft, pulleys and pawls must be examined to see that they are sound and free from excessive wear. Nuts, screws, rivet joints must be checked for tightness and freedom from cracks and fracture.

### Maintenance

Ladder maintenance can be separated into two procedures. The first—repair maintenance, should be performed any time the visual inspection discloses any defect. The repair maintenance procedure must be performed prior to using the ladder again. The second— preventive maintenance procedure reduces the likelihood of ladder damage and injuries. All ladders, regardless of the manufacturer or material of construction, require both procedures to ensure a safe ladder.

- Apart from the regular testing of ladders, the following points on maintenance must be born in mind:
- Ladders of all types must be used regularly at drills in order to keep them in good shape.
- They must not be stored in one position; otherwise warping may occur.
- The wood of the ladder is liable to splinter, as such, any roughness or incipient splinters must be treated and smoothened with sandpaper.
- If the ladder is affected, it must be sent to the workshop for dismantling and recoating with preservative (varnish).
- When the preservative becomes worn, it must be replaced immediately. Otherwise the weather might affect the wood and alter its strength.
- Ladder construction joints must be kept water resistant with close fitting and sound coatings of varnish, to prevent moisture from penetrating the timber.
- Repairs must be carried out by experts, and should not be attempted by those unacquainted with the structure of ladders.

- All moving parts, pulley—-wheel, rotating shaft, etc. must be examined.
- All nuts, bolts, screws, etc. should be checked-up for looseness and tightened where required. The extending line has to be examined for any signs of wear.

## Hook Ladder

This ladder works on an entirely different principle from all other ladders. It must be suitable for suspension with the help of a hook from a window-sill, so that one man can ascend or descend without any assistance. The ladder should not weigh more than 13.2 kg and its total length is usually 4.1 m. On account of their special nature, hook ladders have to meet very strict and detailed requirements, with regard to, e.g.

i. Materials and their finish.
ii. Type of hook, anchorage and safety ring.
iii. Toe pieces to provide a 75 mm clearance from the wall and protection and safety where necessary. For example, metal shoes, tie rods under certain rounds, high tensile wire reinforcement. The requirements vary between ladders made of wood and those of metal.

### Hook Ladder Belts

Firemen must wear a hook ladder belt whenever they use a hook ladder. Its width is about 115 mm and length just over one meter. It is made of best quality webbing-lined leather. It is fixed with two straps with buckle from belt to the right to obviate loose ends of the straps interfering with the spring hook. A large spring hook is fixed to the belt and is used to hook on to the safety ring fitted to the head of the hook ladder. The belt also has a bobbin line pouch, which contains 40 meters plated cord, and possibly an axe pouch.

## Turn-Table Ladders (TTL)

### Terminology

Firemen can carry out a number of different manoeuvres with a turntable ladder and use it for a variety of purpose.

The various actions which can be done with the turntable are:
- Depress
- Elevate
- Extend
- House
- Plumb (to right or left)
- Projection
- Shoot up (to right or left)

In order to avoid any confusion, the fire service has adopted a standard terminology for use when referring to the various turntable operations.

- Depress—to bring down the head of the ladder by reducing the angle of elevation.
- Elevate—to raise the head of the ladder by increasing the angle of elevation.
- Extend—to increase the length of the ladder.

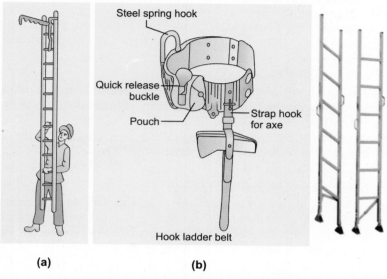

Fig. A2.70: (a) Hook ladder, (b) Ladder belt.

- House—to decrease the length of the ladder.
- Plumb—to keep the center line of the ladder in a vertical plane by removing any tilt to one side when the ladder is extended on a slope. This adds greater stability and obviates side stress.
- Projection—the horizontal distance measured from a vertical line dropped from the head of the ladder to the rim of the turntable.
- Shoot—to extend the ladder with a fireman at its head.
- Turn—to move the head of a. ladder by rotating the turntable

(Some tend to use the expression 'slewing')

Firemen must first acquaint themselves with these and with various parts of a turntable ladder.

## Design of Turn Table Ladders

*Specifications:* Standards' committees have laid down specifications for the design, construction and performance of turntable ladders.

Manufactures may choose to exceed the specifications or fire brigades may always lay down more rigid or additional requirements, perhaps to meet special conditions. Turntable ladders must also conform with current motor vehicle (construction and use regulations) and to road vehicle regulations.

*General description:* A turntable ladder is a self-supporting and power operated extension ladder mounted on a turntable. The ladder assembly is mounted at the rear of a heavy, self-propelled, chassis, approximately above the back axle. The chassis design takes into account that the appliance generally stands laden. Its wheelbase does not go above 5.3 m or its overall width 5.5 m, except at the axle casing. It has at least a 230 mm ground clearance. At the base of the ladder assembly, there is a turntable, which rotates, in a circular track bolted to the chassis. On this is mounted the fulcrum frame, where there is a strong steel structure which supports the rest of the ladder assembly and houses the operating mechanism. At the fulcrum point of this frame, there is a trunnion, on which pivots the swinging or elevating frame. This supports the ladder section and forms a mount on which to operate the ladder. Also fitted to the near side of the fulcrum frame or on a separate console, are control levers for the ladder. There may also be a secondary set of control and indicators in the ladder cage, or capable of being fitted at the head of the ladder, but the main controls at the base will at all times be capable of over-riding them. All turntable ladders have plumbing gear, which keeps the ladder, plumb on gradients up to 7°.

The ladder itself usually consists of a main ladder, secured by a strong pivot-bearing to the swinging frame, and three or four extensions, which extends telescopically.

## Other Features of Appearance

*Cages:* All ladders complying with specifications have a cage at the head of the ladder. This is either a fixed or demountable light rescue cage capable of carrying at least two people or a permanently attached rescue cage capable of carrying at least four persons.

*Monitors:* A monitor is fitted at the head of a ladder or front of its cage. A man at the head of the ladder usually operates this, but it can be operated from the ground. A special length of hose feeds the monitor and a fixed light weight water supply pipe may be provided to have a permanent connection between the monitor and top ladder section.

*Operator's platform:* If the ladder does not have a cage permanently fitted, it will have secured near its head, a hinged platform for the monitor operator. The platform must have handrails and toe guard.

*Rescue gear:* All ladders, without a permanently fitted cage, must have provision for lowering the rope-line in order to affect a rescue. The line used is about 70 m long, fitted with a sling and made up on a cradle.

## Safety Devices in Turntable Ladders

*Automatic safety device:* Many turntable ladders have several automatic safety devices.

*Jacks and axle locks:* These slow down and stop false movements at maximum elevation, and stops extension and depression below the horizontal; and when fully housed, stop extension or depression of the ladder, when the safety stops' safe working limits are reached. They help:

- To prevent damage to the ladder, if it strikes an obstruction.
- To prevent any ladder movement until the jacks are down and on retraction of the jacks until the ladder is housed and depressed on the hand rest.

A turntable ladder requires a solid working base. A total of four jacks fitted in front of and behind each rear wheel help achieve this by being lowered to the ground and taking the excess weight of the ladder and chassis off the tyres. An axle-locking device prevents the chassis rising on the rod springs when a ladder extended on the opposite side reduces the loading.

*Communications:* Turntable ladders have provision for the fireman at the head of the ladder and the operator to communicate in all circumstances.

## Safe Working Indicators

*Inclinometer:* It is very important to maintain the stability of the ladder. Principal factors affecting this are—the length of the extension, the angle of elevation and the loading of a ladder. Therefore, the turntable has a device, which shows its elevation and indicates the maximum permissible extension at different angles for different loadings.

*Other indicators:*
  i. Safe load indicator—this provides both a visible and an audible warning when the maximum permissible load limit is reached.
  ii. Ladder position indicator.
  iii. Plumbing indicator.
  iv. "Rounds inline" indicator
  v. Indicator to show when the ladder sections are fully housed.
  vi. Indicator to show the working of the hydraulic power supply system.

## Operation and Performance of Turntable Ladders

In modern turntable ladders, all the movements except depression are power operated. Depression, elevation, extension and housing are controlled by levers and are infinitely variable throughout the speed range. A turntable ladder has a road turning circle of not more than 20.7 m in either direction. It is designed for speed (at least 80 kmph on a flat road), good road holding, and fast cornering and for giving a comfortable ride to the crew. It is capable of being used at any angle up to 75° but not more than 78° and can also work below the horizontal. A ladder can be turned 360° to right or left at an angle exceeding 7° above the horizontal. The turntable monitor has a range pf 10-15° either side or a vertical range as great as possible and not less than 100°. The minimum pressure of a pump is 7 bars.

*Power take off:* A turntable ladder generally has a power take off which forms part of the road gearbox and transmits adequate power to meet the requirements of the ladder mechanism. It is operated by an engagement lever in the driving compartment of the appliance unless it can be engaged without the use of the clutch.

*Pumps:* Some of the turntable ladders are also provided with a pump, which complies to specifications such as JCDD 29, but must be a separate self-contained unit.

## Acceptance Test

Turntable ladders complying with specifications such as JCDD 36 are subject to acceptance tests, which are laid down by the specification.

## Precaution in use of Ladders

A ladder should never be placed on slanting, oily, slippery or on vibrating footings, unless the ladder is held by another person or securely fastened to prevent it from slipping or twisting. The base of a ladder should not be placed more than 1/4 of the length of the ladder from a wall or supporting surface. Ladders placed near doors or in passageways should be protected against being struck by doors or traffic.

All ladders should be inspected at regular intervals and maintained properly. When a ladder has fallen or has been struck, it should be carefully inspected for possible damage before use. The overlap of sections for extension ladders should be sufficient to prevent collapse of extensions.

All ladders for outdoor use should be given a suitable coating such as clear vanish or linseed oil. Metallic paint should not be used on wood ladders. The use of ladders for other than their intended use and the use of defective ladders is prohibited. Split, broken or otherwise defective ladders should be destroyed or cut to smaller lengths.

While going up or down a ladder, always face the ladder and use both hands for climbing.

When selecting a portable ladder, choose the proper type and size and inspect it for defects. Whenever possible, make sure the top of the straight or extension ladder will be long enough to extend at least three feet above the step-off level.

**Firemen in action**

Fire engines on the way

Ladder operation

Fixing hose to fire engine

Fire fighting

CPR on a victim

**Fig. A2.71:** I fight what you fear–Firemen slogan.

Straight ladders and extension ladders, except for job made ladders, are not to be used unless they have safety feet. Job-made ladders should only be used for a specific purpose intended. If possible, they should also be provided with slip resistant feet.

Never climb an extension or straight ladder that is not adequately tied-off unless someone is holding the ladder to prevent it from moving.

Ensure that the horizontal distance from the foot of a straight or extension ladder to the base of the vertical it rests against is equal to one-fourth the working length of the ladder.

Do not work higher than the second step from the top of a stepladder or the third rung from the top of a straight ladder. If it is necessary to place a ladder in the swing of travel of a doorway, barricade the door and post warning signs and post a watchman to a ladder-climbing safety device when it is provided.

Do not use stepladders as straight ladders.

*Hand Signaling*

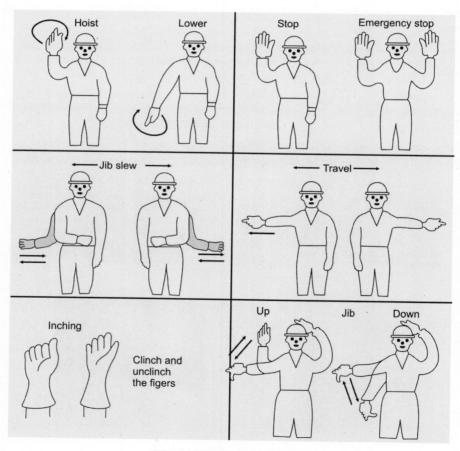

**Fig. A2.72:** Crane signals.

**What to do in the event of a fire**

- Call for help immediately
- Do not enter the house/building, but wait for the firemen
- If somebody is burnt, pour water on the burn and then cover it with a loose clean bandage.
- If burns are large or if the burnt person is a child, send the person to hospital.

## PERSONAL PROTECTION EQUIPMENT (PPE)

In fire fighting, PPE is a necessary requirement, since the hazard is already existent and firemen fighting a fire for suppression, are exposed to the dangers (Figs A2.73 and A2.74).

Protection required:
- Head protection
- Eye and face protection
- Hearing protection
- Respiratory protection
- Arm and hand protection
- Foot and leg protection
- Protective clothing

### Classification of Occupancies

Occupancies classified as light hazard, ordinary hazard and high hazard-A and high hazard-B as per the Indian Factories Act are listed below. This classification is helpful in designing appropriate fire protection system.

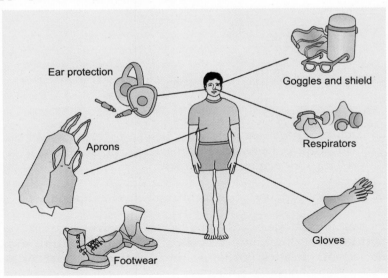

**Fig. A2.73**

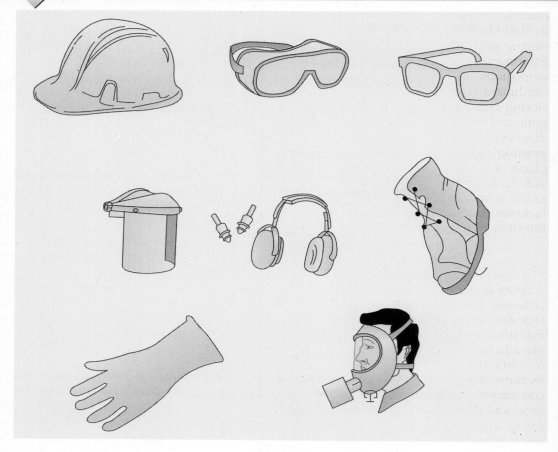

Fig. A2.74

## Light Hazard Occupancies

Abrasive manufacturing premises; aerated water factories; agarbatti manufacturing; aluminium/zinc and copper factories; analytical and/or quality control laboratories; asbestos steam packing and lagging manufacture.

Battery charging/service station; battery manufacturing; breweries;

Brick works; canning factories; cardamom factories; cement factories and/or asbestos products manufacturing; ceramic factories and crockery and stoneware pipe manufacturing; cinema theatres (including preview theatres); clay works; clock and watch manufacturing; club; coffee curing and roasting premise; computer installations (main frame); condensed milk factories, milk pasteurizing plant and dairies; confectionery manufacturing; dwellings; educational and research institutes; electric generating houses (hydel); electric lamps (incandescent and fluorescent) and TV picture tube manufacturing; electric sub-station/distribution station; electro plating work; electric and/or computer equipment assemble and manufacturers; empty containers storage yard; engineering workshops; fruits and vegetables dehydrating/drying factories; fruit products and condiment factories; glass and glass fibre manufacturing; godowns and warehouses storing non-combustible goods; green houses; gold thread factories/gilding factories; gum and/or glue and gelatin manufacturing; hospitals including X-ray and other diagnostic clinics; ice candy and ice-cream manufacturing; ice factories; ink(excluding printing ink) factories; laundries; libraries, mica products manufacturing; office premises; places of worship; pottery works; poultry farm; residential hotels, cafes and restaurants; salt crushing factories and refineries; stables; steel plants (other than gas based); sugar candy manufacturing; sugar factories and refineries; tea blending and tea packing factories; umbrella assembling factories; vermicelli factories; water treatment/water filtration plants and water pump house.

## Ordinary Hazard Occupancies

Airport and other transportation terminal building; arecanut slicing and betel-nut factories; atta and cereal grinding; bakeries; beedi factories; biscuit factories; bobbin factories; book-binders, envelopes and paperbag manufacturers; bulk storage; cable manufacturing; camphor boiling; candle works; carbon paper; typewriter ribbon manufacturers; card board box manufacturing; carpenters, wood-wool and furniture manufacturers; carpet and drugget factories; cashew nut factories; chemical manufacturing; cigar and cigarette factories; coffee grinding premises; coir factories; coir carpets, rugs and tobacco, hides and; skin presses; cold storage premises; cork products manufacturing; dry cleaning, dyeing, laundries; electric generating stations (other than hydel); enamelware factories; filter and wax-paper manufacturing; flour mills; garages; garment makers; ghee factories (other than vegetable); godowns and warehouses (others); grain and/or seeds disintegrating and/or crushing factories; grease manufacturing; hat and topee factories. Hosiery, lace, embroidery and thread factories; incandescent gas mantle manufacturers; industrial gas mfg. including halogenated hydro carbon gases; linoleum factories and man-made yarn/fibre manufacturing (except acrylic); manure and fertilizer works. (Blending, mixing and granulating only); mercantile occupancies (dept. Stores, shopping complexes/malls) mineral oil blending and processing. Museums; oil and leather cloth factories. Oil terminals/depots other than those categorized under high hazard 'a; oxygen plants. Plywood manufacturing/wood veneering factories; paper and cardboard mills; piers, wharves, dockyards. Plastic goods manufacturing; printing press premises; pulverizing and crushing mills; rice mills; rope works; rubber goods manufacturing; rubber tyres and tubes manufacturing; shellac factories; shopping complexes (underground); silk filatures and cocoon stores. Spray painting; soaps and glycerine factories; starch factories; steel plants (gas based); tanneries/leather goods manufacturers; tank farms (other than those categorized under high hazard 'A'.); textile mills; tea factories; telephone exchanges; theatres and auditorium; tobacco (chewing) and pan-masala making; tobacco grinding and crushing; tobacco redrying factories; woolen mills.

*High Hazard Occupancies*

a. Aircraft hangers; aluminium/magnesium powder plants; bituminized paper and/or hessian; cloth manufacturing including tar felt manufacturing.; cotton waste factories; coal and/or coke and/or charcoal ball briquettes manufacturing.; celluloid goods manufacturing; cigarette filter manufacturing; cinema films and tv. Production studios; collieries. Cotton seed cleaning or delinting factories.; distilleries; duplicating and stencil paper manufacturing; fire-works manufacturing; foamed plastics manufacturing and/or converting plants; grass, hay, fodder and bhoosa (chaff); pressing factories; jute mills and jute presses; LPG bottling plants (mini); match factories; man made fibres (acrylic fibres/yarn) making mattress and pillow making.; metal or tin printers (where more than 50% of floor area is occupied as engineering workshop this may be taken as ordinary hazard risk); oil mills; oil extraction plants (other than those forming part of ghee factories and oil refining factors); oil terminals/depots handling flammable liquids having flash point of 32° C and below; paints and varnish factories; printing ink manufacturing; sponge iron plants; surgical cotton manufacture; tank farm storing flammable liquids having flash point of 32° C and below; tarpaulin and canvas proofing factories; turpentine and rosin distilleries; tyre retreading and resoling factories, bottling saw mills; bottling plants having total inventory not exceeding 100 mt of LPG and also battling a total quantity of not exceeding 20 mt of LPG per shift of 8 hrs.

b. Ammonia and urea synthesis plants; CNG compressing and bottling plants; explosives factories; LPG bottling plants (other than mini); petrochemical plants; petroleum refineries.

## NEC Explosive atmosphere classification

Certain locations are hazardous because the atmosphere does or may contain gas, vapor or dust in explosive quantities. The National Electrical Code (NEC) divides these locations into classes and groups according to the type of explosive agent which may be present. Listed are some of the agents in each classification.

**CLASS I**

**Group A:** Acetylene.

**Group B:** Butadiene, ethylene oxide, hydrogen, propylene oxide, manufactured gases containing more than 30% hydrogen by volume.

**Group C:** Acetaldehyde, cyclopropane, diethyl ether, ethylene.

**Group D:** Acetone, acrylonitrile, ammonia, benzene, butane, ethanol, ethylene dichloride, gasoline, hexane, isoprene, methane (natural gas), methanol, naphtha, propane, propylene, styrene, toluene, vinyl acetate, vinyl chloride, xylene.

**CLASS II**

**Group E:** Aluminum, magnesium, and other metal dusts with similar characteristics.

**Group F:** Carbon black, coke, or coal dust.

**Group G:** Flour, starch, or grain dust.

**CLASS III**

Easily ignitable fibers, such as rayon, cotton, sisal, hemp, cocoa fibre, oakum, excelsior, and other materials of similar nature.

## Chemical Hazards

Industrial processes involve use of chemical and hazardous materials, for which safety considerations and controls are highly warranted. The raw materials used in industry could be hazardous due to toxicity inherent in materials. The products, intermediate and finished, by-products (including industrial wastes), can cause serious problems of chemical safety. The storage, handling, manufacture and use of chemicals call for strict surveillance on the part of managers, supervisors and employees.

Chemical hazards can be grouped as:

1. Solids, comprising of combustible/flammable solids, toxic and corrosive solids (including radioactive substances), solids causing spontaneous ignition or violent reactions (in water or air), explosive solids or detonators.
2. Liquids comprising of combustible/ flammable liquids, toxic and corrosives causing explosions.
3. Gases comprising combustible/flammable gases, toxic and corrosive gases, explosive or a mixture of gases.

| A few examples of chemical hazards | | |
|---|---|---|
| Name of hazardous substance | Examples | Remarks/ Effects |
| **SOLIDS** | | |
| (a) Combustible/ flammable | Wood, wool, paper , etc. | Catch fire easily and burn. |
| (b) Toxic solids | Certain compounds of lead, manganese, chromium, mercury, arsenic, cadmium | Lead poisoning, manganese poisoning, chromium poisoning, mercury poisoning, arsenic poisoning, cadmium poisoning. |
| (c) Corrosive solids | Caustic soda, caustic potash | Cause blisters or burns. |
| (d) Radioactive materials | Uranium, plutonium, thorium | Cause radiation hazards |
| (e) Spontaneous ignition or violent reactions | Sodium, lithium, phosphorus | Cause accident |
| (f) Explosives | Gunpowder, coal dust | Causes explosions |
| **LIQUIDS** | | |
| (a) Combustible/ flammable liquids | Petrol, kerosene, methyl alcohol | Catch fire easily and burn |
| (b) Corrosive liquids | Nitric acid, sulphuric acid | Cause burn or blisters |

*(Contd.)*

| | | |
|---|---|---|
| (c) Toxic liquids | Methyl alcohol, benzene, carbon tetrachloride, carbon disulphide | Cause poisoning |
| (d) Explosives (liquid) | Nitroglycerine, trinitro-toluene (TNT) | Cause explosion |

**GASES**

| | | |
|---|---|---|
| (a) Combustible/ flammable | Oxygen, hydrogen, acetylene | Catch fire easily and burn |
| (b) Corrosive gases | Sulphurdioxide, chlorine | Cause discomfort, corrosion, irritation |
| (c) Toxic gases | Carbon monoxide, methyl isocyanate, phosgene | Fatal in nature and cause death |
| (d) Explosives or form explosive mixtures with air | Hydrogen, propane | Causes explosion |

**Evaporation rate:** The rate at which a particular material will vapourize (evaporate) when compared to the rate of vapourization of a known material. The evaporation rate can be useful in evaluating the potential health and fire hazards of a material. The known material is usually normal butyl acetate with an evaporation rate designated as 1.0.

Evaporation rates of other solvents or materials are then classified as:
—FAST evaporating, if greater than 3.0.
Examples: Methyl ethyl ketone (MEK) 3.5, acetone 5.6, hexane 8.3:
—MEDIUM evaporating, if 0.8 to 3.0.
Examples: (95%) Ethyl alcohol = 1.4, naphtha = 1.4.
—SLOW evaporating if less than 0.8.
Examples: Xylene 0.6, isobutyl alcohol = 0.68, normal butyl alcohol = 0.4.

## HIGH-RISE BUILDINGS

High-rise buildings have:

- Multiple floors, one above the other.
- The occupancy concentration is much higher than normal buildings and therefore the fuel load will also be much higher.
- Since fire and smoke, both spread much easily upwards, the probability of uncontrolled movement of fire in the structure is very high.

- Access for fire fighting is very limited, especially to the top floors.
- In case of emergencies, evacuation is very difficult and can get delayed.
- High-rise buildings develop a stack-effect, which causes movement of fire upwards.
- The effect of wind in high-rise buildings is considerable.

| Building classification by height. | |
| --- | --- |
| *Height* | *Classification* |
| 1. 0 to 15 m | Ordinary building |
| 2. 15 to 24 m | High-rise building |
| 3. Above 24 m | Special high-rise building |

## TRANSPORT OF HAZARDOUS CHEMICALS—MOTOR VEHICLES ACT

Due to unprecedented growth of chemical industries, the proportion of hazardous chemicals in total freight traffic is increasing at a rapid rate. Of the carriers that carry hazardous goods, approximately two-thirds of them carry flammable petroleum products including kerosene, petrol, LPG, naphtha, etc. The movement of such substances is more prone to accident than the movement of other goods. When involved in a road accident, the vehicle may cause disastrous consequences like fire, explosion, injuries, in addition to property loss and environmental pollution

The safety codes and safety requirements to be followed in transportation of hazardous materials are laid down in the Central Motor Vehicles Rules.

### Features of the Act

#### Rule 129

Every goods carriage carrying dangerous or hazardous goods shall display a distinct mark of the class label appropriate to the type of dangerous or hazardous goods.

#### Rule 130

The class label shall be so positioned that the size of the label is at an angle of 45° to the vertical and the size of such label shall not be of less than 75 mm square which may be divided into two portions, the upper half portion being reserved for the pictorial symbol and the lower half for the rest.

#### Rule 131

The goods carriage has a valid registration to carry the hazardous goods. The vehicle is equipped with necessary first aid, safety equipment and antidotes as may be necessary. The transporter or owner of the goods carriage has full and adequate information about the dangerous or hazardous goods being transported. The driver of the goods carriage is trained in handling the dangers posed during transport of such goods.

### Rule 132

The goods carriage has valid registration and permit and is safe for the transportation of the said goods. The vehicle is equipped with necessary first aid, safety equipment, tool box and antidotes as may be necessary to contain any accident. The driver should have successfully passed a course connected with the transport of hazardous goods.

### Rule 133

It is the responsibility of the driver to keep all information provided to him in writing, i.e. in the form TREM CARD (Transport Emergency Card). This is to be kept in the drivers cabin and is available at all times while hazardous material related to it is being transported.

### Rule 136

The driver of a goods carriage transporting any dangerous or hazardous goods shall, on the occurrence of an accident involving any dangerous or hazardous goods transported by his carriage, report forthwith to the nearest police station and also inform the owner of the goods carriage or the transporter, regarding the accident.

## Minimum Fire Safety Requirements as per Indian Factories Act-1948

Explosive or inflammable dust, gas, etc.

1. Where in any factory, any manufacturing process produces dust, gas, fume or vapour of such character and to such extent as to be likely to explode on ignition, all practicable measures shall be taken to prevent any such explosion by:

   a. Effective enclosure of the plant or machinery used in the process;
   b. Removal or prevention of the accumulation of such dust, gas, fume or vapour;
   c. Exclusion or effective enclosure of all possible sources of ignition.

2. Where in any factory, the plant or machinery used in a process which involves flammable, combustible, explosive materials, is not so constructed as to withstand the probable pressure which such an explosion as aforesaid would produce, all practicable measures shall be taken-to restrict the spread and effects of the explosion by the provision in the plant or machinery of chokes, baffles, vents or other effective appliances.

## Precautions in Case of Fire

1. In every factory, all practicable measures shall be taken to prevent outbreak of fire and its spread, both internally and externally, and to provide and maintain:

   a. Safe means of escape for all persons in the event of a fire, and
   b. The necessary equipment and facilities for extinguishing fire.

2. Effective measures shall be taken to ensure that in every factory all the workers are familiar with the means of escape in case of fire and have been adequately trained in the routine to be followed in such cases.

| (Front) | (Rear) |
|---|---|

**(Front)**

**PERMT**
**FOR CUTTING AND WELDING**
**WITH PORTABLE GAS OR ARC EQUIPMENT**

Date _____

Building _____

Dept. _____ Floor _____

Work to be done _____

_____

Special precautions _____

_____

Is fire watch required? _____

The location where this work is to be done has been examined, necessary precautions taken, and permission is granted for this work. (See other side.)

Permit expires _____

Signed _____
(Individual responsible for authoizing cutting and welding)

Time started _____ Completed _____

**FINAL CHECK-UP**

Work area and all adjacent areas to which sparks and heat might have spread (including floors above and below and on opposite sides of walls) were inspected 30 minutes after the work was completed and were found firesafe.

Signed _____
(Supervisor or Fife Watcher)

**(Rear)**

**ATTENTION**
Before approving any cutting and welding permit, the fire safety supervisor or his appointee shall inspect the work area and confirm that precautions have been taken to prevent fire in accordance with NFPA 51B.

**PRECAUTIONS**
☐ Sprinklers in service
☐ Cutting and welding equipment in good repair

**WITHIN 35 FT (10.7 M) OF WORK**
☐ Floors swept clean of combustibles
☐ Combustible floors wet down, covered with damp sand. metal, or other shields
☐ No combustible material or flammable liquids
☐ Combustibles and flammable liquids protected with covers, guards, or metal shields
☐ All wall and floor openings covered
☐ Covers suspended beneath work to collect sparks

**WORK ON WALLS OR CEILINGS**
☐ Construction noncombustible and without combustible covering
☐ Combustibles moved away from opposite side of wall

**WORK ON ENCLOSED EQUIPMENT**
(Tanks, containers, ducts, dust collectors, etc.)
☐ Equipment cleaned of all combustibles
☐ Containers purged of flammable vapours

**FIRE WATCH**
☐ To be provided during and 30 minutes after operation
☐ Supplied wfth extinguisher and small hose
☐ Trained in use of equipment and in sounding fire alarm

**FINAL CHECK-UP**
☐ To be made 30 minutes after completion of any operation unless fire watch is provided

Signed. _____
(Supenrisol)

**Example of a hot work permit (source–NFPA)**

## OXIDIZERS

Oxidizers are materials which assist combustion by contributing the required oxygen or oxidizing agent.

*General:* The oxidizers on the following lists are typical for their class. Each oxidizer is undiluted unless a concentration is specified from class 1 to 4.

## Typical Class 1 Oxidizers

All inorganic nitrates (unless otherwise classified); all inorganic nitrites (unless otherwise classified); ammonium persulfate; barium peroxide; calcium peroxide; hydrogen peroxide solutions (greater than 8% up to 27.5%); lead dioxide; lithium hypochlorite (39% or less available chlorine); lithium peroxide; magnesium peroxide; manganese dioxide; nitric acid (40% concentration or less); perchloric acid solutions (less than 50% by weight); potassium dichromate; potassium percarbonate; potassium persulfate; sodium carbonate peroxide; sodium dichloro-s-triazinetrione dehydrate; sodium dichromate; sodium perborate (anhydrous); sodium perborate monohydrate; sodium perborate tetrahydrate; sodium percarbonate; sodium persulfate; strontium peroxide; trichloro-s-triazinetrione (trichloroisocyanuric acid, all forms); zinc peroxide.

## Typical Class 2 Oxidizers

barium bromate; barium chlorate; barium hypochlorite; barium perchlorate: Barium permanganate; 1-bromo-3-chloro-5, 5-dimethylhydantoin (BCDMH).

Calcium chlorate; calcium chlorite; calcium hypochlorite (50% or less by weight); calcium perchlorate; calcium permanganate; chromium trioxide (chromic acid); copper chlorate; halane (1,3-dichloro-5,5-dimethylhydantoin): Hydrogen peroxide (greater than 27.5% up to 52%); lead perchlorate; lithium chlorate; lithium hypochlorite (more than 39% available chlorine); lithium perchlorate; magnesium bromate; magnesium chlorate; magnesium perchlorate; mercurous chlorate; nitric acid (more than 40% but less than 86%); nitrogen tetroxide; perchloric acid solutions (more than 50% but less than 60%); potassium perchlorate; potassium permanganate; potassium peroxide; potassium superoxide; silver peroxide; sodium chlorite (40% or less by weight); sodium perchlorate; sodium perchlorate monohydrate; sodium permanganate; sodium peroxide; strontium chlorate; strontium perchlorate; thallium chlorate; urea; hydrogen peroxide; zinc bromate; zinc chlorate; zinc permanganate.

## Typical Class 3 Oxidizers

Ammonium dichromate; calcium hypochlorite (over 50% by weight); chloric acid (10% maximum concentration); hydrogen peroxide solutions (greater than 52% up to 91%); mono-(trichloro)-tetra-(monopotassium dichloro)-pentas-triazinetrione; nitric acid, fuming (more than 86% concentration); perchloric acid solutions (60% to 72% by weight); potassium bromate; potassium chlorate; potassium dichloro-s-triazinetrione (potassium dichloroisocyanurate); sodium bromate; sodium chlorate; sodium chlorite (over 40% by weight); sodium dichloro-s-triazinetrione (sodium dichloroisocyanurate).

## Typical Class 4 Oxidizers

Ammonium perchlorate (particle size greater than 15 microns); ammonium permanganate; guanidine nitrate; hydrogen peroxide solutions (greater than 91%); tetranitromethane.

NOTE: Ammonium perchlorate less than 15 microns is classified as an explosive and is not mentioned here.

## GENERAL CLASSIFICATION OF BUILDINGS

All buildings are classified, according to the use or the character of occupancy in one of the following groups:

| | |
|---|---|
| Group A | Residential |
| Group B | Educational |
| Group C | Institutional |
| Group D | Assembly |
| Group E | Business |
| Group F | Mercantile |
| Group G | Industrial |
| Group H | Storage |
| Group J | Hazardous |

## Fire Zones

In land development and town planning, fire zones are designated as follows:

- Fire zone no. 1
- Fire zone no. 2, and
- Fire zone no. 3.

*Fire zone no. 1:* This comprises areas having residential (group A), educational (group B), institutional (group C) and assembly (group D), small business (subdivision E-l) and retail mercantile (group F) buildings, or areas which are under development for such occupancies.

*Fire zone no. 2:* This comprises business {group E (part)} and industrial buildings (group G (part)}, except high hazard industrial buildings or areas which are under development for such occupancies.

*Fire zone no. 3:* This comprises areas having high hazard industrial buildings, storage buildings (group H) and buildings for hazardous uses (group J) or areas which are under development for such occupancies.

## Fire Protection for Pallet Storage

Pallets are an important material handling tool for warehousing. However, idle pallets present a significant fire hazard: they provide a source of dry fuel, their frayed edges are subject to easy ignition, and their open construction provides flue spaces through which fire can grow very hot and spread quickly. A pallet fire can severely test a sprinkler system's ability to contain the fire, and may result in severe damage to a building's structure and its contents.

While most pallets are made of wood, the durability, reusability, and recyclability of newer plastic pallets has spurred an increase in the number of plastic pallets in use. While plastic pallets have many advantages over wooden pallets, including the capacity to be customized for specific operations (e.g., pharmaceutical and food industries), their greater flammability requires special attention to fire protection.

During a fire, plastic pallets will release more heat, (approximately twice that of wood), will burn for longer periods, and may melt and pool, contributing to rapid fire spread.

Preferred storage location for idle pallets:

- Store pallets outside and away from the main building, or in a detached, low-value building.
- Adequate clearance is to be provided between outside idle pallet storage and the main building.

If pallets must be stored inside, guidelines are:

- Eliminate or reduce the storage of idle combustible pallets.
- Limit the number of pallets on hand to the amount required for one day's operations.
- Prohibit storage of idle pallets in aisles or between racks.

For wood pallets:

* Store pallets flat and stack no higher than 4 m; never stack them on end.
* Separate each pallet pile (no more than 4 stacks per pile) from other pallet piles by at least 8 feet.
* Separate each pallet pile from other commodities by at least 1 m.

For plastic pallets:

* Store pallets in a cut-off room. Storage should be separated from the remainder of the building by 3-hour-rated fire walls. The room should have at least one exterior wall, and storage should be piled no higher than 4 m.
* When pallets are not within a cut-off room, storage should be piled no higher than 1.5 m. Separate each pallet pile (no more than 4 stacks per pile) from other pallet piles by at least eight feet. Separate each pallet pile from other commodities by at least 2 m.

For both wood and plastic idle pallets:

* It is advisable to install a sprinkler system.

## Combustible Gas Cylinder Storage Rules

* The storage area must be well-ventilated.
* Keep fuel cylinders 6 m (20 ft) or more away from combustibles
* Close valves, ensure valves are protected
* Limit inside storage to 70 m$^3$.
* Store cylinders in the upright position and secured from falling
* Separate oxygen from fuel gas
* Never lift cylinders by the service valve or valve protection (use slings, net, or other approved means)
* Keys, handles, and hand wheels must be present
* Use the proper regulator

## Fire Safety in a Typical Construction Site

People exposed to hazard:

* All workers working in the site.
* Site engineers and supervisors.
* Messengers/visitors entering into the site.

Fire is one of the main hazards in a construction site where numerous activities are being carried out by using different types of equipment, tools and electrical energy.

Fire will not occur without existence of combustible material or exposures to heat or ignition. In a construction site, all the three things namely combustible material (fuel), heat producing activities and ignition sources and atmospheric air are available. The personnel working in the site should work carefully so as not to cause a fire.

Possible injuries from a fire incident:

* Burns (of different degrees)
* Fatality

## Fire Safety Precautions

1. *Wooden shuttering planks:* Wooden planks are used for scaffolding and to make moulds for RCC in the site. Numerous planks will be found stacked, stored, moulded. During unsafe welding or other hot work, these planks may get ignited and catch fire. Personnel doing hot work or welding bar/cutting must get work permit from safety department and carefully perform their work following the safety precaution given in the work permit sheet.

2. *Bamboos:* Nowadays working with bamboos in the construction site has been eliminated and these have been replaced by steel pipes. Hence, using of bamboos should not be allowed in the site.

3. *Oil storage:* For the operation of mobile crane, excavator oil engine pumps, HSD oil may be stored in the 200 lit. capacity metal drums in the site. Keeping this oil may cause fire and explosion. If it is very essentially required, two drums (200 L) may be allowed to be stored within the site at the safe place where heat exposure will be minimum or nil.

4. *Litter and packing materials:* Garbage, paper and other packing materials are frequently found in considerable quantity in a construction site. The litter and other combustible materials should not be allowed to be scattered within the site, since it will encourage a fire.

5. *General stores:* Normally a general stores will be maintained to store the materials required for the next day use or future use. Some of them may be combustible materials. The store keeper should keep an eye on the combustible and explosive materials stored.

6. *Possible heat sources:* Following are the some of the sources emitting heat to cause fire.
   1. Arc welding.
   2. Oxyacetylene gas welding/cutting.
   3. LPG cutting.
   4. Heating of tar or bitumen.
   5. Black smithy oven.
   6. Electrical flash.
   7. Heat from poorly aligned or lubricated or over loaded motor and machines.
   8. Spark/heat from grinding wheels.

All personnel should be given training in fire safety. All should be familiar with the operation of first aid fire fighting hand appliances.

The safety precaution to be taken during: A. welding and cutting operation:

1. Qualified and experienced operators should only the allowed to carry out welding and cutting works.
2. Wherever required, asbestos cloth screen and metal shields should be provided.
3. Beside the welder, the helper also should wear welding goggles, gloves and clothing.
4. The welding equipment used should be in operating condition.
5. All electrical equipment should be properly grounded.
6. Power cables of arc welding should be protected and maintained in good manner.
7. While carrying out welding or cutting work, proper type of fire extinguisher should be kept near by.
8. Inspection should be carried out for fire hazards before starting welding or cutting work.
9. The flammable material kept or stored near by should be well protected.
10. Gas cylinder should always be chained upright.
11. The hoses connected to the gas cylinder should be protected and maintained in good condition.

The safety precaution to be taked during flammable gases and liquids:

1. All containers should be clearly identified.
2. Proper storage practices should be absorbed.
3. Proper storage temperature and protection should be maintained.
4. Fire hazards should be checked.
5. Proper types and required number of extinguishers should be provided near by.
6. Gas cylinder should always be conveyed by means of carts.

## FIRE INSURANCE

Insurance is a loss-prevention measure. Insurance is a means of indemnity against occurrence of a uncertain undesirable event.

Today, the manner of manufacturing and business activity has changed with regard to the scale and complexity of the operations. There are:

- Higher level of automation and use of computers
- Lesser manpower
- Higher volume of storage, throughput and processing of materials

Consequently, the nature of risks has also changed. Earlier insurance companies were charging a uniform premium for insurance of similar activities. Now, insurance companies are employing methodologies such as HAZOP (hazard and operability studies) to determine the premium for the risk. To undertake such studies, insurance companies have to be provided with details about the process, safety procedures and also details about the fire protection system installed.

## Documents to be Submitted for Claiming Fire Insurance

- Incident report of the fire occurrence along with photographs, if any, of the fire incidence–accident report, log entries, etc.
- Details as per the format of the insurance company.
- Copy of the insurance policy.
- Report of the insurance surveyor.
- Report of the fire brigade (if the fire brigade was called).
- Damage and loss assessment report.
- FIR report from the police.

## Insurance Companies Covering Fire Risk in India

- New India Insurance Company.
- Oriental Insurance Company.
- Bajaj Alliance General Insurance.
- ICICI Prudential Insurance Company.
- Reliance General Insurance.

## Fires can Happen!

We all believe that fires rarely happen and even if they do happen, they occur in high risk areas such as depots for inflammable chemicals or petrol pumps, etc. and not in our type of offices and homes. However, statistics and analysis of accidents and fires that have happened in the past–according to data collected by "loss prevention association" clearly reveal that the actual trend is very different.

## Causes Resulting in Insurance Claims

- Fire accidents       80.0%
- Explosion            10.0%
- Arson                 8.29%
- Lightning             0.29%
- Miscellaneous         1.42%

## Insurance Premium

Insurance companies charge an annual premium for insuring a facility. For manufacturing facilities categorized as light to medium hazard, the insurance rate is 0.25% of the value insured. Discount is given if the facility has a well designed and adequate "fire protection" system and the installation is TAC approved.

## Discount in Insurance Premium for Loss Making Manufacturing Companies

This is based on the annual net profit of the business as per the previous year's accounts plus the 'standing charges'. An insured is entitled to return of premium (not exceeding 50% of the premium paid) in case the gross profit earned during the financial year concurrent with the period of insurance is less than the sum insured.

Fire Drill

| Fire drill register | | | | | | | |
|---|---|---|---|---|---|---|---|
| *Sl. no.* | | | | | | | *Date* |
| Type of drill Dry or wet | Serial no. of hydrants operated | Duration of operation of pumps | Remarks/defects observed, if any | Signature of fire marshall | Signature of works manager | Remarks | |

Name and designation of persons attending the drill

1.
2.
3.
4.
5.
6.

Data, Conversions and Constants

**Temperature conversions:**

$$° \text{Fahrenheit} = 9/5° \text{ C} + 32$$
$$° \text{Celsius} = 5/9 \text{ (f–32)}$$
$$° \text{Rankine to } ° \text{ F: } ° \text{ F} + 459.69$$
$$° \text{Kelvin to } ° \text{ C: } ° \text{ C} + 273.16$$

| Standard wire gauges (diameter) | | | | | |
|---|---|---|---|---|---|
| *SWG* | *Inch* | *mm* | *SWG* | *Inch* | *mm* |
| 1 | 0.300 | 7.620 | 14 | 0.082 | 2.032 |
| 2 | 0.276 | 6.910 | 15 | 0.072 | 1.629 |
| 3 | 0.252 | 6.401 | 16 | 0.064 | 1.226 |
| 4 | 0.232 | 5.893 | 17 | 0.056 | 1.442 |
| 5 | 0.212 | 5.385 | 18 | 0.048 | 1.219 |
| 6 | 0.192 | 4.877 | 19 | 0.040 | 1.016 |
| 7 | 0.175 | 4.445 | 20 | 0.036 | 0.914 |
| 8 | 0.160 | 4.064 | 21 | 0.032 | 0.813 |
| 9 | 0.144 | 3.658 | 22 | 0.028 | 0.711 |
| 10 | 0.128 | 3.251 | 23 | 0.024 | 0.610 |
| 11 | 0.116 | 2.946 | 24 | 0.022 | 0.059 |
| 12 | 0.092 | 2.642 | 25 | 0.020 | 0.058 |
| 13 | 0.092 | 2.337 | | | |

| GI wires | | |
|---|---|---|
| SWG | Metres/kg | kg/metre |
| 16 | 42.35 | 0.204 |
| 14 | 27.5 | 0.036 |
| 12 | 22.73 | 0.044 |
| 10 | 18.25 | 0.055 |
| 8 | 9.0 | 0.102 |
| 4 | 4.587 | 0.218 |

| Unit weight of construction materials | |
|---|---|
| Material | kg/cu m |
| Cement ordinary | 1440 |
| Cement concrete | 2240 |
| Sand dry | 1600 |
| Sand wet | 1700–2000 |
| Mild steel | 7840 |
| Cast iron | 7200 |
| Brass | 8550 |
| Lead | 11350 |
| Tar | 1010 |
| Aluminium | 2800 |

| Multiples and submultiples of units: the symbols are used as prefixes along with the units | | |
|---|---|---|
| Factor | Prefix | Symbol |
| $10^{18}$ | exa | E |
| $10^{15}$ | peta | P |
| $10^{12}$ | tera | T |
| $10^{9}$ | giga | G |
| $10^{6}$ | mega | M |
| $10^{3}$ | kilo | k |
| $10^{2}$ | hector | h |
| 10 | deca | da |
| $10^{-1}$ | deci | d |
| $10^{-2}$ | centi | c |
| $10^{-3}$ | milli | m |
| $10^{-6}$ | micro | ì |
| $10^{-9}$ | nano | ç |
| $10^{-12}$ | pico | p |
| $10^{-15}$ | femto | f |
| $10^{-18}$ | atto | a |

## RELIABILITY OF SYSTEMS

## 5S and Maintenance Concepts

### 5S

5S is a basic technique to enhance the appearance of our surroundings, be it at home, office or workplace and make it pleasant and safe and efficient. The concept helps to keep a constant level of cleanliness and orderliness and improve our working efficiency.

The concept of 5S is not new. But the Japanese have systematized the concept and envisioned it in an orderly and regular manner. 5S helps to achieve best results with limited resources.

The 5 steps

| | |
|---|---|
| **SEIRI** | **SORT** |
| **SEITON** | **ORGANIZE** |
| **SEISO** | **SHINE** |
| **SEIKETSU** | **STANDARDIZE** |
| **SHITSUKE** | **SUSTAIN** |

*Seiri:* To remove all the unwanted material from the use area in a systematic way.

*Seiton:* To identify and allocate a place for all the materials and tools being used.

*Seiso:* To keep spic and span everywhere.

*Seiketsu:* Standardizing all the 5S activities.

*Shitsuke:* To maintain and continuously improve.

## MAINTENANCE METHODS

Maintenance is upkeep of machinery, appliances, equipment, etc. or keeping them in proper working condition. Maintenance is required to rectify the detioration of equipment due to normal degradation by wear and tear.

If maintenance is not carried out in a systematic manner, the process or operation is disrupted, the life is shortened and the quality of the output is affected. Such frequent disruptions lead to process and equipment damage and the system reliability reduces. Due to increased hazard levels, this also leads to secondary losses such as accidents, which can be debilitating. The occurrence of catastrophic breakdowns is quite possible.

Maintenance can now be termed as physical asset management. Many of the major physical asset disasters have been due to the root cause of inadequate maintenance. Manufacturing and other systems continue to rise in complexity and are becoming more complicated to maintain and mitigate the safety and environmental consequences of physical asset failures.

## Types of Maintenance Methods

### Reactive or Breakdown Maintenance

Reactive or breakdown maintenance follows the principle of run-to-failure. This method can be used for systems where the consequence of failure or breakdown is not severe.

### Preventive Maintenance

The objective of preventive maintenance is to detect, preclude or mitigate the degradation of a component or system. Preventive maintenance practice has shown a reduction in breakdowns and failure to the extent of 16–18% over the reactive maintenance method.

Features of a preventive maintenance plan:
- Checklist for inspection
- Schedule for inspection
- Replacement plan

High standards of cleanliness are very much a part of preventive maintenance. The practice of concepts such as the Japanese "5S" has yielded very concrete benefits in both operation and maintenance of systems (See "5S" described separately). Methods such as the LLF—"look, listen and feel" are very important for effective implementation of preventive maintenance programmes.

### Predictive Maintenance

Predictive maintenance is based on measurements that detect the onset of a degradation mechanism, so that control action can be initiated before breakdown. This helps to maintain the functional capability. Corrective actions can be initiated at the right time, reduces breakdowns, and increases the safety of the operations. The practice of predictive maintenance requires the availability of diagnostic equipment.

### Reliability Centered Maintenance (RCM)

RCM is the process used to determine the maintenance requirement of any physical asset in its operating context. RCM uses a mix of reactive maintenance (for less critical equipment), predictive and preventive maintenance for high reliability.

Modern large technological systems have to be maintained on predictive and preventive methods, if the hazards due to failure of the system are to be controlled. A combination of 55% reactive, 31% preventive, 12% predictive and 2% other methods has been practised in some of the well-managed enterprises with appreciable success. But a higher qualification, skill and competency level of the operating personnel is necessary to implement the concepts of RCM, predictive and preventive maintenance.

## Human Factor Design

### Manual or Automatic Operation

Industrial plant sizes have grown in size. The operation of such plants is sometimes under extreme conditions and close to the limits of safety. Instrumentation and controls have developed to the extent necessary to operate such systems. Quality, precision, speed and non-ambiguity of operation are now the hallmark of new technological culture. Higher level of control with automation is necessary to mitigate the runaway creation of technological hazards.

### Some Methods to Ensure High Reliability

- Redundancy—by providing more than one means to accomplish a given task/operation.

- De-rating—for assuring that the forces and stress applied are lower than the stresses the parts can normally withstand.

- Fool-proofing—by elimination of error-prone elements in the system. By designing to make human error improbable or impossible.

- Warning mechanisms—by incorporating protective systems such as fire-detection and sound alarms.

---

**Heat released from combustion**

- Heat flux is considered the more appropriate measure by which to examine the radiation effects from a fire.
- A radiant heat flux of 4.7 kw/m² will cause pain on exposed skin, a flux density of 12.6 kw/m² or more may cause secondary fires.
- A flux density of 37.8 kw/m² will cause major damage to a process plant and storage tanks.

---

**Risk managing strategies**

**Avoid risk**—select alternative methods; abandon planned method
**Reduce risk**—change method to reduce risk; add methods to reduce risk
**Plan for contingency**—plan to cope with the risk

---

## FAULT TREE ANALYSIS

A fault tree analysis is a technique for risk assessment. It is a probability based method which helps to identify the risks in an activity or process or system.

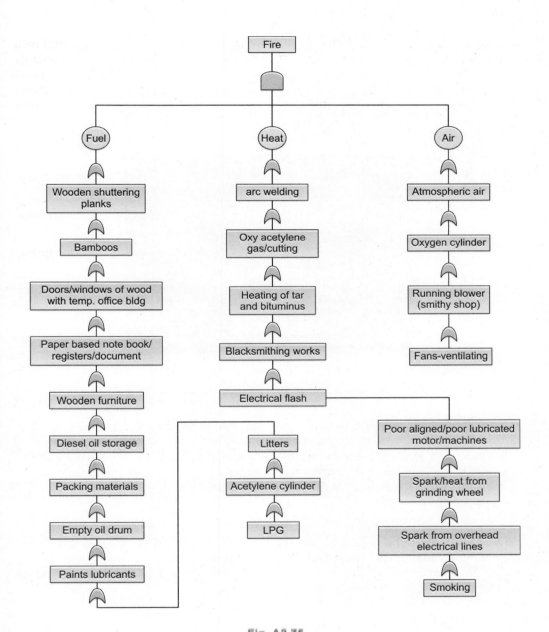

**Fig. A2.75**

### Jet fire

- Most fires involving gas will be associated with a high pressure and labelled as "jet" fires.
- A jet fire is a pressurized stream of combustible gas or atomized liquid that is burning. If such a release is ignited soon after it occurs (i.e. within 2–3 minutes), the result is an intense jet flame.
- The jet fire stabilizes to a point that is close to the source of release, until the release is stopped.
- A jet fire is usually a very localized, but very destructive to anything close to it.

Fig. A2.76: An oil storage tank on fire.

### Flash fire

- If a combustible gas release is not ignited immediately, a vapour plume will form. This will drift and will be dispersed by the ambient winds or natural ventilation. If the gas is ignited at this point, but does not explode, it will result in a flash fire, in which the entire gas cloud burns very rapidly.
- It is unlikely to cause any fatalities, but will damage steel structures.

### Pool fire

- Once a pool of liquid is ignited, gas evaporates rapidly from the pool as it is heated by the radiation and convective heat of the flame.
- Pool fires have some of the characteristics of a vertical jet fire, but their convective heating will be much less.

## RISK ASSESSMENT

Risk assessment involves:

- The systematic identification of hazards associated with a task, before the task is started.

- The systematic identification of the potential consequence caused by the release of the hazard and the measurement of the probability and potential severity.
- The identification and implementation of adequate safeguards which will minimize the risk and reduce the human injury and asset damage.

| Rating of likelihood | | |
|---|---|---|
| *Rating* | *Likelihood* | *Example* |
| 5 | Very likely/frequent | There is high certainty that an accident will happen or is likely to happen frequently. |
| 4 | Likely/probable | The effects of physical hazards, chemical hazards, natural phenomena, human carelessness or other factors could precipitate an accident but which is likely to happen without this additional factor, or the accident will occur several times. |
| 3 | Quite possible/occasional | The accident may happen, if additional factors precipitate it, but is unlikely. |
| 2 | Remote | If the factors were present, this incident or illness might occur, but its probability is low and the risk is minimal. |
| 1 | Unlikely/improbable | There is generally no likelihood of an accident occurring. Only under unexpected conditions could there be a probability of an accident or illness. All reasonable precautions have been taken so far as is reasonably practicable. This should be the normal state of the workplace. |

| Rating of severity | | |
|---|---|---|
| *Rating* | *Likelihood* | *Example* |
| 5 | Catastrophic/very high | Fatal injury. |
| 4 | Critical/high | Serious injury, occupational illness or system damage. Serious injury includes fractures, amputations and hospitalization for more than 24 hrs. |
| 3 | Moderate | Causing injury, occupational illness or disease could keep an individual off work for more than 3 days. |
| 2 | Slight | Causing minor injury that keep an individual off work for not more than 3 days. |
| 1 | Negligible | Less than minor injury that would permit the individual to continue work after first aid treatment. |

| Calculating risk factor number | | | | | |
|---|---|---|---|---|---|
| Likelihood | *Very likely/ frequent* | *Likely/ probable* | *Quite possible/ occasional* | *Remote* | *Unlikely/ improbable* |
| **Severity** | **5** | **4** | **3** | **2** | **1** |
| Catastrophic/very high 5 | 25 | 20 | 15 | 10 | 5 |
| Critical/high 4 | 20 | 16 | 12 | 8 | 4 |
| Moderate 3 | 15 | 12 | 9 | 6 | 3 |
| Slight 2 | 10 | 8 | 6 | 4 | 2 |
| Negligible 1 | 5 | 4 | 3 | 2 | 1 |

| Assessment based on the risk factor number | |
|---|---|
| *Risk level* | *Action and time scale* |
| Intolerable/unacceptable 16–25 | Work should not be started or continued until the risk has been reduced. If it is impossible to reduce risk even with unlimited resources, work has to remain prohibited. |
| Substantial 10–15 | Work should not be started until the risk has been reduced. Considerable resources may have to be allocated to reduce the risk. Where the risk involves work in progress, urgent action should be taken. |
| Moderate/acceptable with review 6–9 | Efforts should be made to reduce the risk, but the costs of prevention should be carefully measured and limited. Risk reduction measures should be implemented within extremely harmful consequences, further assessment may be necessary to establish more precisely the likelihood of harm as a basis for determining the need for improved control of measures. |
| Tolerable 3–5 | No additional controls are required. Consideration may be given to a more cost-effective solution or improvement that imposes no additional cost burden. Monitoring is required. |
| Trivial/acceptable 1–2 | No action is required. |

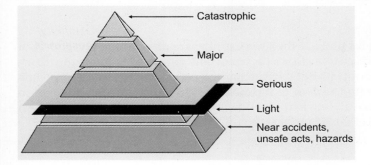

**High risk–intolerable**
**Medium risk–unacceptable**
**Low risk–improve (proactive)**

Fig. A2.77

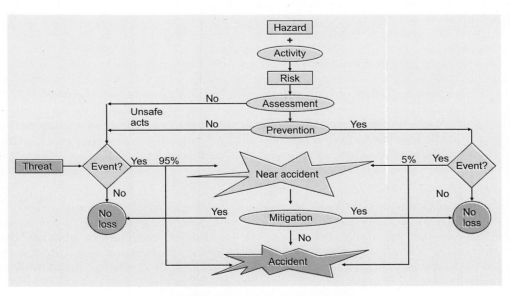

Fig. A2.78: The risk reduction process.

## AIRCRAFT FIRE SAFETY

Preservation of life and prevention of injury should be the prime objective of fire fighting. Saving the aircraft comes second.

An aircraft fire requires immediate, intelligent and carefully planned action. Knowledge of the location of each of the aircraft components and its function is necessary. Fire fighters should be familiar with the design of aircraft and the specific aircraft components and the hazards associated with them. Flammable and fire accelerating materials are present in an aircraft which may be a combination of the following:

- Gasoline, turbine engine lubricating oil, and jet fuel
- Oxygen
- Oils
- Hydraulic fluid

- Anti-icing fluid
- Grease

Of the above, the gasoline and fuel are the greatest fire threats with the possibility of both fire and explosion. Aviation gasoline has a higher octane rating of 115 to 145 (ordinary petrol has an octane rating of about 90), but the fire characteristics of aviation fuel are same as that of ordinary petrol. The fuel system of the aircraft stores and transfers the fuel to the engine or engines. Fuel can leak due to impact, twist or due to poor maintenance. An aircraft contains permanent fuel tanks and also auxiliary and reserve fuel tanks.

### Oxygen System

All aircraft designed for high altitude flying carry oxygen required for breathing of the passengers and crew. Oxygen storage tanks are generally painted green. The oxygen storage pressure may be up to 125 bar in the tank and also in the distribution system. Damage to this system will release the oxygen under high pressure. Oxygen presents two hazards.

Explosion may occur due to rapid heating of the oxygen cylinders or due to oxygen coming in contact with grease or oil coming in contact with the oxygen. The explosion may cause a fire or rupture of fuel and electrical lines. Release of oxygen during a fire will result in acceleration of the burning and great release of heat. Rapid release of oxygen causes fast growth of the fire and resists the action of the fire extinguishing agents. Fire fighters must be familiar with the controls of the oxygen system provided in an aircraft and if possible close the valves, etc to prevent oxygen release.

### Aircraft Electrical System

An aircraft has extensive electrical wiring circuits required for lighting, pumps, hydraulic systems and various other electrical and electronic systems. The principal fire hazard from these is short circuiting and arcing. The electrical system may be powered by a dynamo or the aircraft also has a system of batteries. In a fire incident, the batteries should be disconnected. The ground connection should be disconnected first in order to eliminate the danger of arcing which can ignite fuel vapours.

### Anti-icing Fluid

The anti-icing fluid is used to prevent the formation of ice on propellers, windshield, etc. This fluid may be alcohol or alcohol with 15% glycerine. The fire hazard of the anti-icing fluid is less than that of the fuel, but should be treated with caution in the event of a crash.

### Action in an Airfield

Normally the first information about any accident or impending emergency is received by the control tower. The tower then immediately notifies the fire rescue crew about the situation giving correct status and details about the incident and the rescue crew and other ground staff will respond immediately. In case of impending emergency, the rescue crew will assume standby position but being clear of the runway. An emergency communication system has to be available between the control tower and the ground crew to enable rapid and reliable conveying of messages.

## Aircraft Fire Fighting

The fire fighting crew has to consider:

- The method of fire fighting to be adopted
- The time available to accomplish the fire fighting and rescue
- The cooling techniques to be used.

Safety regulations control aircraft materials and the requirements for automated fire safety systems. Usually these requirements take the form of required tests. The tests measure flammability and the toxicity of smoke.

The cargo holds of most airliners are equipped with "fire bottles" (essentially remote-controlled fire extinguishers) to combat a fire that might occur in the baggage holds, below the passenger cabin.

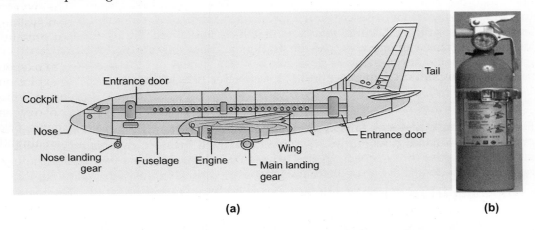

**(a)** **(b)**

**Fig. A2.79:** (a) Parts of an aircraft, (b) Halon 1211 extinguisher for aircraft.

## FIGHTING FOREST FIRES

Forest fires cause enormous loss not only by the destruction of forests but also due to the consequent ecological damage. Hence preventing and controlling the forest fires has assumed great importance.

Forest fires may be caused by nature or by man by his carelessness. The natural elements may be the sun, lightning and the wind. Human elements originate from picnics, revelers, or local residents. Burning of rubbish, automobiles, littering, lumber operations or arson may be major reasons. So, the principal strategy to prevent forest fires should be through enforcement of regulations and education.

Since forest fires can occur over a vast area, detection of the fire is essential to be able to control the fire. Observation patrol squads play a good role in this.

Roads through forest areas act as fire breaks and also limit the vehicles in their movement in the forest. Vehicles with hot exhaust systems parked on litter or dry grass/ leaves in a forest area can initiate a fire.

In forest fires, the part where the fire is progressing fastest is called the head of the fire. There can be several heads in a forest fire depending on the type, abundance and location of the fuel. Fires head generally in the direction of the wind. The stronger the wind, the faster the spread of the fire. It is very important to remember that the wind direction is not fixed and can change abruptly and thereby suddenly exposing the fire fighting personnel to the danger of the fire engulfing them. The fire spread also depends on the topographic condition with a fire generally spreading uphill.

Crown fires are the most difficult to tackle. The hot air rising from the fire causes a partial vacuum. Cold air with a fresh supply of oxygen rushes in at the base. The larger and hotter the fire, stronger is the draught created by the fire. Indirect methods such as creating a fire break may be the only method to control strong forest fires.

Fire fighting personnel and equipment should never approach the head of a fire from upwind and a fire should never be controlled from uphill.

The first step in fighting forest fires is to estimate the extent and condition of the fire such as type, density, flammability of the fuel, head, wind direction and velocity, topography, etc. so that necessary and appropriate actions can be initiated. Water supply in forest areas is difficult. Portable pumps may have to be used. Back-pack are generally used. The persons can use axes, shovels (to dig or throw soil/mud on the fire front). In grass fires, wet cloth or strips of cloth can be used to beat out the fire. For very large forest fires, aircraft are being used today for effective fire fighting.

## EXPLOSIVES AND AMMUNITION FIRES

Fires involving explosives and ammunition are amongst the dreaded because of their potential destruction possibilities by detonation. Such fires are capable of generating violent destructive forces when exposed to heat. Hence it is necessary for the fire-fighting personnel to be totally aware of the characteristics of various explosives and ammunition.

The principal hazard in such fires is:
* Intense heat and rapid fire growth
* Explosion and shock waves
* Flying burning material

In fighting explosive and ammunition fires, fire fighting should be done from a distance greater than 300 metres to avoid unnecessary risk when such materials are involved. In explosive fires, it is prudent to try to confine the fire and prevent it from spreading to surrounding areas and buildings. The materials burn with intense heat and fire fighting personnel and equipment should be adequately protected.

In fires involving phosphorus, persons entering the fire zone must have lifelines attached to themselves to enable them to find their way out through the heavy smoke produced by such fires. Phosphorus will stop burning as long as it is under water, but when exposed again to air, it ignites spontaneously. Fires involving pyrotechnic materials such as magnesium based materials are very difficult to control or extinguish. Water may accelerate the burning and cause explosion which can scatter burning material.

## VEHICLE FIRES

Fires in vehicles may result from various causes. The most frequent causes are:
* Electrical short circuits
* Collisions

- Over heated brakes
- Over heated tyres
- Improper fuelling method
- Fuel leakage
- Careless driving habits

Fig. A2.80

Motor vehicle fire safety depends on:

- Design and construction features
- Method of use of the vehicle
- Vehicle repair and maintenance
- Operational safety procedures

The use of metal body has reduced the fire hazard. But the interior finishing consisting of the upholstery and linings and plastic usage need to have fire-retardant properties.

The fuel tanks' construction, location and security and its vulnerability to damage during a collision are the most important features of a vehicle's fire safety. Fuel tanks are nowadays positioned below the body in the rear of the vehicle. This should provide the occupants more time to escape if the vehicle catches fire. Vibration, corrosion or improper maintenance can detiorate the fuel system design.

The fire safety of the electrical system in a vehicle depends on the design and installation, fusing and maintenance of the system. Fires can originate from the electrical system due to fuel deposits in the engine area or due to ignition of combustible materials such as the seating linings. During a collision of the vehicle, the electrical circuit can get short-circuited.

Proper installation of the exhaust system is important as the hot exhaust gases or incandescent carbon particles from the exhaust can ignite the fuel deposits or nearby combustible materials or wiring cables. The brakes, if overheated can cause the ignition of oil, grease or the brake fluid. Excessive friction can cause underinflated tyres to catch fire.

Vehicle gets overturned or when it has crashed into another vehicle or object trapping the occupants inside, rescue of the occupants is very much necessary to prevent death by injury or fire.

Automobile fires may be extinguished using $CO_2$, halon (or its substitute) or ABC dry powder extinguishers.

# Glossary

**Absolute Temperature:** Temperature scale. Absolute zero (–273.15° C). The temperature at which particles would lose all their kinetic energy.

**Absorption:** A bulk phenomenon or a process by which a gas is taken up by a liquid or solid or in which a liquid is taken up by a solid.

**Accelerant:** Any material which initiates or promotes the spread of a fire, e.g. flammable liquids. The most common accelerants encountered are flammable or combustible liquids. Whether a substance is an accelerant depends not on its chemical structure, but on its use.

**Adsorption:** The retention of solid or liquid particles at the surface of another solid or liquid.

**Air:** A mixture of gases. Atmospheric air is composed of 78% nitrogen, 21% oxygen and 1% are gases including carbon dioxide, argon, helium, etc.

**Air-conditioning:** The process of treating air so as to control simultaneously its temperature, humidity, purity and distribution to meet the requirements of a conditioned space.

**Alarm:** A signal or message from a person or device indicating the existence of a fire, medical emergency, or other situation that requires fire department action.

**Alternative Escape Route:** A path, sufficiently separated by either-direction and space, or by fire resisting construction, to ensure that one is still available should the other be affected by fire.

**Ambient Temperature:** The temperature of the surroundings.

**Anemometer:** An instrument for measuring the velocity of air.

**Anoxia:** Severe lack of oxygen in the blood or the brain.

**Apartment Building:** A building containing three or more dwelling units with independent cooking and bathroom facilities.

**Arson:** Fire originated by malicious intent.

**ASME:** American Society of Mechanical Engineers.

**Assembly Occupancy:** An occupancy: (1) used for a gathering of 50 or more persons for deliberation, worship, entertainment, eating, drinking, amusement, awaiting transportation, or similar uses; or (2) used as a special amusement building, regardless of occupant load.

**Asphyxiant:** A vapour or gas which can cause unconsciousness or death by suffocation (lack of oxygen).

**Asphyxiation:** The suspension of respiration and animation, as the result of the inhalation of substances such as carbon dioxide, methane, nitrogen, etc. when present in atmosphere to an extent sufficient materially to decrease its normal oxygen.

**ASTM:** American Society of Testing and Materials.

**Atmospheric Pressure:** The weight of one sq cm column of the earth's atmosphere. At sea level, this pressure is 1 kg per sq cm (76 cm of a mercury column).

**Auto-door Release:** A device when fitted to self-closing door allows to remain closed under normal conditions but on the actuation of the alarm, releases the door which returns to the closed position.

**Autoignition:** Initiation of combustion by heat but without a spark or flame.

**Autoignition Temperature:** Minimum temperature at which a material will ignite by itself in air and sustain combustion without initiation by an external spark or flame under specified test condition.

**Automatic Fire Detecting and Alarm System:** An arrangement of automatic fire detectors, such as a fuse working at a given temperature, a thermostat or a fluid filled tube or an electronic device, for detecting an outbreak of fire, and sounders and other equipment for automatic transmission and indication of alarm signals without manual

441

intervention. The system also has provision for testing of circuits and, where required for the operation of auxiliary services.

**Automatic Fire Extinguishing System:** Any system designed and installed to detect a fire and subsequently discharge an extinguishing agent without the necessity of human intervention.

**Automatic Sprinkler System:** An arrangement of piping, sprinklers and connected equipment designed to operate automatically by the heat of fire and to discharge water upon that fire and which may also simultaneously give automatic audible alarm.

**Back Draught:** An explosion, of greater or lesser degree, caused by the inrush of fresh air from any source or cause, into a burning building, where combustion has been taking place in a shortage of air.

**Basement:** Any storey that has more than 50% of the total area of the building's perimeter below grade.

**Beam, Fire-resisting:** A structural member with or without any additional protection, capable of satisfying one of the criteria of fire resistance namely, resistance to collapse.

**Blast Effect:** Damage caused by shock waves from an explosion.

**BLEVE (Boiling liquid expanding vapour explosion):** An explosion caused by rapid expansion of flammable gas stored in a container resulting in sudden release of huge quantities of atomized burning liquid which appear as a fire ball and radiating intense heat all around.

**Boil Over:** In a flammable liquid storage tank, it is the discharge with exceptional violence of part of the contents of burning tank.

**Boiling Point:** Minimum temperature at which the vapour pressure of the liquid is equal to atmospheric pressure.

**Building:** Any structure used or intended for supporting or sheltering any use or occupancy.

**Burn:** Injury or damage caused by flame/heat to living beings.

**Burn Back:** Flames going back over an area previously extinguished, may be due to incomplete cover or fire extinguishing media foam degradation.

**Burning:** Normal combustion in which the oxidant is molecular oxygen.

**Burning Behaviour:** All the physical and/or chemical changes that take place when a material, product and/or structure burns or is exposed to fire.

**Burning Rate:** The rate at which combustion proceeds across a fuel. A specialized use of this term, describes the rate at which the surface of a pool or burning liquid recedes. For gasoline, this rate is reported to be approximately 6 mm per minute.

**Burn, to:** To consume or be consumed by rapid oxidation with the production of heat, usually with incandescence or flame, or both.

**Business Occupancy:** An occupancy used for account and record keeping or the transaction of business other than mercantile.

**Calorie:** The amount of energy required to raise the temperature of one gram of water by one degree centigrade. One calorie equals 4.18 joules. One BTU equals 252 calories.

**Calorific Value:** Calorific energy which could be released by the complete combustion of a unit mass of a material.

**Carbon Dioxide:** A colourless, odourless, electrically non-conductive inert gas that is a suitable medium for extinguishing Class B and Class C fires. Liquid carbon dioxide forms dry ice ("snow") when released directly into the atmosphere. Carbon dioxide gas is 1.5 times heavier than air. Carbon dioxide extinguishes fire by reducing the concentrations of oxygen, the vapour phase of the fuel, or both, in the air to the point where combustion stops.

**Cavity Wall:** Built in two thicknesses, separated by an air space, the two thicknesses being connected by occasional ties of metal or brick. Known as hollow wall.

**Ceiling (suspended, fire-resisting):** A ceiling assembly capable of contributing, wholly or in part to the overall fire resistance of the floor above and/or its supporting members.

**CFC:** A chlorofluorocarbon molecule consisting of chlorine, fluorine and carbon atoms.

**Chain Reaction:** A self-propagating chemical reaction in which activation of one molecule leads successfully to activation of many others. Most, perhaps all, combustion reactions are of this kind.

**Change of State:** Condition in which a substance changes from a solid to a liquid or from a liquid to a gas caused by addition of heat or the reverse by removal of heat.

**Charring:** The formation of a light, friable, mainly carbonaceous constituent residual on wood or other organic matter resulting from incomplete combustion and/or devolatilization following exposure to heat.

**Chemical Flash Fire:** The ignition of a flammable and ignitable vapour or gas that produces an outward expanding flame front as those vapours or gases burn. This burning and expanding flame front, a fireball, will release both thermal and kinetic energy to the environment.

**Class 1 Liquid:** A liquid that has a flash point below 37.8° C.

## Classification of Hazards for Portable Fire Extinguishers

*Light (Low) Hazard:* Light hazard occupancies are locations where the total amount of Class A combustible materials, including furnishings, decorations, and contents, is of minor quantity. This can include some buildings or rooms occupied as offices, classrooms, churches, assembly halls, guest room areas of hotels/motels, and so forth. This classification anticipates that the majority of content items are either non-combustible or so arranged that a fire is not likely to spread rapidly. Small amounts of Class B flammables used for duplicating machines, art departments, and so forth, are included, provided that they are kept in closed containers and safely stored.

*Ordinary (Moderate) Hazard:* Ordinary hazard occupancies are locations where the total amount of Class A combustibles and Class B/C flammables are present in greater amounts than expected under light (low) hazard occupancies. These occupancies could consist of dining areas, mercantile shops, and allied storage; light manufacturing, research operations, auto showrooms, parking garages, workshop or support service areas of light (low) hazard occupancies.

*Extra (High) Hazard:* Extra hazard occupancies are locations where the total amount of Class A combustibles and Class B/C flammables present, in storage, production, use, finished product, or combination thereof, is over and above those expected in occupancies classed as ordinary (moderate hazard). These occupancies could consist of woodworking, vehicle repair, aircraft and boat servicing, cooking areas, individual product display showrooms, product convention center displays, and storage and manufacturing processes such as painting, dipping, and coating, including flammable liquid handling.

**Chimney Effect (flue effect):** The upward thrust of convection currents of hot gases through vertical openings. The tendency of air or gas to rise in a duct or other vertical passage when heated.

**Circuit Breaker:** A device designed to open and close a circuit by non-automatic means and to open the circuit automatically on a predetermined over-current without damage to itself when properly applied within its rating.

**Clean Agent:** Electrically non-conducting, volatile, or gaseous fire extinguishant that does not leave a residue upon evaporation.

**Clean Zone:** A defined space in which the concentration of air-borne particles is controlled to specified limits.

**Closed-circuit Self-contained Breathing Apparatus (SCBA):** A recirculation-type SCBA in which the exhaled gas is re-breathed by the wearer after the carbon dioxide has been removed from the exhalation gas and the oxygen content within the system has been restored from sources such as compressed breathing air, chemical oxygen, and liquid oxygen, or compressed gaseous oxygen.

**Closed Container:** A container, so sealed by means of a lid or other device that neither liquid nor vapour will escape from it at ordinary temperatures.

**Code:** A standard that is an extensive compilation of provisions covering broad subject matter or that is suitable for adoption into law independently of other codes and standards.

**Co-efficient of Thermal Expansion:** The proportional increase in length, volume of superficial area of a body per degree rise in temperature.

**Combustion:** Exothermic reaction of a combustible substance with an oxidizer, usually accompanied by flames, and/or glowing and/or emission of smoke.

**Combustible:** Capable of burning. A material that, in the form in which it is used and under the conditions anticipated, will ignite and burn; a material that does not meet the definition of non-combustible or limited-combustible.

**Combustible Aluminum Dust:** Any finely divided aluminum material 420 microns or smaller in diameter that presents a fire or explosion hazard when dispersed and ignited in air.

**Combustible Gas Detectors:** A device used to detect the presence of flammable vapours and gases and to warn when concentrations in air approach the explosive range.

**Combustible Liquid:** A liquid which is capable of forming a flammable vapour/air mixture. All flammable liquids are combustible. Whether a liquid is flammable or combustible depends on its flash point. A liquid that has a closed-cup flash point at or above 37.8° C.

**Combustible Dust:** Any finely divided solid material 420 microns or smaller in diameter (material passing a US no. 40 standard sieve) that presents a fire hazard or a deflagration hazard when dispersed and ignited in air.

**Combustible Fibre:** Any material in a fibrous or shredded form that will readily ignite when heat sources are present.

**Combustible Particulate Solid:** Any combustible solid material comprised of distinct particles or pieces, regardless of size, shape, or chemical composition that generates combustible dusts during handling. Combustible particulate solids include dusts, fibres, fines, chips, chunks, flakes, or mixtures of these.

**Combustible Waste:** Combustible or loose waste materials that are generated by an establishment or process and, being salvageable, are retained for scrap or reprocessing on the premises where generated or transported to a plant for processing. These include, but are not limited to, all combustible fibres, hay, straw, hair, feathers, wood shavings, turnings, all types of paper products, soiled cloth trimmings and cuttings, rubber trimmings and buffing, metal fines, and any mixture of the above items, or any other salvageable combustible waste materials.

**Compartmentation:** The divisions of a building into fire tight compartments by fire-resisting elements of building construction in order to contain a fire within the compartment of origin for a specific period of time.

**Compartment Wall:** A fire-resisting wall used in dividing a building horizontally into separate compartments.

**Compound:** Material formed when elements combine with one another to form a new substance with homogeneous properties.

**Compression:** The reduction of the volume of a gas or vapour by mechanical means.

**Compressed Gas:** Any material or mixture having, when in its container, an absolute pressure exceeding 276 kPa at 21.1° C or, having an absolute pressure exceeding 717 kPa at 54.4° C.

**Compressed Gas Cylinder:** Any portable pressure vessel of 45.4 kg water capacity or less designed to contain a gas or liquid that is authorized for use at gauge pressures over 276 kPa at 21° C .

**Concentration:** The ratio of the amount of one constituent of a homogeneous mixture to the total amount of all constituents in the mixture.

**Conduction:** Passage of heat from one material to another by direct contact.

**Conductive Floor:** Floors which are rendered electrically conductive by integral or applied floor finish for preventing static sparks.

**Conflagration:** A fire which involves not only the building in which it originates, but also other buildings and property over a considerable area adjacent to it.

**Confined Space:** An area large enough and so configured that a member can bodily enter and perform assigned work. An area with limited or restricted means for entry and exit. An area that is not designed for continuous human occupancy. Additionally, a confined space is further defined as having one or more of the following characteristics: (a) The area contains or has a potential to contain a hazardous atmosphere, including an oxygen-deficient atmosphere. (b) The area contains a material with a potential to engulf a member. (c) The area has an internal configuration such that a member could be trapped by inwardly converging walls or a floor that slopes downward and tapers to a small cross-section. (d) The area contains any other recognized serious hazard.

**Containment:** Restricting the spread of fire to surrounding structures or areas.

**Contaminant, Dangerous:** A substance not normally found in a pure state and the presence of which even in small quantity may be dangerous owing to it acting as a catalyst or by itself entering into a potentially dangerous reaction.

**Convection:** Transfer of heat by the movement of molecules in a gas or liquid with the less dense fluid rising. The majority of heat transfer in a fire is by convection.

**Cooling:** A process of fire extinguishment or control by reduction of temperature.

**Corrosive:** As identified by OSHA, a corrosive material is a chemical that causes visible destruction or irreversible alterations in human skin tissue at the site of contact or in the case of leakage will materially damage or even destroy other material.

**Cryogenic Gases:** Substances in a gaseous state which cannot be liquefied by pressure alone and are therefore cooled to a low temperature for storage and temperature in the liquid state.

**Damper, Fire-resisting:** A movable closure within a duct, which on operation is intended to prevent the passage of fire or smoke or gases and which together with its frame, will be capable of satisfying the criteria of fire resistance with respect to collapse and flame penetration.

**Damper, Smoke:** Movable device for smoke control, open or closed in its normal position, which is automatically or manually actuated.

**Danger, Imminent:** Any conditions or practices in any occupancy or structure that pose a danger that could reasonably be expected to cause death, serious physical harm, or serious property loss.

**Dead End:** An area from which escape is possible in one direction only.

**Decomposition:** A chemical reaction in which a compound is split into compounds or elements with simpler molecules. Breakdown of a material or substance (by heat, chemical reaction, electrolysis, decay or other processes) into parts, elements or simpler compounds.

**Dedicated Smoke Control Systems:** Systems that are intended for the purpose of smoke control only. They are separate systems of air moving and distribution equipment that do not function under normal building operating conditions. Upon activation, these systems operate specifically to perform the smoke control function.

**Deflagration:** Propagation of a combustion zone at a velocity that is less than the speed of sound in the unreacted medium. Explosion propagating at subsonic velocity.

**Deliquescent:** a solid substance which absorbs water from the atmosphere to form a solution.

**Deluge Valve:** A type of system actuation valve that is opened by the operation of a detection system installed in the same areas as the spray nozzles or by remote manual operation supplying water to all spray nozzles.

**Density:** The amount of mass in unit volume of a substance, expressed in gram per cc or kg/cu.m.

**Detached Storage:** Storage in a separate building or in an outside area located away from all structures.

**Detonation:** Propagation of a combustion zone at a velocity that is greater than the speed of sound in the unreacted medium. Explosion propagating at supersonic velocity and characterized by a shock valve. Extremely rapid combustion resulting in the generation of a shock wave in the combustible mixture.

**Diffusion:** The process of one gas separating through another gas, e.g. petrol vapour spreads through air by diffusion.

**Discharge Rate:** The rate at which a single file of persons can pass through one unit of exit width (generally accepted as 40 persons per minute).

**Discharge Value:** The maximum number of persons who can effect egress through a given number of units of exit width in a given period of time having regard, in multistorey buildings, to the capacity of the staircase(s).

**Distillation:** A process in which both evaporation and condensation takes place together.

**Door, Fire-check/Fire Door:** Door to prevent or restrict the spread of hot gases and smoke. It must hold-back fire for 20 minutes (integrity) and must not collapse within 30 minutes (stability). Used an entrance doors to flats and apartments sharing a common access area, doors between small garages and houses, doors to habitable rooms and kitchens in buildings of three storeys or more.

**Door, Fire-resisting:** A door which, together but also along with its frame, is capable of satisfying the criteria of with respect to collapse, flame penetration and excessive temperature rise. Such a door may be automatic or self-closing.

**Draught (Draft):** Current of air accelerating towards a fire supplying air for combustion.

**Drenchers:** A device to give protection to the roofs, windows and door openings of a building by interposing a curtain of water between the protected building and a fire.

**Drill:** An exercise involving a credible simulated emergency that requires personnel to perform emergency response operations for the purpose of evaluating the effectiveness of the training and education programs and the competence of personnel in performing required response duties and functions.

**Dry Chemical:** A mixture of finely divided solid particles, usually sodium bicarbonate-, potassium bicarbonate-, or ammonium phosphate-based with added particulate material supplemented by special treatment to provide resistance to packing, and moisture absorption (caking), and to promote proper flow characteristics.

**Duct, Fire-resisting:** A duct which conveys liquid, gas or services through building, and which is capable of satisfying the criteria of fire resistance with respect to collapse, flame penetration and rise of temperature beyond any prescribed value.

**Dust Explosion:** Rapid oxidation type of explosion in a suspension of combustible dust in air.

**Dust-tight Enclosure:** A type of enclosure for electrical equipment located in dusty atmospheres, which prevent ingress of dust within its interior.

**Dwelling Unit:** A single unit, providing complete, independent living facilities for one or more persons, including permanent provisions for living, sleeping, eating, cooking,

**Eductor (Inductor):** A device that uses the venturi principle to introduce a proportionate quantity of foam concentrate into a waterstream.

**Educational Occupancy:** An occupancy used for educational purposes by six or more persons for four or more hours per day or more than 12 hours per week.

**Egress:** A route of travel from any point inside a building to a point outside the building.

**Electron:** A particle having one unit negative charge ($1.60 \times 10^{-19}$ coulombs). Electrons are present in all atoms in shells around the nucleus.

**Element:** A substance which cannot be decomposed by chemical means into simpler substances. There are 113 elements known out of which 92 are naturally occurring and the rest elements were obtained experimentally.

**Emergency:** A fire, explosion, or hazardous condition that poses an immediate threat to the safety of life or damage to property.

**Emergency Lighting:** A provision for lighting in designated parts of premises, for use when normal light fails.

**Emergency Response Operations:** Activities related to emergency incidents, including response to the scene of the incident and specific response duties performed at the scene.

**Escape Chute:** Open slide-like escape used for emergency evacuation.

**Escape Hatch:** An emergency means of providing escape from a room or part of a building in the form of a removable or breakable panel in a wall or floor.

**Escape Route:** A route forming part of the means of escape from any point in a building to a final exit.

**Escape Route, External:** An escape route external to a building, having an adequate degree of fire protection by way of a roof, staircase, balcony, walkway or external court, and terminating at a final exit.

**Escape Route, Protected:** An escape route having an adequate degree of fire protection.

**Evacuation Drill:** Rehearsal of the evacuation procedure involving participation of the occupants of the premises.

**Evacuation Procedure:** A predetermined plan of action designed to achieve the safe evacuation of the occupants of a building to a place of safety.

**Evacuation Time:** The time taken for all occupants of a building or part of a building, on receipt of an evacuation signal, to reach a final exit.

**Exit:** That portion of a means of egress that is separated from all other spaces of a building or structure by construction or equipment as required to provide a protected way of travel to the exit discharge.

**Exit Access:** That portion of a means of egress that leads to an exit.

**Exit, Final:** The terminal point of an escape route beyond which persons are no longer in danger from fire.

**Exit, Fire:** A way out leading to an escape route.

**Exit, Horizontal:** An arrangement which allows alternative egress from a floor area to another floor at or near the same level in an adjoining building or an adjoining part of the same building with adequate fire separation.

**Exit Sign:** Sign which clearly indicates the exit.

**Exit Width, Unit of:** The minimum width required for a single file of persons to pass through an ex (taken as 500 mm width).

**Expellant Gas:** The medium used to discharge dry chemical extinguishant from its container.

**Explosion:** The sudden conversion of chemical energy into kinetic energy with the release of heat, light and mechanical shock.

**Explosion (decomposition type):** Instantaneous decomposition of certain endothermic compounds with evolution of hot gases and extremely rapid rise of surrounding air pressure.

**Explosion (pressure release type):** Rupturing of pressure containers due to abnormally high pressure.

**Explosion (rapid oxidation type):** An extremely rapid oxidation reaction with evolution of light, heat and dynamic energy capable of causing structural or other physical damage.

**Explosion Suppression:** Appliance containing an explosion suppressant which can be expelled by the action of internal pressure. This pressure may be stored pressure or may be obtained by a chemical reaction such as the activation of an explosive or pyrotechnical device.

**Explosion Vent:** An opening in a vessel or building, usually covered by a fragile diaphragm, or a hinged or spring loaded flap, which in the event of an explosion in the vessel or building allows gaseous products to escape. This venting process is also known as 'explosion relief'.

**Explosive Limit:** Flammability limit. The highest or lowest concentration of a flammable gas or vapour in air that will explode or burn readily when ignited. The limit is usually expressed as a volume percent of gas or vapour in air.

**Exposure Hazard:** The risk of fire spreading from a building structure, or other property to an adjacent separate building or structure, or to another part of the same building or structure, by radiated heat across the intervening space. Factors determining the exposure hazard include the width of the intervening space, the heights of the buildings and their construction, types of windows, doors, etc.

**Exothermic Reaction:** A chemical reaction which evolves heat. Combustion reactions are exothermic.

**Fire:** (1) Process of combustion characterized by the emission of heat accompanied by smoke or flame or both. (2) Combustion spreading uncontrolled in time and space.

**Fire Ball:** A spherical mass of flame, which may occur if a large quantity of flammable vapour suddenly ignites in the air as it occurs following BLEVE.

**Fire Break:** An open space separating buildings, stored products or other combustible materials, being itself free of combustible material and designed to restrict spread of fire.

**Fire Classification:** Standardized system of classifying fires in terms of the nature of fuel. These are: Class A-Fire involving solid materials, usually of an organic nature, in which combustion normally takes place with the formation of glowing embers. Class B-fires involving liquids or liquefiable solids. Class C-fires involving gases. Class D-fires involving metals.

**Fire Compartment:** A space within a building that is enclosed by fire barriers on all sides, including the top and bottom.

**Fire Extinguisher (Portable):** A portable device, carried or on wheels and operated by hand, containing an extinguishing agent that can be expelled under pressure for the purpose of suppressing or extinguishing fire.

**Fire Extinguisher (Stored-pressure).** A fire extinguisher in which both the extinguishing material and expellant gas are kept in a single container, and that includes a pressure indicator or gauge.

**Fire Extinguisher (Water-type):** A water-type fire extinguisher contains water-based agents, such as water, AFFF, FFFP, antifreeze, and loaded stream.

**Fire Hazard:** Characteristics of materials, building construction and occupancy which affect the initiation, development and spread of fire. Any situation, process, material, or condition that, on the basis of applicable data, can cause a fire or explosion or provide a ready fuel supply to augment the spread or intensity of the fire or explosion and that poses a threat to life or property.

**Fire Hydrant:** A connection to a water main for the purpose of supplying water to fire hose or other fire protection apparatus.

**Fire Lane:** The road or other means developed to allow access and operational setup for fire fighting and rescue apparatus.

**Fire Lift:** A lift designated to have additional protection that enables it to be used under the direct control of the fire brigade in fighting a fire.

**Fire Load:** Calorific energy, of the whole contents contained in a space, including the facings of the walls, partitions, floors and ceilings.

**Fire Load Density:** Fire load divided by floor-area.

**Fire Point:** The lowest temperature at which a liquid gives off sufficient flammable vapour in air to produce sustained combustion after the removal of the ignition source.

**Fire Prevention:** The concept of elimination of all probable causes of a fire outbreak.

**Fire Propagation Index:** A comparative measure of the contribution to the growth of fire of a combustible.

**Fire Protection:** Design features, systems, equipment, buildings or other structures to reduce danger to persons and property by detecting, extinguishing or containing fires.

**Fire Protection System:** Any fire alarm device or system or fire extinguishing device or system, or their combination, that is designed and installed for detecting, controlling, or extinguishing a fire or otherwise alerting occupants, or the fire department, or both, that a fire has occurred.

**Fire Retardants:** Liquids, solids, or gases that tend to inhibit combustion when applied on, mixed in, or combined with combustible materials.

**Fire Resistance:** Ability of an element of building construction, component for structure to fulfill, for a stated period of time, the required stability, fire integrity and/or thermal insulation and/or other expected duty in a standard fire resistance test.

**Fire Resistance (Criteria of):** Fire resistance is a property of an element of building construction and is the measure of its ability to satisfy for a stated period some or all of the following criteria: (a)resistance to collapse, (b) resistance to flame penetration, and (c) resistance to excessive temperature rise on unexposed face.

**Fire Resistant Construction:** The type of construction in which the structural members including wall, partition, columns, floors and roofs are designed to withstand resistance to fire for a specified period.

**Fire Retardant (Flame Retardant):** A substance or treatment applied to combustible material to increase its ignition temperature, decrease its tendency to propagate flame across its surface and increase its resistance to pyrolysis and destruction by heat. Liquids, solids, or gases that tend to inhibit combustion when applied on, mixed in, or combined with combustible materials.

**Fire Separation:** The distance in metres measured from any other building on the site, or from other site, or from the opposite side of street or other public space to the building for the purpose of preventing the spread of fire.

**Fire Stop:** A physical barrier designed to restrict the spread of fire in cavities within and between elements of building construction.

**Fire Suppression:** The activities involved in controlling and extinguishing fires.

**Fire Tetrahedron:** Fuel, heat, oxygen and a chemical chain reaction.

**Fire Triangle:** Fuel, heat and oxygen.

**Fire Vent:** An opening in the enclosing walls or roof of a building, intended for releasing heat and smoke in the event of fire and automatically or manually opened or both.

**Fire Wall:** A fire resistance rated wall, having protected openings, which restricts the spread of fire and extends continuously from the foundation to at least 1 m above the roof.

**Fire Watch:** The assignment of a person or persons to an area for the express purpose of notifying the fire department and/or building occupants of an emergency, preventing a fire from occurring, extinguishing small fires, or protecting the public from fire or life safety dangers.

**Flame:** A zone of oxidation of gas usually characterized by the liberation of heat and the emission of light.

**Flame Arrestors:** A device fitted to prevent the passage of flames.

**Flame Ionization Detector (FID):** A nearly universal gas chromatographic detector. It responds to almost all organic compounds. An FID does not respond to nitrogen, hydrogen, helium, oxygen, carbon monoxide or water. This detector ionises compounds as they reach the end of the chromatographic column by burning them in an air/hydrogen flame. As the compounds pass through the flame, the conductivity of the flame changes, generating a signal. This is the most commonly used detector in arson debris analysis.

**Flame-proof Enclosure:** An enclosure for electrical machinery or apparatus that will withstand, when covers or other access doors are properly secured, an internal explosion of the flammable gas or vapour which may enter or which may originate inside the enclosure without suffering damage and without communicating the internal flame (or explosion) to the external flammable gas or vapour.

**Flame Propagation Rate:** The velocity with which the combustion front travels through a body of gas, measured as the highest gas velocity at which stable combustion can be maintained, and the velocity at which combustion travels over the surface of a solid or liquid.

**Flame Spread:** The propagation of flame over a surface.

**Flame Spread Rating:** The comparative performance of fire travel over the surface of a material when tested in accordance with the provisions of relevant standard.

**Flammability:** Degree of ease with which a material catches fire and its intensity.

**Flammable:** A combustible material which ignites very easily and either burns very intensely or has a rapid flame spread.

**Flammable Gas:** A gas that is flammable at atmospheric temperature and pressure in a mixture of 13% or less (by volume) with air, or that has a flammable range with air wider than 12%, regardless of the lower limit.

**Flammable Limits:** The minimum and maximum concentration of fuel vapour or gas in a fuel vapour or gas/gaseous oxidant mixture (usually expressed as percent by volume) defining the concentration range (flammable or explosive range) over which propagation of flame will occur on contact with an ignition source.

**Flammable Liquid:** Any liquid having a closed-cup flash point below 37.8° C and having a vapour pressure not exceeding 276 kPa at 37.8° C.

**Flammable Vapour:** A vapour/air mixture of any concentration within the flammability range of that vapour.

**Flash-fire:** A flame or fire of very short duration.

**Flashover:** A stage in the development of a contained fire at which all the combustibles in the enclosures flash into fire simultaneously.

**Flash Point:** The minimum temperature at which a liquid gives off sufficient vapour to produce a flammable vapour-air mixture at the lower limit of flammability.

**Floor, Fire-resisting:** A floor, with or without a ceiling beneath, which, when exposed to fire conditions from below, is capable of satisfying for a stated period of time the criteria of fire resistance with respect to collapse, flame penetration and excessive temperature rise.

**Flue Effect:** The upward thrust of convection currents of hot gases through vertical openings.

**Fuel:** A substance that will produce energy as heat (in useful amounts), it may be gaseous, liquid or solid.

**Fuel Load:** The total quantity of combustible contents of a building, space, or fire area.

**Fuse:** An over-current protective device with a circuit opening fusible part that is heated and severed by the passage of over-current through it.

**Gas:** A gas is a substance which at 50° C has a vapour pressure greater than 300 kPa; or is completely gaseous at 20° C at a standard pressure of 101.3 kPa.

**Gas, Non-flammable:** A class of gases that is non-flammable, generally non-reactive.

**Gas, Liquefied:** A gas which when packaged for transport is partially liquid at 20° C.

**Glazing, Fire-resisting:** Glazing capable of satisfying for a stated period of time the criteria of fire resistance with respect to collapse and flame penetration.

**Hazard:** A condition or set of circumstances that presents a specific injury or adverse health potential.

## Hazard of Contents

**Hazard, High:** High hazard contents shall be classified as those that are likely to burn with extreme rapidity or from which explosions are likely. These are contents that are liable to burn with extreme rapidity or from which poisonous fumes or explosions are to be feared in the event of a fire.

**Hazard, Low:** Hazard contents shall be classified as those of such low combustibility that no self-propagating fire therein can occur.

**Hazard, Ordinary:** Ordinary hazard contents shall be classified as those that are likely to burn with moderate rapidity or to give off a considerable volume of smoke.

**Health Care Occupancy:** An occupancy used for purposes of medical or other treatment or care of four or more persons where such occupants are mostly incapable of self-preservation due to age, physical or mental disability, or because of security measures not under the occupants' control.

**Health Hazard:** A chemical for which there is statistically significant evidence based on at least one study conducted in accordance with established scientific principles that acute or chronic health effects may occur in exposed persons.

**Heat:** A mode of energy associated with and proportional to molecular motion that may be transferred from one body to another by conduction, convection or radiation.

**Heat Release Rate (HRR):** The rate at which heat energy is generated by burning.

**High-rise Building:** A building in which a 'stack effect' is created by a fire and in which fire fighting cannot be carried out from outside with the fire appliances available with the fire services. Generally, all buildings more than 15 m in height shall be considered as high-rise buildings.

**Highly Volatile Liquid:** A liquid with a boiling point of less than 20° C.

**Housekeeping:** State of maintenance and cleanliness of an occupancy, affecting the frequency and growth of fires.

**Hospital:** A building or portion thereof used on a 24-hour basis for the medical, psychiatric, obstetrical, or surgical care of four or more inpatients.

**Hotel:** A building or groups of buildings under the same management in which there are sleeping accommodations for more than 16 persons and primarily used by transients for lodging with or without meals.

**Hot Work:** Work involving flames or temperatures likely to be sufficiently high to cause ignition of flammable material.

**Hydrostatic Testing:** Pressure testing of the extinguisher to verify its strength against unwanted rupture.

**Hyperbolic Chambers:** Enclosures within which air pressure is much higher than under normal atmospheric conditions.

**Ignition:** The means by which burning is started.

**Ignitability:** Degree of ease with which a material can be ignited.

**Ignition Energy, Minimum:** The minimum energy required to ignite a flammable mixture; usually the minimum energy of an electric spark or arc expressed in joules.

**Ignition Temperature:** The lowest temperature of a substance at which sustained combustion can be initiated.

**Incendiaries:** Substances or mixtures of substances consisting of a fuel and an oxidizer used to initiate a fire.

**Incendiarism:** Intentional and culpable generation of fire.

**Incipient Stage:** Refers to the severity of a fire where the progression is in the early stage and has not developed beyond that which can be extinguished using portable fire extinguishers.

**Industrial Occupancy:** Includes factories that manufacture products of all kinds and properties devoted to operations such as processing, assembling, mixing, and packaging, finishing or decorating, and repairing.

**Inerting:** Filling an enclosed space with an inert, gas to prevent formation of explosive vapour-air mixture.

**Inhibition:** A process of fire extinguishment by the use of an agent which interrupts the chemical reactions in the flame.

**Landing:** A level space at the top or bottom of a staircase and/or at each floor level.

**Law of Conservation of Mass:** Put forth by Lavoisier (1774) – states that matter cannot be created or destroyed. When a chemical change occurs, the total mass of the products is the same as the total mass of the reacting substances.

**Lean Mixture:** A fuel and oxidizer mixture having less than the stoichiometric concentration of fuel.

**Lightning Arrestor:** A device for the protection of a structure from damage by a lightning discharge or other accidental electrical surge.

**Lightning conductor:** A metal strip connected to earth at its lower end, and its upper end terminated in one or more sharp points where it is attached to the highest part of a building or structure. By electrostatic induction, it will tend to neutralize a charged cloud in its neighbourhood and the discharge will pass directly to earth through the conductor.

**Lining Material:** Any material used for lining walls, ceiling or floors of building for insulation, decoration or other purposes.

**Linking Balcony:** An arrangement which provides access to an adjacent protected area, via a balcony.

**Liquefied Natural Gas (LNG):** A fluid in the liquid state that is composed predominantly of methane and that can contain minor quantities of ethane, propane, nitrogen, or other components normally found in natural gas.

**Liquefied Petroleum Gas (LP-Gas):** Any material having a vapour pressure not exceeding that allowed for commercial propane composed predominantly of the following hydrocarbons, either by themselves or as mixtures: propane, propylene, butane (normal butane or isobutane), and butylenes.

**Lobby, Fire Fighting Access:** A protected lobby and permanently ventilated of specified dimensions, suitable for use as a means of access for fire fighting purposes.

**Lobby, Protected:** A lobby forming part or whole of the horizontal component of a protected escape route.

**Lobby Ventilated:** A protected lobby provided with means of ventilation to the open air for use when required.

**Lodging or Rooming House:** A building or portion thereof that does not qualify as a one- or two-family dwelling, that provides sleeping accommodations for a total of 16 or fewer people on a transient or permanent basis, without personal care services, with or without meals, but without separate cooking facilities for individual occupants.

**Lower Limit of Flammability:** The lowest percentage concentration by volume of flammable vapour (gas) mixed with air which will burn with a flame.

**Material, Compatible:** A material that, when in contact with an oxidizer, will not react with the oxidizer or promote or initiate its decomposition.

**Material, Incompatible:** A material that, when in contact with an oxidizer, can cause hazardous reactions or can promote or initiate decomposition of the oxidizer.

**Maintenance:** Work, including, but not limited to, repair, replacement, and service, performed to ensure that equipment operates properly.

**Means of Egress:** A continuous and unobstructed way of travel from any point in a building or structure to a public way consisting of three separate and distinct parts: (1) the exit access, (2) the exit, and (3) the exit discharge.

**Means of Escape:** A way out of a building or structure that does not conform to the strict definition of means of egress but does provide an alternative way out.

**Melting Point:** The temperature at which a solid substance changes to a liquid state. For mixtures, the melting range may be given.

**Mercantile Occupancy:** An occupancy used for the display and sale of merchandise.

**Mercantile Occupancy, Sub-classification of:** Mercantile occupancies shall be sub-classified as follows.

(a) **Class A.** All mercantile occupancies having an aggregate gross area of more than 2800 m$^2$ (30000 sq ft) or using more than three levels, excluding mezzanines, for sales purposes. (b) **Class B.** All mercantile occupancies of more than 280 m$^2$ (3000 sq ft) but not more than 2800 m$^2$ (30000 sq ft) aggregate gross area, or using floors above or below the street floor level for sales purposes. (c) **Class C.** All mercantile occupancies of not more than 280 m$^2$ (3000 sq ft) gross area used for sales purposes on one storey only, excluding mezzanines.

**Mezzanine:** A part-floor in between two other floors of a building.

**Material Safety Data Sheet (MSDS):** Written or printed material concerning a hazardous material that is prepared in accordance with the provisions of regulations.

**Mushroom Effect:** A horizontal spread of hot gases at ceiling or roof level due to the vertical restriction of convection currents.

**Naked Lights:** Open flames or fires, exposed incandescent material, or any other unconfined source of ignition.

**National Fire Protection Association NFPA USA:** NFPA is an international voluntary membership organization to promote and improve fire protection and prevention and establish safeguards against the loss of life and property by fire.

**Non-combustible Material:** Not capable of undergoing combustion under normal atmospheric pressure and oxygen concentration.

A material that, in the form in which it is used and under the conditions anticipated, will not ignite, burn, support combustion, or release flammable vapours when subjected to fire or heat. Materials that are reported as passing ASTM E136, *Standard test method for behavior of materials in a vertical tube furnace at 750° C*, shall be considered noncombustible materials.

**Occupancy:** Purpose for which a building, or part of a building is used, or intended to be used.

**Occupation Density:** Number of persons per square metre of the usable floor area of a room for a given activity. Used to calculate in particular the number and the width of the exits of a room or space.

**Occupant Load Factor:** A factor used in calculating the population density when planning means of escape from a building or part of a building.

**Organic Peroxide:** Any organic compound having a double oxygen or peroxy (-O-O-) group in its chemical structure.

**OSHA:** The Occupational Safety Health Administration of the US Department of Labour.

**Oxidation:** Originally, oxidation meant a chemical reaction in which oxygen combines with another substance. The usage of the word has been broadened to include any reaction in which electrons are transferred. The substance which gains electrons is the oxidizing agent.

**Oxidizer:** A chemical or substance other than a blasting agent or explosive that initiates or promotes combustion in other materials, thereby causing fire of itself or through the release of oxygen or other gases. Any material that readily yields oxygen or other oxidizing gas, or that readily reacts to promote or initiate combustion of combustible materials.

**Partition, Fire-resisting:** A partition either load-bearing or non-load-bearing capable of satisfying the criteria of fire resistance with respect to collapse, flame penetration and excessive temperature rise.

**Parting Wall:** Generally called a separating wall now. A wall common to two buildings or two pieces of land.

**Permissible Exposure Limit (PEL):** An exposure limit established by a regulatory authority. May be a time-weighted average (TWA) limit, STEL or a maximum exposure limit.

**Percentage Volatile:** (percent volatile by volume)—the percentage of a liquid or solid (by volume) that will evaporate at an ambient temperature of 21° C. Example-butane, petrol are 100% volatile; they will evaporate completely over a period of time.

**Permit:** A document issued by the authority having jurisdiction for the purpose of authorizing performance of a specified activity.

**Peroxide Forming Chemical:** A chemical that, when exposed to air, will form explosive peroxides that are shock, pressure, or heat sensitive.

**Petrochemicals:** Flammable chemicals which are derived (in whole or part) from petroleum or natural gas constituents.

**Physical hazard:** A combustible liquid, a compressed gas, explosive, flammable, an organic peroxide, an oxidizer, pyrophoric, unstable (reactive) or water reactive.

**Place of Safety:** A place in which persons are not in danger from fire.

**Population Density:** The number of persons in a given area for whom means of escape shall be provided as determined by the functional use(s) of the building or floor.

**PPM:** Parts per million. A measurement of concentration such as 1 microgram per gram

**Pressure:** The force per unit of area. Expressed in the International System of Units (SI), in pascal (Pa) or newton per square meter ($N/m^2$).

**Pressurization:** A method of keeping escape route in high-rise buildings clear of smoke by increasing the air pressure in staircases and lobbies.

**Process:** The manufacturing, handling, blending, conversion, purification, recovery, separation, synthesis, or use, or any combination, of any commodity or material.

**Products of Combustion:** Total gaseous, particulate and aerosol effluents from a fire or pyrolysis.

**Protected Area:** Area giving an adequate degree of fire resisting enclosure from other areas and from which there is alternative means of escape.

**Purging:** Freeing an enclosed space from flammable or toxic vapours/gases by blowing air or inert gas.

**Pyrolysis:** Irreversible chemical decomposition of a material due to an increase in temperature without oxidation. The transformation of a substance into one or more other substances by heat alone without oxidation.

**Pyrophoric Gas:** A gas that will spontaneously ignite in air at or below a temperature of 54.4° C.

**Radiation (Heat):** Transfer of heat through a gas or vacuum other than by heating of the intervening space. (1) Transfer of heat through electromagnetic waves from hot to cold. (2) Electromagnetic waves of energy having frequency and wavelength. The shorter wavelengths (higher frequencies) are more energetic. The electromagnetic spectrum is comprised of: (a) cosmic rays, (b) gamma rays, (c) X-rays, (d) ultraviolet rays, (e) visible light rays, (f) infrared, (g) microwaves and (h) radio-waves.

**Ramp:** An inclined plane taking the place of steps. (2) The curved portion connecting the horizontal part of a handrail or moulding to a raking or inclined part of the same.

**Reaction:** A chemical transformation or change; the interaction of two or more substances to form different substances.

**Reaction Rate:** A measure of the amount of reactant which disappears in a chemical reaction in unit time.

**Reactivity:** Description of the tendency of a substance to undergo chemical reaction with the release of energy.

**Reactivity, Air:** Property possessed by certain chemicals of causing dangerous reactions when exposed to air.

**Reactivity, Water:** Property possessed by certain chemicals of causing dangerous reactions when coming into contact with water.

**Recharging:** The replacement of the extinguishing agent (also includes the expellant for certain types of fire extinguishers).

**Reducing agent:** In a reduction reaction (which always occurs simultaneously with an oxidation reaction), the reducing agent is the substance or chemical which combines with oxygen or loses electrons in the reaction.

**Refuge Area:** Area above ground level to which occupants can gain access from a room or building and await rescue.

**Reliability:** The probability a system performs a specified function or mission under given conditions for a prescribed time.

**Riser Dry:** A vertical water main inside a building, not normally connected to a water main or an automatic stationary pump, with an inlet or inlets at street level, through which water can be pumped by fire service pumps to hydrant outlets or hose reels at various floors.

**Roof (external fire exposure, resistance, etc.):** The ability of a roof deck and covering to resist both penetration by external fire and flame spread over the external surfaces.

**Roof Screen or Roof Curtain Boards:** A vertical screen or substantial non-combustible material.

**Risk:** The exposure to chance of injury or illness or loss.

**Risk Level:** The probability or expected frequency of the event, multiplied by the expected magnitude of exposure and the potential for harm.

**Risk Management:** The process of planning, organizing, directing, and controlling the resources and activities in order to minimize detrimental effects.

**Risk Potential:** The possible harm, as opposed to the actual; that which may, but has not yet, come into being; that which is latent, unrealized.

**Roof Venting:** A system of vents which will open automatically in the event of a fire and allow the escape of smoke and hot gases.

**Room, Access:** A room which forms the only escape route from an inner room.

**SCBA:** Acronym for self-contained breathing apparatus.

**Self-contained Breathing Apparatus (SCBA):** A respirator worn by the user that supplies a respirable atmosphere that is either carried in or generated by the apparatus and is independent of the ambient environment.

**Secondary Fire:** A fire which has started some distance from the seat of the original fire, but is due to the latter.

**Self-closing:** Equipped with an approved device that ensures closing after opening.

**Self-extinguishing, Incapable of:** A material undergoing sustained combustion after removal of the external source of heat.

**Self-expelling Fire Extinguisher:** A fire extinguisher in which the agents have sufficient vapour pressure at normal operating temperatures to expel themselves.

**Self-heating:** An exothermic reaction occurring without the application of external heat.

**Shaft, Fire-resisting:** A space bounded by fire-resisting elements of building construction and intended for the passage of persons, services or things.

**Shutter, Fire-resisting:** Shutter which, together with its frame; is capable of satisfying the criteria of fire resistance with respect to collapse and flame penetration.

**Sleeping Risk:** A form of occupancy or a part of a building in which bedroom or dormitory accommodation predominates, such as in hotels, boarding schools, hospitals and similar establishments.

**Smoke:** The visible aerosol that results from incomplete combustion. Visible suspension in atmosphere of solid and/or liquid particles resulting from combustion or pyrolysis.

**Smoke Barrier:** A continuous membrane or a membrane with discontinuities created by protected openings, where such membrane is designed and constructed to restrict the movement of smoke.

**Smoke Compartment:** A space within a building enclosed by smoke barriers on all sides, including the top and bottom.

**Smoke Density:** The proportion of solid matter present in the smoke, measured on various arbitrary scales.

**Smoke Detector:** A device that detects visible or invisible particles of combustion.

**Smoke Exhaust:** An opening or a fire resisting shaft or duct provided in a building to act as an outlet, usually from a basement, for smoke and hot gases produced by an outbreak of fire.

**Smoke Shaft:** Shaft provided to remove smoke in the event of fire.

**Smoke Vent:** Opening in the enclosing walls or roof of a building, intended to release heat and smoke in the event of fire, automatically and/or manually opened.

**Smoking:** The carrying or use of lighted pipe, cigar, cigarette, tobacco, or any other type of smoking substance.

**Smoking Area:** A designated area where smoking is permitted within premises where smoking is otherwise generally prohibited.

**Smothering:** A process of fire extinguishment by the limitation or reduction of oxygen.

**Smouldering:** Slow combustion of material without visible light and generally evidenced by smoke and an increase in temperature.

**Solubility in Water:** A term expressing the percentage of a material by weight that will dissolve in water at ambient temperature. This information is useful for determining fire-extinguishing agent and method for a material. Terms used to express solubility are—negligible ..less than 0.1%, slight ...0.1 to1.0%, moderate ... 1 to 10%, appreciable .... more than 10%, complete .... soluble in all proportions.

**Soot:** Finely divided particles, mainly carbon, produced and deposited during the incomplete combustion of organic material.

**Spark, Electric:** Instantaneous electrical response leading to changes of temperature. Discharge between bodies at different electrical potentials accompanied by heat and light.

**Spark, Fire:** A small incandescent particle. Also known as spontaneous combustion. Initially, a slow, exothermic reaction at ambient temperatures, liberating heat, if undissipated, accumulates at an increasing rate and may lead to spontaneous ignition of any combustibles present. Spontaneous ignition occurs sometimes in haystacks, coal piles, warm moist cotton waste, and in stacks of rags coated with drying oils such as cottonseed or linseed oil.

**Specific Gravity:** The weight of a material compared to the weight of an equal volume of water; an expression of the density (or heaviness) of the material. Insoluble materials with specific gravity of less than 1.0 will float in water, insoluble materials with a specific gravity greater than 1.0 will sink Most flammable liquids (not all) have specific gravity less than 1.0 (an important consideration for designing fire suppression).

**Spontaneous Combustion:** A biological or chemical reaction which produces its own heat resulting in combustion.

**Spontaneous Heating:** A kind of heating internally developed by a body due to bacteriological and/or chemical reaction without drawing off heat from its surroundings. Also known as spontaneous combustion. Initially, a slow, exothermic reaction at ambient temperatures, liberating heat, if undissipated, accumulates at an increasing rate and may lead to spontaneous ignition of any combustibles present. Spontaneous ignition occurs sometimes in haystacks, coal piles, warm moist cotton waste, and in stacks of rags coated with drying oils such as cottonseed or linseed oil.

**Stability:** An expression of the ability of a material to remain unchanged. A material is stable, if it remains in the same form under expected and reasonable condition of storage or use.

**Stack Pressure:** Pressure difference caused by a temperature difference creating an air movement within a duct, chimney or enclosure.

**Staircase, Enclosed:** A staircase physically separated, e.g. by walls, partitions, screens, etc. from the floors of a building through which it passes, and which does not form part of a protected escape route.

**Staircase, Fire-fighting:** A staircase, designated for use by the fire services in obtaining access into a building for fire fighting purposes and provided with fire fighting access lobbies.

**Staircase, Open:** A staircase not separated in any way from the floors through which it passes.

**Staircase, Protected:** A staircase, protected from the remainder of a building by fire-resisting construction, accessible only through self-closing, fire resisting doors, and forming the vertical component of a protected escape route.

**Stairway Lobby Approach:** Protected stairway separated from the accommodation space in a building by protected lobbies.

**Standard:** A document, the main text of which contains only mandatory provisions using the word "shall" to indicate requirements and which is in a form generally suitable for mandatory reference by another standard or code or for adoption into law.

**Standard Operating Procedure:** A written procedure that establishes a standard course of action.

**Standpipe System:** An arrangement of piping, valves, hose connections, and allied equipment installed in a building or structure, with the hose connections located in such a manner that water can be discharged in streams or spray patterns through attached hose and nozzles, for the purpose of extinguishing a fire, thereby protecting a building or structure and its contents in addition to protecting the occupants. This is accomplished by means of connections to water supply systems or by means of pumps, tanks, and other equipment necessary to provide an adequate supply of water to the hose connections.

**Starvation:** A process of fire extinguishment by the limitation or reduction of fuel.

**States of Matter:** The three physical conditions in which substances exist: solid, liquid and gas.

**Static Electricity:** Electricity, generated as a result of friction between two non-conducting substances.

**Storage: Types of Storage**

> *Cut-off Storage:* Cut-off storage refers to storage in the same building or inside area, but physically separated from incompatible materials by partitions or walls.

> *Detached Storage:* Detached storage refers to storage in either an open outside area or a separate building containing no incompatible materials and located away from all other structures.

> *Segregated Storage:* Segregated storage refers to storage in the same room or inside area, but physically separated by distance from incompatible materials. Sills, curbs, intervening storage of non-hazardous compatible materials, and aisles shall be permitted to be used as aids in maintaining spacing.

**Storage Occupancy:** An occupancy used primarily for the storage or sheltering of goods, merchandise, products, vehicles, or animals.

**STP (Standard Temperature and Pressure):** A temperature of 70° F (21° C) and a pressure of 1 atmosphere (14.7 psi or 760 mm Hg).

**Structural Fire Protection:** Features in the layout and/or construction of a building intended to reduce the effects of a fire in a building.

**Surface Spread of Flame Classification:** The classes of combustible building materials according to the rate at which flames spread over their surfaces.

**System:** Several items of equipment assembled, grouped, or otherwise interconnected for the accomplishment of a purpose or function.

**Temporary Wiring:** Approved wiring for power and lighting during a period of construction, remodelling, maintenance, repair, or demolition, and decorative lighting, carnival power and lighting, and similar purposes.

**Thermocouple:** A junction of wires of dissimilar metals used for measuring temperature.

**Thermal Barrier:** A material that limits the average temperature rise of the unexposed surface to not more than 120° C for a specified fire exposure complying with the standard time-temperature curve.

**Thermostat:** An automatic temperature control device.

**Thresh-hold Limit Value (TLV):** Refers to airborne concentrations of substances and represents conditions under which it is believed that nearly all workers are protected while repeatedly exposed for an 8-hr day, 5 days a week (expressed as parts per million (ppm) for gases and vapours and as milligrams per cubic meter (mg/m$^3$) for fumes, mists, and dusts).

**Time-temperature Curve:** A graph that shows the increase in temperature of a fire as a function of time, beginning with ignition and ending with burnout or extinguishment.

**Tinder:** Material which can be ignited by an ordinary lighted match, and materials such as wood, cardboard, paper, textiles, etc.

**Toxic:** Harmful to living organisms, substances liable either to cause death or serious injury or to harm human health, if swallowed or inhaled or by skin contact.

**Toxicity:** The nature and extent of adverse effects of a substance on a living organism.

**Travel Distance:** The distance to be traveled from any point in a building to a protected escape route, or final exit.

**Ultimate Load:** The maximum load which a structure is designed to withstand.

**Unstable:** A material that readily undergoes decomposition or other unwanted chemical change during normal handling or storage.

**Upper Limit of Flammability:** The highest percentage concentration by volume of flammable vapour (gas) mixed with air above which no combustion can occur.

**UFL or UEL (upper explosive limit or upper flammable limit):** The highest concentration (highest percentage of the substance in the air) that will produce a flash or fire when an ignition source (heat, arc or flame) is present. At higher concentrations, the mixture is too rich to burn.

**Valency:** The relative combining capacity of an atom or group of atoms with that of the standard hydrogen atom.

**Vapour:** A gas obtained by the vapourization of a liquid or solid.

**Vapourization:** The process by which a liquid or solid is changed into a gas or vapour by heat. The physical change of going from a solid or a liquid into a gaseous state.

**Vapour Density:** The weight of a vapour or gas compared to the weight of an equal volume of air; an expression of the density of the vapour or gas. Materials lighter than air have vapour densities less than 1.0 (examples are hydrogen, methane). Materials heavier than air have vapour densities greater than 1.0 (examples are ethane, chlorine butane sulphur dioxide). All vapours and gases tend to mix with air, but the lighter materials will tend to rise and dissipate (unless confined), heavier gases tend to concentrate in low places where they may create fire hazards.

**Vapour Pressure:** The pressure exerted by a saturated vapour above its own liquid layer in a closed container.

**Vent:** A passageway used to convey flue gases from gas utilization equipment or their vent connectors to the outside atmosphere.

**Vent, Emergency:** Opening fitted with easily rupturable shutter of diaphragm fitted on equipment ducts or buildings to relieve pressure of explosions.

**Ventilation:** Is either a mechanical system or natural method of exchanging air within a space. Mechanical systems can include fans and blowers. Natural ventilation is caused by wind pressure and infiltration through a structure.

**Venting, Fire:** The process of inducing heat and smoke to leave a building as quickly as possible by such paths so that lateral spread is checked, fire fighting operations are facilitated and a minimum fire damage is caused.

**Vent, Smoke (roof):** Automatic or manually closing openings on the roof of a building to vent smoke and hot gases of a fire.

**Vertical Opening:** Any aperture through floors in buildings, such as lifts, ducting, stairs, services.

These openings can act as channels for the vertical spread of fire and smoke.

**Vestibule:** Small lobby or enclosed space.

**Volatile:** Prone to rapid evaporation. Both combustible and non-combustible materials may be volatile.

**Volatility:** The tendency of a liquid to vapourize.

**Wall, Fire-resisting:** A wall, either load division into bearing or non-load-bearing, capable of satisfying the criteria of fire resistance with respect to collapse flame penetration and excessive temperature rise.

**Water-miscible Liquid:** A liquid that mixes in all proportions with water without the use of chemical additives, such as emulsifying agents.

**Written Notice:** A notification in writing indicating an instruction or message.

**Wetting Agent:** A chemical compound that, when added to water in proper quantities, materially reduces its surface tension, increases its penetrating and spreading abilities.

# List of Codes and Standards

IS: 949-1967 Specification for emergency tender for fire brigade use and rescue tender for general purposes.

IS: 950-1980 Specification for water tender, type B for fire brigade use.

IS: 952-1969 Specification for fog nozzle for fire brigade use.

IS: 954-1974 Functional requirements for carbon dioxide tender for firebrigade use.

IS: 955-1980 Specification for dry powder tender for fire brigade use.

IS: 957-1967 Specification for control van for fire brigade use.

IS: 1239 Mild steel wrought iron pipes.

IS: 1536 Centrifugally cast (spun) iron pipes.

IS: 1537 Vertically cast iron pipes.

IS: 1642-1960 Code of practice for fire safety of buildings (general): Materials and details of construction.

IS: 1645-1960 Code of practice for fire safety of buildings (general): Chimneys, flues, flue pipes and hearths.

IS: 1646-1982 Code of practice for fire safety of buildings (general): Electrical installations.

IS: 1647-1960 Code of practice for fire safety of buildings (general): Non-electric lighting equipment, oil and gas heaters and burners of small capacity.

IS: 1649-1962 Code of practice for design and construction of flues and chimneys for domestic heating appliances.

IS: 1941 (Part I)-1976 Functional requirements for electric motor sirens: Part I AC 3 phase 50 Hz, 415 V type.

IS: 1984 Specification for extended branch pipe for fire fighting.

IS: 1984 Specification for portable fire extinguisher BCF type.

IS: 1984 Functional requirements of dry powder tender 2 000 kg capacity.

IS: 1984 Specification for bromochloro-difluoromethane (Halon).

IS: 2097-1983 Specification for foam making branch.

IS: 2171-1976 Specification for portable fire extinguishers, dry powder type.

IS: 2175-1977 Specification for heat sensitive fire detectors for use in automatic electric fire alarm system.

IS: 2189-1976 Code of practice for installation of automatic fire alarm system using heat sensitive type fire detectors.

1S: 2190-1979 Code of practice for selection, installation and maintenance of portable first aid fire appliances.

IS: 2217-1982 Recommendations for providing first aid fire fighting arrangements in public buildings.

IS: 2298-1977 Specification for single barrel stirrup pump for fire fighting purposes.

IS: 2546-1974 Specification for galvanized mild steel fire bucket.

IS: 2696-1974 Functional requirements for 125-1/min light fire engine.

IS: 2726-1964 Code of practice for fire safety of industrial buildings: Cotton ginning and pressing (including cotton seed delintering factories).

IS: 2745-1983 Specification for firemen's helmets.

IS: 2871-1983 Specification for branch pipe, universal, for fire fighting purposes.

IS: 2878-1976 Specification for portable fire extinguishers, carbon dioxide type.

IS: 2930-1980 Specification for hose laying tender for fire brigade use.

IS: 3034-1981 Code of practice for fire safety of industrial buildings: Electrical generating and distributing stations.

IS: 3058-1965 Code of practice for fire safety of industrial buildings: Viscose rayon yarn and/or staple fibre plants.

IS: 3079-1965 Code of practice for fire safety of industrial buildings: Cotton textile mills.

IS: 3582-1984 Specification for basket strainers for fire fighting purposes (cylindrical type).

IS: 3594-1967 Code of practice for fire safety of industrial buildings: General storage and warehousing including cold storages.

IS: 3595-1967 Code of practice for fire safety of industrial buildings: Coal pulverizers.

IS: 3614 (Part I)-1966 Specification for fire check doors: Part I plate, metal covered and rolling type.

IS: 3808-1979 Method of test for non-combustibility of building materials.

IS: 3809-1979 Fire resistance test of structures.

IS: 3836-1979 Code of practice for fire safety of industrial buildings: Jute mills.

IS: 4081 Blasting and related drilling operations.

IS: 4082 Recommendations for storage of construction materials at site.

IS: 4209-1966 Code of safety for chemical laboratories.

IS: 4226-1967 Code of practice for fire safety of industrial buildings: Aluminium powder factories.

IS: 4308-1982 Specification for dry powder for fire fighting.

IS: 4571-1977 Specification for aluminium extension ladders for fire brigade use.

IS: 4643-1984 Specification for suction wrenches for fire brigade use.

IS: 4861-1984 Specification for dry powder for fighting fires in burning metals IS: 4927-1968 Specification for unlined flax canvas hose for fire fighting.

IS: 4878-1968 Byelaws for construction of cinema buildings.

IS: 4886-1968 Code of Practice for fire safety of industrial buildings: Tea factories.

IS: 4928-1968 Specification for quick closing clack valve for centrifugal pump outlet.

IS: 4947-1977 Specification for gas cartridge for fire extinguishers.

IS: 4963-1968 Recommendations for buildings and facilities for the physically handicapped.

IS: 4989-1974 Specification for foam compound for producing mechanical foam for fire fighting (first revision).

IS: 5131-1969 Specification for dividing breeching with control, for fire brigade use.

IS: 5216 Part 1, 2–Recommendations on safety procedures and practices in electrical work.

IS: 5290-1983 Specification for landing valve.

IS: 5486-1969 Specification for quick release knife.

IS: 5490 Specification for refills for portable fire extinguishers and chemical fire engines.

IS: 5490 (Part I)-1977 Part I For soda acid portable fire extinguishers.

IS: 5490 (Part II)-1977 Part II For foam type portable fire extinguishers.

IS: 5490 (Part III)-1979 Part III For soda acid chemical fire engines.

IS: 5490 (Part IV)-1979 Part IV For foam chemical fire engines.

IS: 5505-1969 Specification for multi-edged rescue axe (non-wedging).

IS: 5506-1979 Specification for 50-1 capacity chemical fire engine, soda acid type.

IS: 5507-1979 Specification for 50-1capacity chemical fire engine, foam type.

IS: 5612 Specification for hose-clamp and hose-bandages for fire brigade use.

IS: 5612 (Part I)-1977 Part I hose clamps.

IS: 5612 (Part II)-1977 Part II hose bandages.

IS: 5714-1981 Specification for hydrant, stand-pipe for fire fighting.

IS: 5916 Construction using hot bituminous materials.

IS: 6026-1970 Specification for hand operated sirens.

IS: 6067-1983 Functional requirements for water tender, Type 'X' for fire brigade use.

IS: 6234-1971 Specification for portable fire extinguishers, water type (constant air pressure).

IS 6329-1971 Code of practice for fire safety of industrial buildings: Saw mills and wood works.

IS: 6382-1984 Code of practice for design and installation of fixed carbon dioxide fire extinguishing system.

IS: 6994 Safety gloves–leather and cotton gloves.

IS: 7181 Horizontally cast iron pipes.

IS: 8090-1976 Specification for coupling branch pipe, nozzle, used in hose reel tubing for fire fighting.

IS: 8096-1976 Specification for fire beater.

IS: 8149-1976 Functional requirements for twin $CO_2$ fire extinguishers (trolley mounted).

IS: 8423-1977 Specification for controlled percolating hose for fire fighting.

IS: 8442-1977 Functional requirements for stand post type water monitor fire fighting.

IS: 8757-1978 Glossary of terms associated with fire safety.

IS: 8758 Recommendation for fire precautionary measures in construction of temporary structures and pandals.

IS: 9109-1979 Code of practice for fire safety of industrial buildings: Paint and Varnish factories.

IS: 9972-1981 Specification for automatic sprinkler heads.

IS: 10204-1982 Specification for portable fire extinguisher mechanical foam type.

IS: 10460-1983 Functional requirements for small foam tender for fire brigade use.

IS: 10474-1983 Specification for 150 liter capacity chemical fire engine, foam type.

IS: 10658-1983 Specification for higher capacity dry powder fire extinguisher (trolley mounted).

IS: 10667 Safety equipment for protection of foot and leg.

IS: 13416 Recommendations for preventive measures against hazards at work places.

## National Building Code of India–1983

*Standards issued by NFPA–National Fire Protection Association of USA:*

NFPA–1: 2000 Fire prevention code.

NFPA–10: 2002 Standard for portable fire extinguishers.

NFPA–11: 2002 Standard for low, medium and high expansion foam.

NFPA –12: 2000 Standard for carbon dioxide extinguishing systems.

NFPA–13: 2002 Standard for installation of sprinkler systems.

NFPA –14: 2003 Standard for installation of standpipe and hose systems.

NFPA–15: 2001 Standard for water spray fixed systems for fire protection.

NFPA–16: 1999 Standard for installation of deluge foam water sprinkler system.

NFPA–17: 2002 Standard for dry chemical extinguishing system.

NFPA –18: 1995 Standard on wetting agents.

NFPA–20: 1999 Installation of stationary pumps for fire protection.

NFPA –22: 2003 Standard for water tanks for private fire protection.

NFPA –24: 2002 Standard for installation of private fire service mains and their appurtenances.

NFPA–25: 1998 Standard for Inspection, testing and maintenance of water based fire protection system.

NFPA–30: 2000 Flammable and combustible liquids code.

NFPA–31: 2001 Standard for installation of oil burning equipment.

NFPA–32: 2000 Standard for dry cleaning plants.

NFPA–33: 2000 Standard for spray application using flammable or combustible materials.

NFPA –34: 1995 Standard for dipping and coating processes using flammable or combustible liquids.

NFPA–36: 1997 Standard for Solvent Extraction Plants.

NFPA–40: 1997 Standard for the storage and handling of cellulose nitrate motion picture film.

NFPA–45: 2000 Standard for fire protection for labs using chemicals.

NFPA–50: 2001 Standard for bulk oxygen systems.

NFPA–51B: 2003 Standard for prevention during cutting, welding and other Hot-work.

NFPA –52: 1998 Standard for CNG vehicular fuels.

NFPA–53: 1999 Recommendations for work in oxygen enriched environments.

NFPA–54: 1999National fuel gas code.

NFPA–55: 1998 Standard for storage, use and handling of compressed and liquefied gases in portable cylinders.

NFPA–57: 1999 Liquefied natural gas (LNG) vehicular fuel systems Code.

NFPA–58: 2000 LPG code.

NFPA–59: 1998 Standard for the storage and handling of liquefied petroleum gases at utility gas plants.

NFPA–61: 1999 Standard for the prevention of fires and dust explosions in agricultural and food products facilities.

NFPA–69: 1997 Standard on explosion prevention systems.

NFPA–70: 2005 National electric code.

NFPA–72: 2002 National fire alarm code.

NFPA–77 Recommended practice on static electricity.

NFPA–79: 1997 Electrical standard for industrial machinery.

NFPA–80: 1999 Standard for fire doors and fire windows.

NFPA–82: 1999 Standard on incinerators and waste and linen handling systems and equipment.

NFPA–85: 2001 Boiler and combustion systems hazard code.

NFPA–86: 1999 Standard for ovens and furnaces.

NFPA–88: 1998 A, standard for parking structures.

NFPA–88B: 1997 Standard for repair garages.

NFPA–90A: 1999 Standard for the installation of air-conditioning and ventilating systems.

NFPA–90B: 1999 Standard for the installation of warm air heating and air-conditioning systems.

NFPA–91: 1999 Standard for exhaust systems for air conveying of vapours, gases, mists, and noncombustible particulate solids.

NFPA–96: 1998 Standard for ventilation control and fire protection of commercial cooking operations.

NFPA–99: 1999 Standard for health care facilities.

NFPA–101: 2000 Life safety code.

NFPA–101: 2000 National life safety code.

NFPA–101B: 1999 Code for means of egress from buildings and structures.

NFPA–110: 2002 Code for emergency and standby power systems.

NFPA–111: 1996 Standard on stored electrical energy emergency and standby power systems.

NFPA–120: 1999 Standard for coal pre-paration plants.

NFPA–204: 2002 Standard for smoke and heat venting.

NFPA–211: 2000 Standard for chimneys, fire-places, vents, and solid, fuel-burning appliances.

NFPA–220: 1999 Standard on types of building construction.

NFPA–221: 1997 Standard for fire walls and fire barrier walls.

NFPA–230: 1999 Standard for the fire protection of storage.

NFPA–231D: 1998 Standard for storage of rubber tires.

NFPA–232: 1995 Standard for the protection of records.

NFPA–241: 1996 Standard for safeguarding construction, alteration, and demolition oper-ations.

NFPA–251: 1999 Standard methods of tests of fire endurance of building, construction and materials.

NFPA–252: 1999 Standard methods of fire tests of door assemblies.

NFPA–255: 2000 Standard method of test of surface burning characteristics of building materials.

NFPA–256: 1998 Standard methods of fire tests of roof coverings.

NFPA–257: 2000 Standard on fire test for window and glass block assemblies.

NFPA–260: 1998 Standard methods of tests and classification system for cigarette ignition resistance of components of upholstered furniture.

NFPA–303: 1995 Fire protection standard for marinas and boatyards.

NFPA–307: 1995 Standard for the construction and fire protection of marine terminals, piers, and wharves.

NFPA–312: 1995 Standard for fire protection of vessels during construction, repair, and lay-up.

NFPA–318: 1998 Standard for the protection of cleanrooms.

NFPA–385: 2000 Standard for tank vehicles for flammable and combustible liquids.

NFPA–407: 1996 Standard for aircraft fuel servicing.

NFPA–409: 1995 Standard on aircraft hangars.

NFPA–41: 1999 Standard on aircraft maintenance.

NFPA–414: Standard for aircraft rescue and fire fighting vehicles.

NFPA–415: 1997 Standard on airport terminal buildings, fueling ramp drainage, and loading walkways.

NFPA–418: 1995 Standard for heliports.

NFPA–422: 1999 Guide for aircraft accident Response.

NFPA–424: Guide for airport/community emergency planning.

NFPA–430: 2000 Code for the storage of liquid and solid oxidizers.

NFPA–432: 1997 Code for the storage of organic peroxide formulations.

NFPA–434: 1998 Code for the storage of pesticides.

NFPA–471: 2002 Recommended practice for responding to hazardous materials incidents.

NFPA–480: 1998 Standard for the storage, handling, and processing of magnesium solids and powders.

NFPA–481: 1995 Standard for the production, processing, handling, and storage of titanium.

NFPA–482: 1996 Standard for the production, processing, handling, and storage of zirconium.

NFPA–485: 1999 Standard for the storage, handling, processing, and use of lithium metal.

NFPA–490: 1998 Code for the storage of ammonium nitrate.

NFPA–495: 1996 Explosive materials code.

NFPA–498: 1996 Standard for safe havens and interchange lots for vehicles transporting explosives.

NFPA–501: 1999 Standard on manufactured housing.

NFPA–501A: 1999 Standard for fire safety criteria for manufactured home installations, sites, and communities.

NFPA–505: 1999 Fire safety standard for powered industrial trucks including type designations, areas of use, conversions, maintenance, and operation.

NFPA–600: 2001 Standard for industrial fire brigades.

NFPA–601: 2000 Standard for security Services in fire loss prevention.

NFPA–650: 1998 Standard for pneumatic conveying systems for handling combustible particulate solids.

NFPA–651: 1998 Standard for the machining and finishing of aluminum and the production and handling of aluminum powders.

NFPA–654: 1997 Standard for the prevention of fire and dust explosions from the manufacturing, processing, and handling of combustible particulate solids.

NFPA–655: 1993 Standard for prevention of sulfur fires and explosions.

NFPA–664: 1998 Standard for the prevention of fires and explosions in wood processing and woodworking facilities.

NFPA–701: 1999 Standard methods of fire tests for flame propagation of textiles and films.

NFPA–704: 1996 Standard system for the identification of the hazards of materials for emergency response.

NFPA–780: 1997 Standard for installation of lightning protection systems.

NFPA–901: 2001Standard classifications for incident reporting and fire protection data.

NFPA–906: 1998 Guide for fire incident field notes.

NFPA–909: 1997 Standard for the protection of cultural resources, including museums, libraries, places of worship, and historic properties.

NFPA–921: 2004 Guide for fire and explosion investigations.

NFPA–1122: 1997 Code for model rocketry.

NFPA–1123: 1995 Code for fireworks display.

NFPA–1124: 1998 Code for the manufacture, transportation, and storage of fireworks and pyrotechnic articles.

NFPA–1125: 1995 Code for the manufacture of model rocket and high power rocket motors.

NFPA–1126: 1996 Standard for the use of Pyrotechnics before a proximate audience.

NFPA–1127: 1998 Code for high power rocketry.

NFPA–1142: 1999 Standard on water supplies for suburban and rural fire fighting.

NFPA–1150: 1999 Standard for fire fighting foam.

NFPA–1403: 2002 Standard for live fire training.

NFPA–1561: 2000 Standard on emergency service incident management system.

NFPA–1600: 2004 Standard on disaster and emergency management.

NFPA–1901 Standard for automotive fire apparatus.

NFPA–1962 Standard for the care, use, and service testing of fire hose including couplings and nozzles.

NFPA–1963: 1998 Standard for fire hose connections.

NFPA–2001: 2000 Standard on clean agent fire extinguishing systems.

NFPA–8503: 1997 Standard for pulverized fuel systems.

## British Standards

BS: 336 Fire hose couplings and ancillary equipment.

BS: 349 Identification of contents of industrial gas containers.

BS: 476 Fire tests on building materials and structures.

BS: 750 Underground fire hydrants and dimensions of surface box openings.

BS: 764 Automatic change-over contactors for emergency lighting systems.

BS: 889 Flameproof electric light fittings.

BS: 1259 Intrinsically safe electrical apparatus and circuits for use in explosive atmospheres.

BS: 1635 Graphical symbols and abbreviations for fire protection drawings.

BS: 1771 Outdoor uniform cloths for fire service and other staff.

BS: 1945 Fireguards for heating appliances. (Gas, electric and oil burning).

BS: 2052 Ropes made from coir, hemp, manila and sisal.

BS: 2560 Specification for exit signs (internally illuminated).

BS: 2963 Methods of test for the flammability of fabrics.

BS: 3116 Automatic fire alarms in buildings.

BS: 3119 Method of test for flameproof materials.

BS: 3120 Performance requirements of flameproof materials for clothing and other purposes.

BS: 3121 Performance requirements of fabrics described as of low flammability.

BS: 3169 Rubber reel hose for fire fighting purposes.

BS: 3187 Electrically conducting rubber floors.

BS: 3248 Space guards for solid fuel fires.

BS: 3251 Indicator plates for fire hydrants and emergency water supplies.

BS: 3300 Kerosine (paraffin) unflued space heaters for domestic use.

BS: 3367 Fire brigade rescue lines.

BS: 3566 Lightweight salvage sheets for fire service use.

BS: 3791 Clothing for protection against intense heat for short periods.

BS: 3864 Firemen's helmets.

BS: 4422 Glossary of terms associated with fire.

BS: 4547 Classification of fires.

BS: 4667 Breathing apparatus.

BS: 5041 Fire hydrant systems equipment.

BS: 5053 Methods of tests for cordage.

BS: 5173 Methods of tests for hoses.

BS: 5266 Emergency lighting of premises.

BS: 5306 Code of practice for fire extinguishing installations and equipment on premises.

BS: 5378 Specification for safety colours and safety signs.

BS: 5423 Specification for portable fire extinguishers.

BS: 5445 Components of automatic fire alarm systems.

BS: 5446 Components of automatic fire alarm systems for residential purposes.

BS: 5499 Fire safety signs, notices and graphic symbols.

BS: 5502 Fire protection of agricultural buildings.

BS: 5588 Code of practice for fire precautions in the design of buildings.

BS: 5839 Fire detection and alarm systems in buildings.

Pt I Code of practice for installation and servicing.

Pt 2 Specification for manual call points.

BS: 6266 CP for fire protection for electronic date processing installations.

BS: 6165 Specification for small disposable fire extinguishers of the aerosol type.

BS: 6643 Recharging fire extinguishers.

## Standards Issued by ASTM–American Society of Testing and Materials

*List of statutory acts and rules (EHS)*

The Indian Explosives Act 1884 and Rules 1983.

The Motor Vehicle Act 1988 and Central Motor Vehicles Rules 1989.

The Factories Act and Factory Rules 1948.

The Petroleum Act.

The Workmen Compensation Act.

The Gas Cylinders Rules 1981 and the Static and Mobile Pressure Vessel (Unfired) Rules.

The Indian Electricity Act 1901 and Rules 1956.

The Indian Boiler Act 1923 and Regulations 1950.

The Environment (Protection) Act 1986.

The Water (Prevention and Control of Pollution) Act.

The Water (Prevention and Control of Pollution) Cess Act 1974 and 1975.

The Air (Prevention and Control of Pollution) Act.

The Mines and Minerals (Regulation and Development) Act.

The Atomic Energy Act.

The Radiation Protection Rules.

The Indian Fisheries Act.

The Indian Forest Act.

The Wildlife (Protection) Act.

The Hazardous Wastes (Management and Handling) Rules.

The Manufacturing, Storage and Import of Hazardous Chemicals Rules 1989.

The Public Liability Act 1991 Amended 1992.

Building and other Construction Workers (Regulation of Employment and Condition of Service) Act 1996.

The format of Form 34 to be filed by all Industrial establishments is given below

**THE FACTORIES ACT, 1948 FORM 34**
*(Prescribed under Rule 125)*
**ANNUAL RETURN**
**For the year ending 31st December**

1. Registration number of factory: _____

2. Name of factory: _____

3. Name of occupier: _____

4. Name of the manager: _____

5. District: _____

6. Full postal address of factory: _____

    _____

7. Name of industry: _____

    Number of workers and particulars of employment: _____

8. Number of days worked in the year: _____

9. Number of mandays worked during the year: _____

    (a) Men: _____

    (b) Women: _____

    (c) Children: _____

10. Average number of workers employed daily (see explanatory note)

    (a) Adults        (i) Men: _____

                        (ii) Women: _____

    (b) Adolescents (i) Male: _____

                        (ii) Female: _____

    (c) Children      (i) Boys: _____

                        (ii) Girls: _____

11. Total number of man-hours worked including over-time

    (a) Men: _____

    (b) Women: _____

    (c) Children: _____

12. Average number of hours worked per week (see explanatory note)

    (a) Men: _____

    (b) Women: _____

    (c) Children: _____

13. (a) Does the factory carry out any process or operation declared as dangerous under Section 87? (See Rule 120)

1. If so, give the following information:

Name of the dangerous processes or operation

Average number of persons employed daily in each of the processes or operation given in column 1 carried on

| 1 | 2 |
|---|---|
| (i) _____ | _____ |
| (ii) _____ | _____ |
| (iii) _____ | _____ |

### Leave with Wages

14. Total number of workers employed during the year.

    (a) Men: _____

    (b) Women: _____

    (c) Children: _____

15. Number of workers who were entitled to annual leave with wages during the year

    (a) Men: _____

    (b) Women: _____

    (c) Children: _____

16. Number of workers who were granted leave during the year.

    (d) Men: _____

    (e) Women: _____

    (f) Children: _____

17. (a) Number of workers who were discharged, or dismissed from the service or quit employment, or were superannuated, or who died while in service during the year: _____

    (b) Number of such workers in respect of whom wages in lieu of leave were paid: _____

### Safety Officers

18. (a) Number of safety officers required to be appointed as per notification under Section 40-B _____

    (b) Number of safety officer appointed: _____

### Ambulance

19. Is there an ambulance room provided in the factory as required in Section 45? _____

20. (a) Is there a canteen provided in the facto ry as required under Section 46? _____

    (b) Is the canteen provided managed/run

        (i) Departmentally, or _____

        (ii) Through a contractor? _____

### Shelters or Rest Rooms and Lunch Rooms

21. (a) Are there adequate and suitable shelters or rest rooms provided in the factory as required under Section 47? _____

    (b) Are there adequate and suitable lunch rooms provided in the factory as required under Section 47? _____

### Crèches

22. Is there a crèche provided in the factory as required under Section 48* _____

### Welfare Officers

23. (a) Number of welfare officers to be appointed as required under Section-49 _____
    (b) Number of welfare officers to be appointed as required under Section-49 _____

### Accidents

24. (a) Total number of accidents (see explanatory note)

    (i) Fatal _____

    (ii) Non-fatal _____

    (b) Accidents in which workers returned to work during the year to which this return relates:

    (i) Accidents (workers injured) occurring during the year in which injured workers returned to work during the same year;

    (aa) Number of accidents _____

    (bb) Mandays lost due to accidents _____

    (ii) Accidents (workers injured) occurring in the previous year in which injured workers returned to work during the year to which this return relates.

    (aa) Number of accidents _____

    (bb) Mandays lost due to accidents _____

    (c) Accidents (workers injured) occurring during the year in which injured workers did not return to work during the year to which this returns relates:

    (i) Number of accidents _____

    (ii) Mandays lost due to accidents _____

### Suggestion Scheme

25. (a) Is a suggestion scheme in operation in the factory _____

    (b) If so, the number of suggestion _____

    (i) Received during the year _____

    (ii) Accepted during the year _____

    (c) Amount awarded in cash prizes during the year

    (i) Total amount awarded _____

    (ii) Value of the maximum cash prize awarded _____

    (iii) Value of the minimum cash prize awarded _____

---

\* *The term "ordinarily employed" as used in Section 48 of the Factories Act, 1948 would mean "Total number of persons employed in all shifts. This should be over 50% of the number of Working days in the. establishment"*

Certified that the information furnished above is, to the best of my knowledge and belief, correct.

*Signature of the Manager:* _____

Date: _____

THIS RETURNS SHOULD BE SENT TO THE PRESCRIBED STATE AUTHORITY BY 31 JANUARY OF THE SUCCEEDING YEAR

*Explanatory Notes:*

1. The average number of workers employed daily should be calculated by dividing the aggregate number of attendances, attendance by temporary as well as permanent employed should be counted, and all employees (including apprentices) should be included, whether they are employed directly or under contractors. Attendance on separate shifts (e.g. night and day shifts) should be counted separately. Days on which the factory was closed for whatever cause, and day on which the manufacturing processes were not carried on should not be treated as working days. However, if more than 40% of workers employed (on previous day) attend to repair maintenance or other such work on closed days, such days should be treated as working days. Partial attendance for less than half a shift on a working day should be ignored, while attendance for half a shift or more on such day should be treated as full attendance.
2. For seasonal factories, the average number of workers employed during the working season and the off season should be given separately. Similarly the number of days worked and average number of man hours worked per week during the working and off season should be given separately.
3. The average number of hours worked per week means the total actual hours worked by all workers during the year excluding the rest intervals what including overtime work divided by the product of average number of workers employed daily in the factory during the year and. In case the factory has not worked for the whole year, the number of weeks during which the factory worked should be used in place of the figure 52.
4. Every person killed or injured should be treated as one separate accident. If in one occurrence, 6 persons were injured or killed it should be counted as six accidents.
5. In item 24 (a), the number of accidents which took place during the year should be given. In case of non-fatal accident only those accidents which prevented workers from working from 48 hours or more, immediately following the accidents should be indicated.
6. In item 8, the information may be furnished as the number of days the factory worked during the year.

Accident details to be furnished

# FORM–15
## HALF YEARLY/ANNUAL RETURN
### [Rule 94 of Factories Act.]

1. Name of the factory: _____

2. Average number of workers employed daily } Male _____

   } Female _____

3. Normal hours per week: _____

4. Number of days worked in a year: _____

5. Does the factory come under Section 87: YES/NO _____
   (dangerous operations)

6. Average daily number of workers employed in: _____
   (dangerous operations)

7. Number of accidents reported involving: _____

1. Fatalities _____

2. Serious occurrences (without injury to persons) _____

3. Minor loss of time (accident causing disability of more than 48 hours) _____

4. Permanent partial disability _____

5. Permanent total disability _____

6. Causative factors of accidents: _____

1. Machinery _____

2. Handling of materials _____

3. Chemicals _____

4. Hand tools _____

5. Fall of persons _____

6. Fall of objects _____

7. Striking against/struck _____

8. Explosion or fire _____

9. Misc. agencies _____

Date: _____

Signature of the Manager: _____

Name: _____

Note: A detailed report on accidents be prepared on the lines suggested in IS 3786 entitled "Industrial accidents, classification and computation of injuries and accidents".

# References

1. Frank P. Lees and M.L. Ang (Eds), *Safety Cases–within the Control of Industrial Major Accident Hazards (CIMAH) Regulations 1984*, Butterworths, 1989.

2. Michael Atchia and Shawna Tropp (Eds): *Environmental Management Issues and Solutions*, United Nations Environment Programme (UNEP), John Wiley & Sons, 1995.

3. Colin S. Todd: *Croner's guide to Fire Safety*, Croner Publications, Surrey, UK.

4. British Medical Association: *Health and Environmental Impact Assessment–An integrated approach*. Earth Scan Publications, London, 1998.

5. Marc L. Janssens: *Evaluating Computer Fire Models, Fire Protection Engineering*, 2002.

6. Barry Kirwan: *A Guide to Practical Human Reliability Assessment*, Taylor & Francis Ltd, London, 1994.

7. R.S.F Schilling (Ed): *Occupational Health Practice*, 2/e, Butterworth's, UK, 1981.

8. McGuire and White: *Liquefied Gas Handling Principles on ships and in terminals*, 2/e, Witherby & Company, London.

9. Allan Gilpin: *Environmental Impact Assessment, cutting edge for the twenty-first century*. Cambridge University Press, 1995.

10. Mohinder Nayyar (Ed): *Piping handbook*, McGraw-Hill, New York.

11. Mark S. Sanders and Ernest J. McCormick: *Human factors in engineering and design*, McGraw-Hill.

12. Martyn S. ray: *Elements of engineering Design: an Integrated Approach*. Prentice Hall International.

13. Charles O. Smith: *Introduction to Reliability in Design*, McGraw-Hill.

14. L.S. Srinath: *Concepts in Reliability Engineering*, Affiliated East-West Press.

15. Edward Hawthorne: *Management of Technology*.

16. George E. dieter: *Engineering Design*.

17. Trevor Kletz: *Learning from Accidents*. Butterworth–Heinemann Ltd, Oxford, U.K, 2nd edition 1994.

18. F.P Lees: *Loss prevention in the process industries, Hazard Identification, Assessment and control*, Volumes 1 and 2, Butterworth's, London, 1983.

19. Gwendolyn Holmes, Ramnarain Singh and Louis Theodore: *Handbook of Environmental Management and Technology*,. John Wiley & sons, N.Y. 1993.

20. Jeremy Strants: *A Manager's Guide to Health and Safety at Work*, Fifth edition, Kogan Page Limited, 120, Pentonville Road, London, 1997.

21. Sue Cox and Tom Cox: *Safety, Systems and People*, Butterworth and Heinemann, 1996.

22. The *Assessment and Control of Major Hazards*. Institution of Chemical Engineers, UK, 1985.

23. *First Aid for Motorists*, published by The British School of Motoring Ltd, London, 1997.

24. *Human factors in Safety* – critical Systems.

25. W. Bartknecht: *Dust explosions*.

26. Neil Schultz: Fire and Flammability handbook, Van-Nostrand Reinhold, NY, 1985.

27. *Fire and Smoke: Understanding the Hazards*, Commission on Life Sciences, 1986.

28. *TAC Manual* (India).

29. *National Building code of India*, 1983.

30. *Fundamentals of Process Safety,* Vic Marshall and Steve Ruhemann. Institution of chemical Engineers, Rugby, UK, 2001.

31. Perry: *Chemical Engineer's Handbook.*

32. Derek James: *Fire Prevention Handbook,* Butterworths, London, 1986.

33. T.Z. Harmathy: *Fire Safety Design and concrete,* Longman Scientific and technical, U.K,1993.

34. *NFPA fire protection handbook: Automatic sprinkler system handbook NFPA SFPE handbook of fire protection engineering.* Society of Fire Protection Engineers.

35. Mcguire and White: *Liquefied Gas Handling Principles on ships and in terminals,* 2/e, Witherby & Company, London.

36. Paul Stollard and John Abrahams: *Fire from First Principles'–A design guide to building fire safety,* E & FN SPON, London, Third edition, 1999.

37. J.A. Purkiss: *Fire safety Engineering-Design of Structures,* Butterworth Heinemann, Oxford, U.K, 1996.

38. M.B.Wood: *Fire Precautions in Computer Installations,* NCC Publications, England, 1986.

# Index